AF889458

METHODS IN MOLECULAR BIOLOGY

Series Editor
John M. Walker
School of Life and Medical Sciences
University of Hertfordshire
Hatfield, Hertfordshire, AL10 9AB, UK

For further volumes:
http://www.springer.com/series/7651

Bacterial Multidrug Exporters

Methods and Protocols

Edited by

Akihito Yamaguchi

Institute of Scientific and Industrial Research, Osaka University, Ibaraki, Osaka, Japan

Kunihiko Nishino

Institute of Scientific and Industrial Research, Osaka University, Ibaraki, Osaka, Japan;
Graduate School of Pharmaceutical Sciences, Osaka University, Suita, Osaka, Japan

Humana Press

Editors
Akihito Yamaguchi
Institute of Scientific and Industrial Research, Osaka University
Ibaraki, Osaka, Japan

Kunihiko Nishino
Institute of Scientific and Industrial Research, Osaka University
Ibaraki, Osaka, Japan

Graduate School of Pharmaceutical Sciences, Osaka University
Suita, Osaka, Japan

ISSN 1064-3745 ISSN 1940-6029 (electronic)
Methods in Molecular Biology
ISBN 978-1-4939-7452-8 ISBN 978-1-4939-7454-2 (eBook)
https://doi.org/10.1007/978-1-4939-7454-2

Library of Congress Control Number: 2017959058

© Springer Science+Business Media, LLC 2018
This work is subject to copyright. All rights are reserved by the Publisher, whether the whole or part of the material is concerned, specifically the rights of translation, reprinting, reuse of illustrations, recitation, broadcasting, reproduction on microfilms or in any other physical way, and transmission or information storage and retrieval, electronic adaptation, computer software, or by similar or dissimilar methodology now known or hereafter developed.
The use of general descriptive names, registered names, trademarks, service marks, etc. in this publication does not imply, even in the absence of a specific statement, that such names are exempt from the relevant protective laws and regulations and therefore free for general use.
The publisher, the authors and the editors are safe to assume that the advice and information in this book are believed to be true and accurate at the date of publication. Neither the publisher nor the authors or the editors give a warranty, express or implied, with respect to the material contained herein or for any errors or omissions that may have been made. The publisher remains neutral with regard to jurisdictional claims in published maps and institutional affiliations.

Printed on acid-free paper

This Humana Press imprint is published by Springer Nature
The registered company is Springer Science+Business Media, LLC
The registered company address is: 233 Spring Street, New York, NY 10013, U.S.A.

Preface

Multidrug exporters are intrinsic membrane proteins widely distributed in bacteria. They play as cellular self-defense mechanisms and have some other physiological roles. They contribute to bacterial tolerance against antibiotics. When they are over-expressed, they cause multidrug resistance. Multidrug-resistant pathogens have caused great difficulties in modern chemotherapy. There have been no clinically useful drugs against bacterial multidrug exporters. They can be classified into three major families: ABC, MFS, and RND. ABC-type exporters are ATP-hydrolysis-coupled transporters, which are major multidrug exporters in mammalian models, but in microorganisms, ABC-type exporters play a minor role in drug resistance. MFS-type exporters are drug/proton antiporters, which are major drug exporters in multidrug resistant Gram-positive pathogens. MATE-type is a subfamily of MFS-type transporters. RND-type exporters are multidrug exporters characteristic in Gram-negative bacteria, which consist of a tripartite complex passing through cytoplasmic membrane, periplasm, and the outer membrane. The most characteristic common properties of multidrug exporters are their extremely broad substrate specificity. For example, major multidrug exporter AcrAB-TolC in *E. coli* exports both aromatic and aliphatic compounds including cationic, anionic, twitter ionic, and neutral compounds. However, it has own specificity. It exports oxacillin but does not export carbenicillin. Clarifying the mechanism of multidrug exporters, specifically as they export such a wide range of drugs and toxic compounds, is one of major challenges posed to our modern science.

Elucidation of multidrug efflux mechanisms has been greatly advanced mainly by structure determination of bacterial multidrug exporters during the last decade. Prior to the 2000s, no transporters' molecular structures were solved. Now, the crystal structures of more than ten multidrug exporters including all three types (RND, MFS, and ABC) have been solved. This success is supported by the advancement of the technology to express a large amount of tagged membrane proteins, development of detergents to solubilize membrane proteins, large-scale purification using affinity chromatography, as well as development of the method for crystallization including facilitator of crystallization such as DARPin and monobody. As a result, we have been able to understand the outline of the structural basis of multidrug recognition and multidrug efflux mechanisms.

In the first part of this book, we present protocols to introduce marvelous success in determining multidrug exporter structures during the last decade. Chapters 1 and 2 show the protocols for determination of the high-resolution structures and the ligand-binding structures of RND-type multidrug exporter AcrB, which are the pioneering works for the structural study of bacterial multidrug exporters. They revealed that the multidrug efflux is mediated by the functional rotation mechanism of an asymmetric homo-trimer, and the structural basis of the multidrug recognition is a multisite drug binding. RND-type transporters are tripartite complexes composed of cell membrane transporters, outer membrane channels, and adaptor proteins. Chapter 3 describes the crystal structure of another interesting type multidrug exporter, MATE-type, bound with its inhibitors. It also describes the novel artificial and systematic transporter-inhibitor construction method. Chapter 4 shows the first crystal structure of the complex of a cell membrane transporter and an adaptor protein two-part complex. Unlikely to the conventional sense, an adaptor protein is not just an adaptor but the hexamer of the adaptor proteins makes a periplasmic channel that connects between a cell membrane transporter and an outer membrane channel. Chapter 5 shows the first whole structure of a tripartite complex of the RND-type exporter by cryo-EM images. Chapter 6 shows another strong tool for structure determination of a membrane transporter using NMR spectroscopy. Chapter 7 describes one of the powerful tools for crystallization of a membrane transporter.

In order to understand how to work the molecular structures of multidrug exporters to transport drugs, biochemical and bioengineering analysis is absolutely necessary. In the second section, we will show a few important examples of the numerous biochemical and genetic studies of exporters. With respect to the RND-type transporters, reconstitution studies for whole tripartite complex are tremendously difficult because the complex penetrates through two membranes. Chapter 8 shows the breakthrough achievement of the reconstitution of the tripartite complex and the measurement of multidrug transport by RND-type transporters through artificial membranes. Chapter 9 proves the functional rotation mechanism of drug export mediated by RND-type exporters using artificial covalently linked trimers of cell membrane transporter. Chapter 10 shows the intra-protein drug translocation pathway by introducing site-directed mutagenesis.

In Part III, we show the computational analysis of how to work the exporter structure and how to predict a novel efflux pump. Chapter 11 shows an excellent example of the molecular dynamic simulations of the RND-type exporter. Chapter 12 shows a transcriptomic approach to identify novel efflux pumps.

Multidrug exporters play a variety of physiological roles under various expression controls. In Part IV, we show how to regulate the exporter expression and biomedical roles of exporters. Chapters 13 and 14 show the condition and the mechanism of expression regulation of the bacterial multidrug exporters. Chapter 15 shows the identification of the expression regulators. Chapter 16 shows high-throughput screening of multidrug efflux system useful for identification of the physiological role of exporters.

Finally, in the Part V, we show the advanced technologies useful for future works of multidrug exporters. Chapter 17 shows single-molecule analysis of membrane transporter activity using artificial membranes stretched on a microchamber. Chapter 18 shows a single-cell efflux assay method using femtoliter droplet arrays. Chapter 19 shows reconstitution and active transport assay using liposomes including V-ATPase for a proton-motive force supplier.

The field of multidrug exporter studies is a fast developing field of science. I hope this collection of protocols will contribute to the work of many researchers studying multidrug exporters. I would like to thank all authors for sharing their valuable experience and insights with the research community at large. I would like to thank the series editor Dr. John Walker for help with reviewing the book.

Ibaraki, Osaka, Japan

Akihito Yamaguchi
Kunihiko Nishino

Contents

Preface ... *v*
Contributors ... *ix*

Part I Structural Analysis of Bacterial Multidrug Exporters

1 High-Resolution Crystallographic Analysis of AcrB Using Designed Ankyrin Repeat Proteins (DARPins) ... 3
Heng Keat Tam, Viveka Nand Malviya, and Klaas M. Pos

2 Crystallographic Analysis of Drug and Inhibitor-Binding Structure of RND-Type Multidrug Exporter AcrB in Physiologically Relevant Asymmetric Crystals ... 25
Ryosuke Nakashima, Keisuke Sakurai, and Akihito Yamaguchi

3 Crystallographic Analysis of MATE-Type Multidrug Exporter with Its Inhibitors ... 37
Tsukasa Kusakizako, Yoshiki Tanaka, Christopher J. Hipolito, Hiroaki Suga, and Osamu Nureki

4 Crystallographic Analysis of the CusBA Heavy-Metal Efflux Complex of *Escherichia coli* ... 59
Jared A. Delmar and Edward W. Yu

5 Purification of AcrAB-TolC Multidrug Efflux Pump for Cryo-EM Analysis ... 71
Dijun Du, Zhao Wang, Wah Chiu, and Ben F. Luisi

6 NMR Spectroscopy Approach to Study the Structure, Orientation, and Mechanism of the Multidrug Exporter EmrE ... 83
Maureen Leninger and Nathaniel J. Traaseth

7 Generation of Conformation-Specific Antibody Fragments for Crystallization of the Multidrug Resistance Transporter MdfA ... 97
Frank Jaenecke, Yoshiko Nakada-Nakura, Kumar Nagarathinam, Satoshi Ogasawara, Kehong Liu, Yunhon Hotta, So Iwata, Norimichi Nomura, and Mikio Tanabe

Part II Biochemical and Bioengineering Analysis of Bacterial Multidrug Exporters

8 Biochemical Reconstitution and Characterization of Multicomponent Drug Efflux Transporters ... 113
Martin Picard, Elena B. Tikhonova, Isabelle Broutin, Shuo Lu, Alice Verchère, and Helen I. Zgurskaya

9 Covalently Linked Trimers of RND (Resistance-Nodulation-Division) Efflux Transporters to Study Their Mechanism of Action: *Escherichia coli* AcrB Multidrug Exporter as an Example ... 147
Hiroshi Nikaido

10 Determining Ligand Path Through a Major Drug Transporter, AcrB, in *Escherichia coli* 167
Fasahath Husain and Hiroshi Nikaido

PART III COMPUTATIONAL ANALYSIS OF BACTERIAL MULTIDRUG EXPORTERS

11 Molecular Modeling of Multidrug Properties of Resistance Nodulation Division (RND) Transporters 179
Pierpaolo Cacciotto, Venkata K. Ramaswamy, Giuliano Malloci, Paolo Ruggerone, and Attilio V. Vargiu

12 A Transcriptomic Approach to Identify Novel Drug Efflux Pumps in Bacteria 221
Liping Li, Sasha G. Tetu, Ian T. Paulsen, and Karl A. Hassan

PART IV BIOMEDICAL APPROACH FOR BACTERIAL MULTIDRUG EXPORTERS

13 Regulation of the Expression of Bacterial Multidrug Exporters by Two-Component Signal Transduction Systems 239
Kunihiko Nishino

14 Study of the Expression of Bacterial Multidrug Efflux Pumps in Anaerobic Conditions 253
Jingjing Sun, Ziqing Deng, Danny Ka Chun Fung, and Aixin Yan

15 Identification of a *Staphylococcus aureus* Efflux Pump Regulator Using a DNA–Protein Affinity Technique 269
Que Chi Truong-Bolduc and David C. Hooper

16 High-Throughput Flow Cytometry Screening of Multidrug Efflux Systems 293
Mark K. Haynes, Matthew Garcia, Ryan Peters, Anna Waller, Pietro Tedesco, Oleg Ursu, Cristian G. Bologa, Radleigh G. Santos, Clemencia Pinilla, Terry H. Wu, Julie A. Lovchik, Tudor I. Oprea, Larry A. Sklar, and George P. Tegos

PART V ADVANCED TECHNOLOGIES EXPECTED FOR APPLICATION TO MULTIDRUG EFFLUX TRANSPORT STUDIES

17 Single-Molecule Analysis of Membrane Transporter Activity by Means of a Microsystem 321
Rikiya Watanabe, Naoki Soga, Shin-ya Ohdate, and Hiroyuki Noji

18 Large-Scale Femtoliter Droplet Array for Single Cell Efflux Assay of Bacteria 331
Ryota Iino, Shouichi Sakakihara, Yoshimi Matsumoto, and Kunihiko Nishino

19 Reconstitution and Transport Analysis of Eukaryotic Transporters in the Post-Genomic Era 343
Hiroshi Omote and Yoshinori Moriyama

Index *353*

Contributors

CRISTIAN G. BOLOGA • *Division of Translational Informatics, Department of Internal Medicine, University of New Mexico School of Medicine, Albuquerque, NM, USA*

ISABELLE BROUTIN • *Laboratoire de Cristallographie et RMN Biologiques, UMR 8015, CNRS, Université Paris Descartes, Faculté de Pharmacie de Paris, Paris, France*

PIERPAOLO CACCIOTTO • *Department of Physics, University of Cagliari, Monserrato, CA, Italy*

WAH CHIU • *National Center for Macromolecular Imaging, Verna and Marrs McLean Department of Biochemistry and Molecular Biology, Baylor College of Medicine, Houston, TX, USA*

JARED A. DELMAR • *Department of Physics and Astronomy, Iowa State University, Ames, IA, USA*

ZIQING DENG • *School of Biological Sciences, The University of Hong Kong, Pokfulam, Hong Kong SAR*

DIJUN DU • *Department of Biochemistry, University of Cambridge, Cambridge, UK*

DANNY KA CHUN FUNG • *School of Biological Sciences, The University of Hong Kong, Pokfulam, Hong Kong SAR; Department of Bacteriology, University of Wisconsin-Madison, Madison, WI, USA*

MATTHEW GARCIA • *Center for Molecular Discovery, University of New Mexico School of Medicine, Albuquerque, NM, USA; Department of Pathology, University of New Mexico School of Medicine, Albuquerque, NM, USA*

KARL A. HASSAN • *Department of Chemistry and Biomolecular Sciences, Macquarie University, North Ryde, NSW, Australia*

MARK K. HAYNES • *Center for Molecular Discovery, University of New Mexico School of Medicine, Albuquerque, NM, USA; Department of Pathology, University of New Mexico School of Medicine, Albuquerque, NM, USA*

CHRISTOPHER J. HIPOLITO • *Department of Chemistry, Graduate School of Science, The University of Tokyo, Bunkyo-ku, Tokyo, Japan; Faculty of Medicine, University of Tsukuba, Tsukuba, Japan*

DAVID C. HOOPER • *Division of Infectious Diseases and Medical Services, Massachusetts General Hospital, Harvard Medical School, Boston, MA, USA*

YUNHON HOTTA • *Department of Cell Biology, Graduate School of Medicine, Kyoto University, Kyoto, Japan*

FASAHATH HUSAIN • *Doctor Evidence, LLC, Santa Monica, CA, USA; Department of Molecular and Cell Biology, University of California, Berkeley, CA, USA; Veterans Affairs, Los Angeles, CA, USA*

RYOTA IINO • *Okazaki Institute for Integrative Bioscience, Institute for Molecular Science, National Institutes of Natural Sciences, Okazaki, Aichi, Japan; Department of Functional Molecular Science, School of Physical Sciences, The Graduate University for Advanced Studies (SOKENDAI), Kanagawa, Japan*

SO IWATA • *Department of Cell Biology, Graduate School of Medicine, Kyoto University, Kyoto, Kyoto, Japan; Research Acceleration Program, Membrane Protein Crystallography Project, Japan Science and Technology Agency, Sakyo-ku, Kyoto, Japan; ERATO, Iwata*

Human Receptor Crystallography Project, Japan Science and Technology Agency, Sakyo-ku, Kyoto, Japan; RIKEN, SPring-8 Center, Sayo, Hyogo, Japan

FRANK JAENECKE • *HALOmem, Membrane Protein Biochemistry, Martin-Luther-University Halle-Wittenberg, Halle (Saale), Germany*

TSUKASA KUSAKIZAKO • *Department of Biological Sciences, Graduate School of Science, The University of Tokyo, Bunkyo-ku, Tokyo, Japan*

MAUREEN LENINGER • *Department of Chemistry, New York University, New York, NY, USA*

LIPING LI • *Department of Chemistry and Biomolecular Sciences, Macquarie University, Sydney, NSW, Australia*

KEHONG LIU • *Department of Cell Biology, Graduate School of Medicine, Kyoto University, Kyoto, Kyoto, Japan*

JULIE A. LOVCHIK • *Center for Infectious Disease and Immunity, Department of Internal Medicine, University of New Mexico School of Medicine, Albuquerque, NM, USA*

SHUO LU • *Department of Chemistry and Biochemistry, University of Oklahoma, Norman, OK, USA*

BEN F. LUISI • *Department of Biochemistry, University of Cambridge, Cambridge, UK*

GIULIANO MALLOCI • *Department of Physics, University of Cagliari, Monserrato, CA, Italy*

VIVEKA NAND MALVIYA • *Institute of Biochemistry, Goethe University Frankfurt, Frankfurt, Germany*

YOSHIMI MATSUMOTO • *Department of Biomolecular Science and Regulation, Institute of Scientific and Industrial Research, Osaka University, Osaka, Japan*

YOSHINORI MORIYAMA • *Department of Membrane Biochemistry, Okayama University Graduate School of Medicine, Dentistry and Pharmaceutical Sciences, Okayama University, Okayama, Japan*

KUMAR NAGARATHINAM • *HALOmem, Membrane Protein Biochemistry, Martin-Luther-University Halle-Wittenberg, Halle (Saale), Germany*

YOSHIKO NAKADA-NAKURA • *Department of Cell Biology, Graduate School of Medicine, Kyoto University, Kyoto, Kyoto, Japan; Research Acceleration Program, Membrane Protein Crystallography Project, Japan Science and Technology Agency, Sakyo-ku, Kyoto, Japan*

RYOSUKE NAKASHIMA • *Department of Cell Membrane Biology, Institute of Scientific and Industrial Research, Osaka University, Ibaraki, Osaka, Japan*

HIROSHI NIKAIDO • *Department of Molecular and Cell Biology, University of California, Berkeley, Berkeley, CA, USA*

KUNIHIKO NISHINO • *Institute of Scientific and Industrial Research, Osaka University, Ibaraki, Osaka, Japan; Graduate School of Pharmaceutical Sciences, Osaka University, Suita, Osaka, Japan*

HIROYUKI NOJI • *Department of Applied Chemistry, Graduate School of Engineering, The University of Tokyo, Bunkyo-ku, Tokyo, Japan*

NORIMICHI NOMURA • *Department of Cell Biology, Graduate School of Medicine, Kyoto University, Kyoto, Kyoto, Japan; Research Acceleration Program, Membrane Protein Crystallography Project, Japan Science and Technology Agency, Kyoto, Kyoto, Japan; ERATO, Iwata Human Receptor Crystallography Project, Japan Science and Technology Agency, Sakyo-ku, Kyoto, Japan*

OSAMU NUREKI • *Department of Biological Sciences, Graduate School of Science, The University of Tokyo, Bunkyo-ku, Tokyo, Japan*

SATOSHI OGASAWARA • *Department of Cell Biology, Graduate School of Medicine, Kyoto University, Kyoto, Kyoto, Japan; Department of Chemistry, Graduate School of Science, Chiba University, Chiba, Chiba, Japan*

SHIN-YA OHDATE • *Department of Applied Chemistry, Graduate School of Engineering, The University of Tokyo, Tokyo, Japan*
HIROSHI OMOTE • *Department of Membrane Biochemistry, Okayama University Graduate School of Medicine, Dentistry and Pharmaceutical Sciences, Okayama University, Okayama, Japan*
TUDOR I. OPREA • *Division of Translational Informatics, Department of Internal Medicine, University of New Mexico School of Medicine, Albuquerque, NM, USA*
IAN T. PAULSEN • *Department of Chemistry and Biomolecular Sciences, Macquarie University, Sydney, NSW, Australia*
RYAN PETERS • *Center for Infectious Disease and Immunity, Department of Internal Medicine, University of New Mexico School of Medicine, Albuquerque, NM, USA*
MARTIN PICARD • *Laboratoire de Biologie Physico-chimique des Protéines Membranaires, UMR7099. IBPC, Université Paris Diderot, Paris, France*
CLEMENCIA PINILLA • *Torrey Pines Institute for Molecular Studies, San Diego, CA, USA*
KLAAS M. POS • *Institute of Biochemistry, Goethe University Frankfurt, Frankfurt, Germany*
VENKATA K. RAMASWAMY • *Department of Physics, University of Cagliari, Monserrato, CA, Italy*
PAOLO RUGGERONE • *Department of Physics, University of Cagliari, Monserrato, CA, Italy*
SHOUICHI SAKAKIHARA • *Technical Division, Institute of Scientific and Industrial Research, Osaka University, Osaka, Japan*
KEISUKE SAKURAI • *Department of Cell Membrane Biology, Institute of Scientific and Industrial Research, Osaka University, Ibaraki, Osaka, Japan*
RADLEIGH G. SANTOS • *Torrey Pines Institute for Molecular Studies, Port St Lucie, FL, USA*
LARRY A. SKLAR • *Center for Molecular Discovery, University of New Mexico School of Medicine, Albuquerque, NM, USA; Department of Pathology, University of New Mexico School of Medicine, Albuquerque, NM, USA*
NAOKI SOGA • *Department of Applied Chemistry, Graduate School of Engineering, The University of Tokyo, Bunkyo-ku, Tokyo, Japan*
HIROAKI SUGA • *Department of Chemistry, Graduate School of Science, The University of Tokyo, Bunkyo-ku, Tokyo, Japan*
JINGJING SUN • *School of Biological Sciences, The University of Hong Kong, Pokfulam, Hong Kong SAR*
HENG KEAT TAM • *Institute of Biochemistry, Goethe University Frankfurt, Frankfurt, Germany*
MIKIO TANABE • *HALOmem, Membrane Protein Biochemistry, Martin-Luther-University Halle-Wittenberg, Halle (Saale), Germany; Structural Biology Research Center, Photon Factory, Institute of Materials Structure Science, High Energy Accelerator Research Organization (KEK), Tsukuba, Ibaraki, Japan*
YOSHIKI TANAKA • *Department of Systems Biology, Graduate School of Biological Sciences, Nara Institute of Science and Technology, Nara, Japan*
PIETRO TEDESCO • *Institute of Protein Biochemistry, National Research Council, Naples, Italy; Department of Chemical Sciences and School of Biotechnological Sciences, University of Naples, Naples, Italy*
GEORGE P. TEGOS • *Department of Dermatology, Harvard Medical School, Boston, MA, USA; Wellman Center for Photomedicine, Massachusetts General Hospital, Boston, MA, USA*
SASHA G. TETU • *Department of Chemistry and Biomolecular Sciences, Macquarie University, Sydney, NSW, Australia*

ELENA B. TIKHONOVA • *Department of Cell Biology and Biochemistry, Texas Tech University Health Science Center, Lubbock, TX, USA*
NATHANIEL J. TRAASETH • *Department of Chemistry, New York University, New York, NY, USA*
QUE CHI TRUONG-BOLDUC • *Division of Infectious Diseases and Medical Services, Massachusetts General Hospital, Harvard Medical School, Boston, MA, USA*
OLEG URSU • *Division of Translational Informatics, Department of Internal Medicine, University of New Mexico School of Medicine, Albuquerque, NM, USA*
ATTILIO V. VARGIU • *Department of Physics, University of Cagliari, Monserrato, CA, Italy*
ALICE VERCHÈRE • *Laboratoire de Cristallographie et RMN Biologiques, UMR 8015, CNRS, Université Paris Descartes, Faculté de Pharmacie de Paris, Paris, France*
ANNA WALLER • *Center for Molecular Discovery, University of New Mexico School of Medicine, Albuquerque, NM, USA; Department of Pathology, University of New Mexico School of Medicine, Albuquerque, NM, USA*
ZHAO WANG • *National Center for Macromolecular Imaging, Verna and Marrs McLean Department of Biochemistry and Molecular Biology, Baylor College of Medicine, Houston, TX, USA*
RIKIYA WATANABE • *Department of Applied Chemistry, Graduate School of Engineering, The University of Tokyo, Bunkyo-ku, Tokyo, Japan; PRESTO, JST, Tokyo, Japan*
TERRY H. WU • *Center for Infectious Disease and Immunity, Department of Internal Medicine, University of New Mexico School of Medicine, Albuquerque, NM, USA*
AKIHITO YAMAGUCHI • *Department of Cell Membrane Biology, Institute of Scientific and Industrial Research, Osaka University, Ibaraki, Osaka, Japan*
AIXIN YAN • *School of Biological Sciences, The University of Hong Kong, Pokfulam, Hong Kong SAR; Institute of Scientific and Industrial Research, Osaka University, Ibaraki, Osaka, Japan*
EDWARD W. YU • *Department of Physics and Astronomy, Iowa State University, Ames, IA, USA; Department of Chemistry, Iowa State University, Ames, IA, USA*
HELEN I. ZGURSKAYA • *Department of Chemistry and Biochemistry, University of Oklahoma, Norman, OK, USA*

Part I

Structural Analysis of Bacterial Multidrug Exporters

Chapter 1

High-Resolution Crystallographic Analysis of AcrB Using Designed Ankyrin Repeat Proteins (DARPins)

Heng Keat Tam, Viveka Nand Malviya, and Klaas M. Pos

Abstract

X-ray crystallography is still the most prominent technique in use to decipher the 3D structures of membrane proteins. For successful crystallization, sample quality is the most important parameter that should be addressed. In almost every case, highly pure, monodisperse, and stable protein sample is a prerequisite. Vapor diffusion is in general the method of choice for obtaining crystals. Here, we discuss a detailed protocol for overproduction and purification of the inner-membrane multidrug transporter AcrB and of DARPins, which are used for crystallization of the AcrB/DARPin complex, resulting in high-resolution diffraction and subsequent structure determination.

Key words AcrB, Resistance Nodulation cell Division, Multidrug resistance, Antibiotic resistance, DARPin, Protein crystallography

1 Introduction

AcrB is the inner membrane component of the Resistance Nodulation cell Division superfamily and part of a large three-component multidrug efflux transporter with homologues in all the three domains of life [1, 2]. In Gram-negative bacteria, the three-component efflux pump AcrAB-TolC provides resistance against various antibiotics [3–5]. The first X-ray structure of AcrB was solved in the year 2002 at 3.5 Å resolution, and displayed this homotrimeric membrane protein in asymmetric state [6]. This membrane protein shows a fascinating architecture with almost 60% of its amino acid residues localized in the periplasm. This periplasmic part and in particular a region designated as porter domain, is central to drug binding and release [6–10]. In the year 2006 and 2007, other X-ray structures (resolutions up to 2.8 Å) of AcrB were obtained, where the three subunits were shown to be present in three different conformations; loose (L), tight (T) and open (O) [7–9], and the conformational differences are particularly pronounced in the porter domain [10]. The trimer has been

Akihito Yamaguchi and Kunihiko Nishino (eds.), *Bacterial Multidrug Exporters: Methods and Protocols*, Methods in Molecular Biology, vol. 1700, https://doi.org/10.1007/978-1-4939-7454-2_1, © Springer Science+Business Media, LLC 2018

suggested to transport drugs at the expense of the proton motive force according to a functional rotating, peristaltic pump mechanism [7–9, 11–13]. The structure and dynamics of this protein have been described extensively by several groups and have been summarized recently [14].

This chapter describes more practical aspects of crystallization of this protein, resulting in high-resolution diffraction datasets and structure determination. For membrane proteins, the diffraction resolution often prohibits detailed analysis and pursuit of function on the molecular level. The first membrane protein for which a crystallization chaperone (Fv antibody fragment) has been described was the cytochrome *c* oxidase from *Paracoccus denitrificans* [15]. Nowadays, the use of crystallization chaperones is, in addition to protein engineering and in meso crystallization, the method of choice to tackle the crystallization of difficult-to-crystallize targets such as G-protein-coupled receptors [16, 17]. Sennhauser et al. [9] described the use of AcrB as membrane protein of choice to select for specific binders from a library of designed ankyrin repeat proteins (DARPins). The enormous improvement of diffraction quality of AcrB/DARPins co-complexes (up to 1.9 Å) led to unprecedented insights into drug binding [18] and the drug/H^+-coupling mechanism [10]. The protein supposed to function with rotating mechanism or peristaltic pump mechanism [7–13].

DARPins are derived ankyrin repeat proteins, which have been raised against a number of target proteins using in vitro selection methods such as ribosome display and phage display and hence are dubbed "Designed Ankyrin Repeat Proteins" [19–21]. Selected DARPins often display high specificity (sometimes conformational selective) and affinity for their binding partners, and have been used for cocrystallization [22, 23]. AcrB has been used as a proof-of-principle for membrane proteins to be amenable for DARPin selection and successful co-crystallization, leading to crystals with improved diffraction quality compared to the apo-AcrB [9, 10, 18]. The AcrB/DARPin co-crystal structure displayed two identical DARPin molecules bound specifically to the periplasmic part of the AcrB trimer at either the L or T conformation, but not to the O conformer [9, 10, 23, 24]. Recently, a symmetric AcrZ/AcrB co-crystal structure (all protomers in the L state) in complex with DARPins has been reported [25].

This protocol describes the purification of AcrB from *E. coli* and DARPins (specifically the 1108_19 variant [9]) by affinity and size-exclusion chromatography. Subsequently, these protein complexes were used for crystallization and the crystals were subjected to synchrotron X-rays resulting in high-resolution diffraction data collection and AcrB/DARPin structure elucidation.

2 Materials

Prepare all solutions using ultrapure water (milliQ grade) and molecular biology grade reagents. Prepare and store all reagents at 4 °C (unless indicated otherwise).

2.1 DARPin Clone 1108_19 Production

1. *Escherichia coli* XL1 Blue competent cells (50 μL aliquots) in 1.5 mL reaction tube. Liquid nitrogen (flash) frozen and store at −80 °C.
2. An expression plasmid pQE (Qiagen GmbH, Hilden, Germany) harbouring the DARPin 1108_19 gene [9], designated pQE-DARPin. Store at −20 °C.
3. Sterile-filtered ampicillin stock solution (100 mg/mL) in water. Store at −20 °C.
4. Sterile-filtered isopropyl-β-D-thiogalactopyranoside (IPTG) stock solution (1 M) in water. Store at −20 °C.
5. 200 mL LB medium for *E. coli* pre-culture for DARPin production: Weigh 2 g tryptone, 1 g yeast extract and 2 g NaCl, in a 250 mL graduated cylinder containing 150 mL water while stirring using a stir bar on a magnetic stirrer. Mix until dissolved. Add water to a volume of 200 mL and pour LB medium in a 500 mL baffled flask. Sterilize by autoclaving at 121 °C for 20 min. Cool down the medium until hand-warm and add 200 μL ampicillin (100 mg/mL stock, final concentration 100 μg/mL) into 200 mL LB medium under sterile conditions.
6. 50% glucose solution: Weigh 10 g glucose and dissolve it into 17 mL of water. Add water to 20 mL in a graduated cylinder and pass through a 0.2 μm sterile syringe filter into a pre-autoclaved Schott bottle.
7. 1 L LB liquid medium + 1% glucose for DARPin production: Weigh 10 g tryptone, 5 g yeast extract and 10 g NaCl, in a 1 L graduated cylinder. Add water to a volume of 800 mL and mix until dissolved. Adjust the final volume to 1000 mL with water before transferring the LB liquid medium in a 5 L baffled flask. Sterilize by autoclaving at 121 °C for 20 min. Cool down the medium until hand-warm. Before inoculation with *E. coli* XL1 Blue cells harboring pQE-DARPin, add 20 mL of sterile 50% glucose solution (**step 6**) and 1 mL of ampicillin (100 mg/mL) to the LB medium.
8. LB agar plate supplemented with 100 μg/mL ampicillin: Prepare the LB agar medium: Weigh 2 g tryptone, 1 g yeast extract, 2 g NaCl and 3 g agar in a 250 mL graduated cylinder. Add water to a volume of 200 mL and mix until dissolved. Pour the LB agar medium in a 250 mL Schott bottle. Sterilize by

autoclaving at 121 °C for 20 min. Cool the LB agar medium to 55 °C. Add 200 μL of 100 mg/mL ampicillin into the medium and mix. Pour the LB agar medium with ampicillin into sterile petri dishes (92 × 16 mm, 20 mL each dish) and leave until solidified. Store the LB agar plates at 4 °C until further use.

9. Ice bath (Polystyrene box, 25 cm × 25 cm × 15 cm).
10. Thermomixer Pro 35 wells (CellMedia GmbH & Co. KG, Gutenborn, Germany).
11. Drigalski spatula.
12. 37 °C incubator.
13. Inoculating loop, 1 μL.
14. Orbital shaker.
15. Centrifuge (Sorvall® Evolution™RC, Thermo Fisher Scientific, Waltham, MA, USA) with Fiberlite F8-6x1000y rotor (Thermo Fisher Scientific, Waltham, MA, USA).
16. Liquid nitrogen (*see* **Note 1**).
17. 0.2 μm membrane filter (cellulose acetate filter, 0.2 μm pore size, 47 mm diameter).
18. Petri dishes (92 × 16 mm, polystyrene).
19. 0.2 μm sterile syringe filter (cellulose acetate membrane, 0.2 μm pore size).

2.2 DARPin Purification

1. DARPin Buffer A: Mix 50 mL of 1 M Tris–HCl, pH 7.5, and 80 mL of 5 M NaCl in a 1 L graduated cylinder. Adjust to 1 L with water and mix. Filter the buffer with 0.2 μm membrane filter and degas with vacuum pump for 10 min. Store at 4 °C.
2. DARPin Buffer B: Mix 50 mL of 1 M Tris–HCl, pH 7.5, 80 mL of 5 M NaCl, 100 mL of 100% glycerol and 20 mL of 1 M imidazole in a 1 L graduated cylinder. Adjust to 1 L with water and mix. Filter the buffer with 0.2 μm membrane filter and degas with vacuum pump for 10 min. Store at 4 °C.
3. DARPin Buffer C: Mix 50 mL of 1 M Tris–HCl, pH 7.5, 80 mL of 5 M NaCl, 100 mL of 100% glycerol and 250 mL of 1 M imidazole in a 1 L graduated cylinder. Adjust to 1 L with water and mix. Filter the buffer with 0.2 μm membrane filter and degas using a vacuum pump for 10 min. Store at 4 °C.
4. DNase I.
5. 0.2 M Phenylmethanesulfonyl fluoride (PMSF): Weigh 35 mg of PMSF and dissolve it in 1 mL of 100% ethanol. Do not store PMSF solution, always use freshly prepared one.
6. Water. Filter water with 0.2 μm membrane filter and degas it with vacuum pump for 10 min. Store at 4 °C.
7. Magnetic stirrer.

8. Cell disruptor (Constant System Inc., Northants, United Kingdom).
9. Ultracentrifuge (Beckman Optima L-70, Beckman Coulter Inc. Brea, CA, USA) with Type 45 Ti fixed angle rotor (Beckman Coulter Inc. Brea, CA, USA).
10. 0.2 μm membrane filter (cellulose acetate filter, 0.2 μm pore size, 47 mm diameter).
11. 150 mL superloop (GE Healthcare Europe GmbH, Freiburg, Germany).
12. 5 mL HisTrap HP column (GE Healthcare Europe GmbH, Freiburg, Germany).
13. Äkta Prime system (GE Healthcare Europe GmbH, Freiburg, Germany).
14. Vacuum pump.
15. 1.5 mL reaction tubes.
16. Pierce™ BCA Protein Assay Kit (Thermo Fisher Scientific, Rockford, IL, USA).
17. 0.22 μm filter unit (Hydrophilic Polyvinylidene Fluoride (PVDF) filter, 0.22 μm pore size, 4 mm diameter, ethylene oxide sterilized).

2.3 AcrB Protein Production

1. *E. coli* C43 (DE3) *ΔacrAB* competent cells (50 μL aliquots) in 1.5 mL reaction tube. Liquid nitrogen (flash) frozen and store at −80 °C.
2. An expression plasmid pET24a(+) (Novagen, EMD Millipore, Darmstadt, Germany) borne *acrB* gene, pET24-$acrB_{\mathrm{His}}$.
3. Kanamycin stock solution (50 mg/mL) in water. Stored at −20 °C.
4. Isopropyl-β-D-thiogalactopyranoside (IPTG) stock solution (1 M) in water. Stored at −20 °C.
5. 100% glycerol: Sterilize by autoclaving at 121 °C for 20 min. Store at room temperature.
6. 1 M Potassium phosphate solution: Weigh 4.62 g KH_2PO_4 and 32.9 g $K_2HPO_4 \cdot 3H_2O$, in a 250 mL graduated cylinder. Add water to a volume of 150 mL and mix until dissolved. Add water to 200 mL. Sterilize by autoclaving at 121 °C for 20 min. Store at room temperature.
7. 200 mL LB liquid medium for *E. coli* pre-culture for AcrB production: Weigh 2 g tryptone, 1 g yeast extract and 2 g NaCl in a 250 mL graduated cylinder containing 150 mL water while stirring using a stir bar on a magnetic stirrer. Mix until dissolved. Add water to a volume of 200 mL and pour the LB liquid medium in a 500 mL baffled flask. Sterilize by

autoclaving at 121 °C for 20 min. Cool down the medium until hand-warm and add 200 μL kanamycin (50 mg/mL, final concentration 50 μg/mL) into 200 mL LB medium under sterile condition.

8. 1 L TB medium for AcrB production: Prepare Terrific broth by weighing 12 g tryptone and 24 g yeast extract in a 1 L graduated cylinder containing 800 mL water while stirring using a stir bar on a magnetic stirrer. Mix until dissolved. Add water to a volume of 900 mL and pour the Terrific broth in 5 L baffled flask. Sterilize by autoclaving at 121 °C for 20 min. Before inoculation with *E. coli* C43 (DE3) *ΔacrAB* harboring pET24-$acrB_{His}$, add 100 mL sterile 1 M potassium phosphate solution, 4 mL sterile 100% glycerol and 1 mL kanamycin (50 mg/mL, final concentration 50 μg/mL) into 900 mL Terrific broth.
9. LB agar plate supplemented with 50 μg/mL kanamycin: Prepare the LB agar medium: Weigh 2 g tryptone, 1 g yeast extract, 2 g NaCl and 3 g agar in a 250 mL graduated cylinder. Add water to a volume of 200 mL and mix until dissolved. Pour the LB agar medium in a 250 mL Schott bottle. Sterilize by autoclaving at 121 °C for 20 min. Cool the LB agar medium to 55 °C. Add 200 μL kanamycin (50 mg/mL, final concentration 50 μg/mL) into the medium and mix it. Pour the LB agar medium with kanamycin into sterile petri dishes (92 × 16 mm, 20 mL each dish) and leave until solidified. Store the LB agar plates at 4 °C until further use.
10. Ice bath (Polystyrene box, 25 cm × 25 cm × 15 cm).
11. Thermomixer Pro 35 wells (CellMedia GmbH & Co. KG, Gutenborn, Germany).
12. Glass spreader.
13. 37 °C incubator.
14. Inoculating loop, 1 μL.
15. Orbital shaker.
16. Centrifuge (Sorvall® Evolution™RC, Thermo Fisher Scientific, Waltham, MA, USA) with Fiberlite F8-6x1000y rotor (Thermo Fisher Scientific, Waltham, MA, USA).
17. Liquid nitrogen (*see* **Note 1**).
18. Petri dishes (92 × 16 mm, polystyrene).

2.4 Suspension, Lysis, and Membrane Preparation of AcrB Produced E. coli C43 (DE3) ΔacrAB Cells

1. AcrB Cell Suspension Buffer: Mix 20 mL of 1 M Tris–HCl, pH 8.0, 100 mL of 5 M NaCl and 4 mL of 0.5 M $MgCl_2 \cdot 6H_2O$ in a 1 L graduated cylinder. Adjust to 1 L with water and mix. Filter the buffer with 0.2 μm membrane filter. Store at 4 °C.

2. AcrB Membrane Suspension Buffer: Mix 2 mL of 1 M Tris–HCl, pH 8.0 and 10 mL of 5 M NaCl in a 100 mL graduated cylinder. Adjust to 1 L with water and mix. Filter the buffer with 0.2 μm membrane filter. Store at 4 °C.
3. DNase I.
4. 0.2 M Phenylmethanesulfonyl fluoride (PMSF).
5. Magnetic stirrer.
6. Cell disruptor (Constant System Inc., Northants, United Kingdom).
7. Centrifuge (Sorvall® Evolution™RC, Thermo Fisher Scientific, Waltham, MA, USA) with Fiberlite F8-6x1000y rotor (Thermo Fisher Scientific, Waltham, MA, USA).
8. Ultracentrifuge (Beckman Optima L-70, Beckman Coulter Inc. Brea, CA, USA) with Type 45 Ti fixed angle rotor (Beckman Coulter Inc. Brea, CA, USA).
9. 0.2 μm membrane filter (cellulose acetate filter, 0.2 μm pore size, 47 mm diameter).

2.5 Membrane Solubilization and Purification of AcrB Protein by Ni^{2+}-NTA Affinity Chromatography

1. 20% n-dodecyl-β-D-maltopyranoside (DDM, GLYCON Biochemicals, Luckenwalde, Germany): Weigh 2 g of DDM in a 10 mL graduated cylinder. Add water up to a volume of 8 mL and mix until dissolved, adjust to 10 mL and store at −20 °C.
2. AcrB Buffer A: Mix 5 mL of 1 M Tris–HCl, pH 7.5, 7.5 mL of 5 M NaCl, 25 mL of 100% glycerol and 5 mL of 1 M imidazole in a 250 mL graduated cylinder. Adjust to 250 mL with water and mix. Filter the buffer with 0.2 μm membrane filter and degas with vacuum pump for 10 min. Store at 4 °C. Before protein purification, add 250 μL of 20% DDM and mix.
3. AcrB Wash Buffer I: Mix 2 mL of 1 M Tris–HCl, pH 7.5, 3 mL of 5 M NaCl, 10 mL of 100% glycerol and 8 mL of 1 M imidazole in a 100 mL graduated cylinder. Adjust to 100 mL with water and mix. Filter the buffer with 0.2 μm membrane filter and degas with vacuum pump for 10 min. Store at 4 °C.
4. AcrB Wash Buffer II: Mix 2 mL of 1 M Tris–HCl, pH 7.5, 3 mL of 5 M NaCl, 10 mL of 100% glycerol and 11 mL of 1 M imidazole in a 100 mL graduated cylinder. Adjust to 100 mL with water and mix. Filter the buffer with 0.2 μm membrane filter and degas with vacuum pump for 10 min. Store at 4 °C. Before protein purification, add 100 μL of 20% DDM and mix.
5. AcrB Elution Buffer: Mix 2 mL of 1 M Tris–HCl, pH 7.5, 3 mL of 5 M NaCl, 10 mL of 100% glycerol and 22 mL of 1 M imidazole in a 100 mL graduated cylinder. Adjust to 100 mL with water and mix. Filter the buffer with 0.2 μm membrane

filter and degas with vacuum pump for 10 min. Store at 4 °C. Before protein purification, add 100 μL of 20% DDM and mix.

6. Ultracentrifuge (Beckman Optima L-70, Beckman Coulter Inc. Brea, CA, USA) with Type 45 Ti fixed angle rotor (Beckman Coulter Inc. Brea, CA, USA).
7. 0.2 μm membrane filter (cellulose acetate filter, 0.2 μm pore size, 47 mm diameter).
8. DDM (GLYCON Biochemicals, Luckenwalde, Germany).
9. 50 mL superloop (GE Healthcare Europe GmbH, Freiburg, Germany).
10. 2 × 1 mL HisTrap HP column (GE Healthcare Europe GmbH, Freiburg, Germany).
11. Äkta Prime system (GE Healthcare Europe GmbH, Freiburg, Germany).
12. Vacuum pump.
13. 0.22 μm filter unit (Hydrophilic Polyvinylidene Fluoride (PVDF) filter, 0.22 μm pore size, 4 mm diameter, ethylene oxide sterilized).

2.6 Purification of AcrB/DARPin Complex by Gel Filtration Chromatography

1. 20% n-dodecyl-β-D-maltopyranoside (DDM, GLYCON Biochemicals, Luckenwalde, Germany): Weigh 2 g of DDM in a 10 mL graduated cylinder. Add water up to a volume of 8 mL and mix until dissolved, adjust to 10 mL with water and store at −20 °C.
2. AcrB GF Buffer: Mix 3 mL of 1 M Tris–HCl, pH 7.5, 4.5 mL of 5 M NaCl in a 150 mL graduated cylinder. Adjust to 150 mL with water and mix. Filter the buffer with 0.2 μm membrane filter and degas with vacuum pump for 10 min. Store at 4 °C. Before protein purification, add 225 μL of 20% DDM and mix.
3. Amicon Ultra-15 Centrifugal Filter Ultracel® 100 K (Merck Millipore, Cork, Ireland).
4. 1 mL syringe.
5. 0.22 μm filter unit (Hydrophilic Polyvinylidene Fluoride (PVDF) filter, 0.22 μm pore size, 4 mm diameter, ethylene oxide sterilized).
6. Superose 6 10/300 GL column (GE Healthcare Europe GmbH, Freiburg, Germany).
7. Äkta Purifier system (GE Healthcare Europe GmbH, Freiburg, Germany).
8. Pierce™ BCA Protein Assay Kit (Thermo Fisher Scientific, Rockford, IL, USA).
9. Bench-top centrifuge (Heraeus Megafuge 1.0R, Thermo Electron Corporation, Osterode, Germany).

2.7 Crystallization of AcrB/DARPin Complex by Hanging-Drop Vapor Diffusion Method

1. 1 M N-(2-acetamido)-iminodiacetic acid (ADA), pH 6.6: Weigh 9.5 g ADA in a 50 mL graduated cylinder. Add water to a volume of 40 mL. Mix using a magnetic stirrer until dissolved and adjust the pH with 10 M NaOH. Adjust to 50 mL with water. Filter the solution with 0.2 μm sterile syringe filter. Store at room temperature.
2. 50% glycerol: Filter the solution with 0.2 μm sterile syringe filter. Store at room temperature.
3. 50% Poly(ethylene glycol) average molecular weight 4000 (PEG4000): Weigh 15 g PEG4000 in a 50 mL graduated cylinder. Add water to a volume of 20 mL and mix until dissolved, adjust to 30 mL with water. Filter the solution with 0.2 μm sterile syringe filter. Store at room temperature (*see* **Note 2**).
4. 2 M Ammonium sulfate: Weigh 26.43 g $(NH_4)_2SO_4$ in a 100 mL graduated cylinder. Add water to a volume of 80 mL and mix until dissolved, adjust to 100 mL with water. Filter the solution with 0.2 μm membrane filter. Store at room temperature.
5. MilliQ water: Filter water with 0.2 μm sterile syringe filter. Store at room temperature.
6. Linbro 24-well plates (Hampton Research, CA, USA): Place some silicon grease over the rim of the well before sealing with siliconized circular cover slides (*see* **Note 3**).
7. Siliconized circular cover slides with 22 mm in diameter (Hampton Research, CA, USA).
8. Light microscope (Zoom stereomicroscope SMZ1500) (Nikon Corporation, Kanagawa, Japan).
9. 0.2 μm sterile syringe filter (cellulose acetate membrane, 0.2 μm pore size).
10. 1.5 mL reaction tubes.

2.8 AcrB/DARPin Complex Crystal Harvesting

1. 1 M N-(2-acetamido)-iminodiacetic acid (ADA), pH 6.6: Weigh 9.5 g ADA in a 50 mL graduated cylinder. Add water to a volume of 40 mL. Mix using a magnetic stirrer until dissolved and adjust the pH with 10 M NaOH. Adjust to 50 mL with water. Filter the solution with 0.2 μm sterile syringe filter. Store at room temperature.
2. 70% glycerol: Filter the solution with 0.2 μm sterile syringe filter. Store at room temperature.
3. 50% Poly(ethylene glycol) average molecular weight 4000 (PEG4000): Weigh 15 g PEG4000 in a 50 mL graduated cylinder. Add water to a volume of 20 mL and mix until dissolved, adjust to 30 mL with water. Filter the solution with

0.2 μm sterile syringe filter. Store at room temperature (*see* **Note 2**).

4. 2 M Ammonium sulfate: Weigh 26.43 g $(NH_4)_2SO_4$ in a 100 mL graduated cylinder. Add water to a volume of 80 mL and mix until dissolved, adjust to 100 mL with water. Filter the solution with 0.2 μm sterile syringe filter. Store at room temperature.
5. MilliQ water: Filter the water with 0.2 μm sterile syringe filter.
6. 3% n-dodecyl-β-D-maltopyranoside (DDM, GLYCON Biochemicals, Luckenwalde, Germany): Store at −20 °C.
7. 5% Cryo Solution: Mix 2.5 μL of 1 M ADA, pH 6.6, 3.6 μL of 70% glycerol, 12.0 μL of 50% PEG4000, 4.5 μL of 2 M ammonium sulfate, 0.5 μL of 3% DDM and 26.4 μL of MilliQ. Mix the solution and store at room temperature.
8. 15% Cryo Solution: Mix 2.5 μL of 1 M ADA, pH 6.6, 10.7 μL of 70% glycerol, 12.0 μL of 50% PEG4000, 4.5 μL of 2 M ammonium sulfate, 0.5 μL of 3% DDM and 19.3 μL of MilliQ. Mix the solution and store at room temperature.
9. 28% Cryo Solution: Mix 2.5 μL of 1 M ADA, pH 6.6, 20.0 μL of 70% glycerol, 12.0 μL of 50% PEG4000, 4.5 μL of 2 M ammonium sulfate, 0.5 μL of 3% DDM and 10.0 μL of MilliQ. Mix the solution and store at room temperature.
10. Magnetic cryo loops of diameter 0.1–0.2 mm or 0.2–0.3 mm and vials (Molecular Dimensions Limited, Suffolk, UK) (*see* **Note 4**) (Fig. 1).
11. Siliconized circular cover slides with 22 mm in diameter (Hampton Research, CA, USA).
12. Magnetic cryo wand (Molecular Dimensions Limited, Suffolk, UK) (Fig. 1).
13. Vial tongs (Molecular Dimensions Limited, Suffolk, UK) (Fig. 1).
14. Liquid nitrogen foam dewar (Molecular Dimensions Limited, Suffolk, UK).
15. Cryo puck (Molecular Dimensions Limited, Suffolk, UK) (Fig. 1).
16. Liquid nitrogen storage dewar (Molecular Dimensions Limited, Suffolk, UK).
17. 0.2 μm sterile syringe filter (cellulose acetate membrane, 0.2 μm pore size).
18. Liquid nitrogen (*see* **Note 1**).

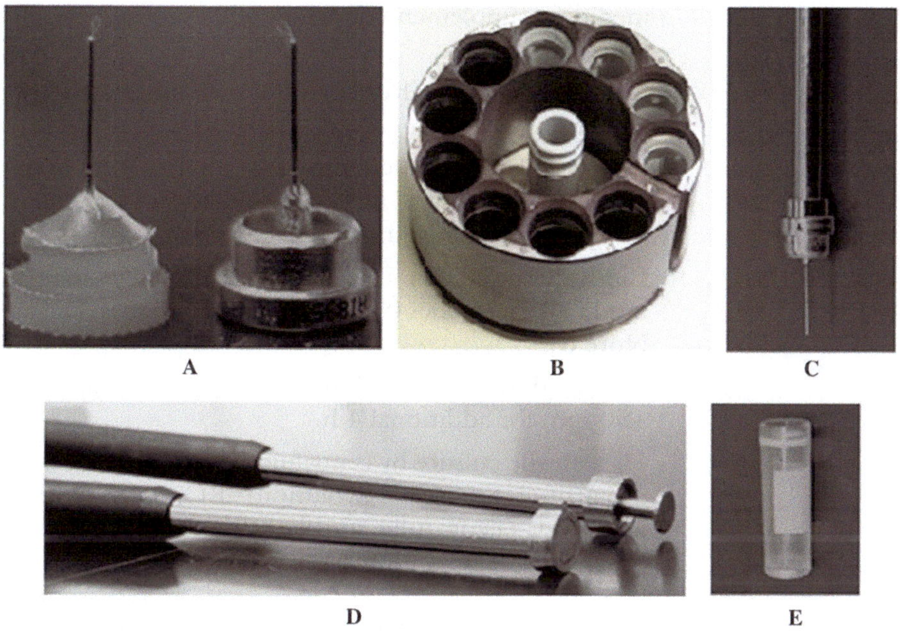

Fig. 1 Tools required for crystal harvesting. (**a**) Cryo loops. (**b**) Cryo puck: first four wells contain empty cryo vials (*see* **e**) in which the cryo loops with crystals will be placed with the help of cryo wand. (**c**) Cryo wand (rod from the top of the image) together with cap base and cryo loop (capped thin rod with non-visible nylon loop pointing down in the image) after harvesting crystal. (**d**) Empty cryo wands. Cryo wands contain an ejection pin (upper wand) to release the magnetic base cap into the cryo vial (*see* **b**). (**e**) Empty cryo vial

3 Methods

3.1 Overproduction of DARPin Clone 1108_19

1. All steps should be performed under strict sterile conditions.
2. Add 0.5 μL of expression plasmid pQE-DARPin (concentration of 100 ng/μL) into 50 μL of *E. coli* XL1 Blue competent cells in a reaction tube. Incubate the cells on ice for 20 min.
3. Heat-shock the competent cells by placing the reaction tube at 42 °C for 40 s. Transfer the reaction tube on ice and incubate for 5 min.
4. Add 200 μL of LB liquid medium into the reaction tube and cultivate the cells at 37 °C for 40 min with 750 rpm on a thermomixer.
5. Transfer the cells on LB agar plate supplemented with 100 μg/mL ampicillin and spread with a glass spreader. Incubate the agar plate at 37 °C overnight in a 37 °C incubator. Take out the plates in the morning and store them at 4 °C until further use.
6. Prepare pre-culture by picking single colony of *E. coli* XL1 Blue cells harboring pQE-DARPin from the LB agar plate with a sterile inoculating loop and inoculate into 200 mL LB liquid

medium supplemented with 100 μg/mL ampicillin (*see* **Note 5**).

7. Cultivate the pre-culture at 37 °C overnight with 180 rpm in a shaker.
8. Inoculate 10 mL pre-culture of *E. coli* XL1 Blue cells harboring pQE-DARPin into 1 L LB medium supplemented with 100 μg/mL ampicillin and 1% glucose.
9. Cultivate the culture at 37 °C with 150 rpm in an orbital shaker until OD_{600} reaches 0.7. Measure OD_{600} after 2, 3 and 4 h (*see* **Note 6**).
10. Add 1 mL of 1 M IPTG into the culture and grow at 37 °C at 150 rpm for additional 4 h.
11. Harvest the culture by centrifugation at 5000 x *g* for 15 min in a centrifuge (Sorvall® Evolution™RC). Collect the cell pellet in 50 mL falcon tube and flash freeze using liquid nitrogen. Store at −80 °C until further use.

3.2 Cell Lysis and Purification of DARPin Clone 1108_19

1. Suspend 1 g of cells with 4 mL of ice-cold DARPin Buffer A at 4 °C by stirring with a stirrer (*see* **Note 7**).
2. Add 1–2 mg of DNase I and 0.2 mM PMSF in final concentration to the cell suspension before cell lysis.
3. Set up the cell disruptor (Constant System Inc.) according to the manufacturer's protocol. Set the pressure to 22 kPsi and wash the cell disruptor with 200 mL water and 400 mL DARPin Buffer A. Lyse the cells by passing the cell suspension through the cell disruptor two times.
4. Pour the cell lysate into an ultracentrifuge tube and centrifuge at 137,000 × *g* for 1 h using Beckman Optima L-70 ultracentrifuge. Collect the supernatant and filter with 0.22 μm filter unit.
5. Pour the supernatant into a 150 mL superloop (GE Healthcare) (*see* **Note 8**).
6. Set up the Äkta Prime System and assemble the 5 mL HisTrap HP affinity column on it (*see* **Note 9**).
7. Wash the 5 mL HisTrap HP affinity column with 5 column volume of water with a flow rate of 2.5 mL/min and equilibrate it with 5 column volume of DARPin Buffer A with a flow rate of 2.5 mL/min.
8. Inject the supernatant on the HisTrap HP affinity column with a flow rate of 2.5 mL/min. Wash the column with 30 column volume of DARPin Buffer B with a flow rate of 2.5 mL/min after the injection of sample is finished.
9. Wash the column containing bound DARPin protein with 15 column volume of Buffer C with a flow rate of 2.5 mL/min.

10. Elute the DARPin protein with 10 column volume of DARPin Buffer D with a flow rate of 2.5 mL/min. Pool all the fractions containing DARPin protein.
11. Determine DARPin protein concentration by BCA method (Pierce™ BCA Protein Assay Kit) (*see* **Note 10**).
12. Aliquot 1 mL DARPin protein into 1.5 mL reaction tubes and flash freeze all the protein aliquots with liquid nitrogen. Store them at −80 °C until further use.

3.3 Overproduction of AcrB Protein

1. Carry out all the procedure under strict sterile conditions.
2. Add 0.5 μL of 100 ng/μL of expression plasmid pET24-$acrB_{\mathrm{His}}$ into 50 μL of *E. coli* C43(DE3) *ΔacrAB* competent cells. Incubate the cells on ice for 20 min.
3. Heat-shock the competent cell by placing the reaction tube at 42 °C for 40 s. Transfer the tube on ice and incubate for 5 min.
4. Add 200 μL LB liquid medium into the cell and cultivate the cell at 37 °C for 40 min with 750 rpm on a thermomixer.
5. Transfer the cells on LB agar plate supplemented with 50 μg/mL kanamycin and spread with a glass spreader. Incubate the plate at 37 °C for overnight in a 37 °C incubator. Take out the plate next morning and store them at 4 °C until further use.
6. Prepare pre-culture by picking one colony of *E. coli* C43 (DE3) *ΔacrAB* harboring pET24-$acrB_{\mathrm{His}}$ from the LB agar plate with a sterile inoculating loop and inoculate the colony into 200 mL LB liquid medium supplemented with 50 μg/mL kanamycin.
7. Cultivate the pre-culture at 37 °C overnight at 180 rpm in a shaker.
8. Inoculate 4 mL of pre-culture into 1 L TB liquid medium supplemented with 50 μg/mL kanamycin.
9. Cultivate the culture at 37 °C with 150 rpm in an orbital shaker until OD_{600} reaches 0.8. Measure the OD_{600} of the culture after 2 and 3 h. Cool the culture on ice for 30 min when the OD_{600} reaches 0.8. Cool the orbital shaker to 20 °C.
10. Add 1 mL of 1 M IPTG into the culture and grow at 20 °C at 130 rpm for additional 16 h.
11. Harvest the culture by centrifugation at 5000 x *g* for 15 min in a centrifuge (Sorvall® Evolution™RC) and collect the pellet (*see* **Note 11**).

3.4 Cell Lysis and Membrane Preparation of E. coli C43(DE3) Δ acrAB Overproducing AcrB Protein

1. Suspend 1 g of cells with 2 mL of ice-cold AcrB Cell Suspension Buffer at 4 °C by stirring with a stirrer (*see* **Note 7**).
2. Add 1–2 mg of DNase I and 0.2 mM PMSF in final concentration to the cell suspension before cell lysis.
3. Set up the cell disruptor according to the manufacturer's protocol. Set the pressure to 22 kPsi and wash it with 200 mL

water and 400 mL AcrB Cell Suspension Buffer. Lyse the cells by passing the cell suspension through the disruptor two times.

4. Pour the cell lysate into a centrifuge tube. Centrifuge it at 25,000 × *g* for 20 min with Sorvall® Evolution™RC centrifuge. Collect and pour the supernatant carefully into a beaker without disturbing the cell debris pellet.
5. Weigh each of the ultracentrifuge tubes (without the lid) and note down their weights. Pour the supernatant into the ultracentrifuge tube and centrifuge the supernatant at 137,000 × *g* for 2 h with Beckman Optima L-70 ultracentrifuge. Weigh the tubes together with the membrane pellet and subtract the weight of empty tubes. Estimate the amount of membrane prepared. Discard the supernatant carefully without disturbing the membrane pellet.
6. Suspend 1 g of membrane with 4 mL of AcrB Membrane Suspension Buffer. Initially, add 10 mL of AcrB membrane suspension buffer into the ultracentrifuge tube containing membrane pellet. Suspend by a 10 mL glass or plastic pipette with an automatic dispensing pipette. Add additional 5–10 mL fresh membrane suspension buffer if the suspension is too concentrated (*see* **Note 12**).
7. Aliquot 10.8 mL membrane suspension into 15 mL plastic falcon tube. Flash freeze the membrane suspension with liquid nitrogen and store them at −80 °C until further use.

3.5 Membrane Solubilization and Purification of AcrB Protein by Ni^{2+}-Affinity Chromatography

1. Carry out all the procedure at 4 °C unless otherwise specified.
2. Thaw the membrane suspension and 20% DDM stock solution in a water bath at room temperature.
3. Pour 10.8 mL membrane suspension into a fresh 50 mL beaker. Put a magnetic stirrer into the beaker. Mix 17.7 mL Buffer A and 246 μL of 1 M imidazole solution into the membrane suspension. Add 1.5 mL of 20% DDM into the membrane suspension. Stir the suspension for 1 h (*see* **Note 13**).
4. Pour the membrane suspension into an ultracentrifuge tube. Centrifuge at 137,000 × *g* for 30 min using Beckman Optima L-70 ultracentrifuge.
5. Filter the supernatant containing AcrB protein with 0.22 μm filter unit. Pour the filtrate into a 50 mL superloop (*see* **Note 8**).
6. Set-up the Äkta Prime System and assemble the 1 mL HisTrap HP affinity column on it (*see* **Note 9**).
7. Wash the 1 mL HisTrap HP affinity column with 5 column volume of degassed water with a flow rate of 1.0 mL/min and

equilibrate it with 5 column volume of Buffer A with a flow rate of 1.0 mL/min.

8. Inject the membrane protein solution onto the affinity column with a flow rate of 0.6 mL/min. Wash it with 15 column volume of Buffer A with a flow rate of 1.0 mL/min after the injection is complete.
9. Wash the affinity column containing bound AcrB protein with 15 column volume of Buffer B with a flow rate of 1.0 mL/min. Afterwards, wash it with 15 column volume of Buffer C with a flow rate of 1.0 mL/min.
10. Elute the AcrB protein with 10 column volume of Buffer D with a flow rate of 1.0 mL/min. Collect each of the protein fractions with a volume of 1.0 mL. Pool all the fractions containing AcrB protein into a 15 mL falcon tube for subsequent process.
11. Determine AcrB protein concentration by BCA method (*see* **Note 10**).

3.6 Purification of AcrB/DARPin Complex by Gel Filtration Chromatography

1. Carry out all the procedure at 4 °C unless otherwise specified.
2. Mix 0.31 mg DAPRin into 1 mg AcrB protein. Incubate the AcrB/DARPin complex solution on ice for 10 min.
3. Pipette the AcrB-DAPRin protein solution in an Amicon Ultra-15 Centrifugal Filter Ultracel® 100 K and centrifuge at 2800 × *g* using a bench-top centrifuge (Heraeus Megafuge 1.0R) until the protein solution is concentrated to 300 μL (*see* **Note 14**).
4. Set-up the Superose 6 10/300 GL column on Äkta Purifier System (Check the compatibility of Superose 6 10/300 GL column with any Äkta purification system according to manufacturer's recommendation). Wash the column with 1 column volume of water with a flow rate of 0.1 mL/ min and equilibrate with 1 column volume of AcrB GF Buffer with a flow rate of 0.3 mL/min.
5. Filter the AcrB/DARPin complex protein solution with 0.22 μm filter unit and inject the protein solution on the Superose 6 10/300 GL column according to manufacturer's recommendation and let the program run. Use AcrB GF Buffer throughout the purification of AcrB/DARPin complex protein on Superose 6 10/300 GL column with a flow rate of 0.3 mL/ min. Collect each of the protein fractions with a volume of 0.3 mL.
6. After gel filtration, pool all the fractions containing AcrB/ DARPin complex protein solution in an Amicon Ultra-15 Centrifugal Filter Ultracel® 100 K and centrifuge at 2800 × *g* using a bench-top centrifuge (Heraeus Megafuge 1.0R) until

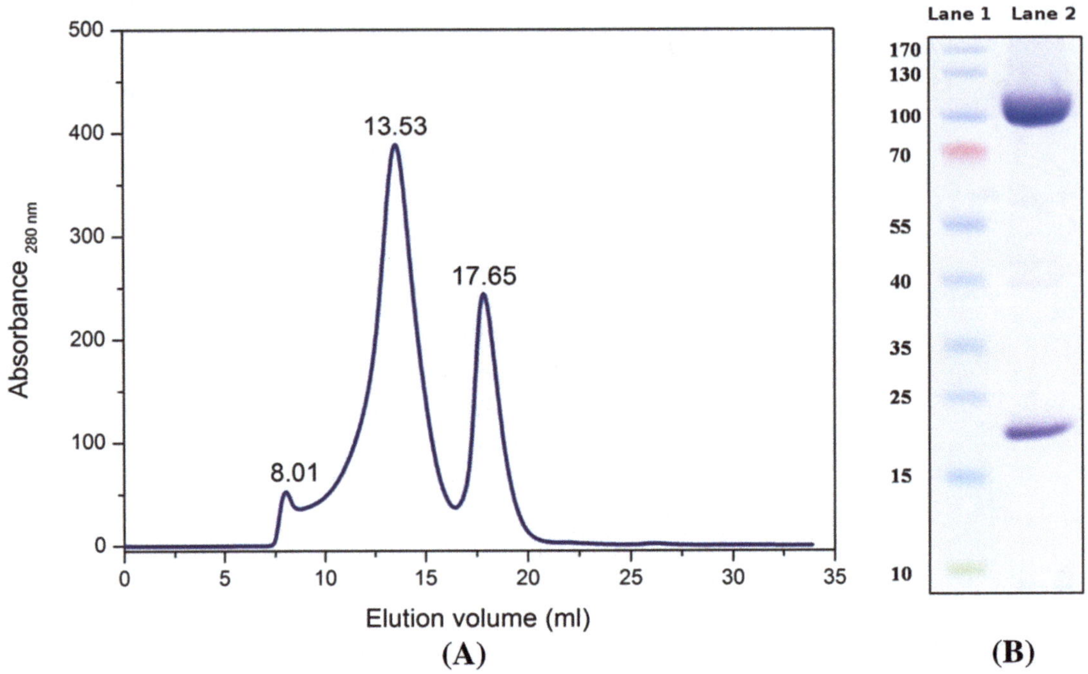

Fig. 2 Purification of AcrB/DARPin complex. (**a**) Gel filtration profile of the AcrB/DARPin complex sample using Superose 6 10/300 GL column on an Äkta Prime System (with AcrB GF Buffer, 0.3 mL/min). The initial peak at 8.01 mL elution volume accounts for aggregated protein fraction, the peak at 13.53 mL elution volume accounts for AcrB/DARPin complex and the peak at 17.65 mL elution volume accounts for free DARPin molecules. Elution fractions between 12.33 and 14.67 mL were collected (peak at 13.53 mL) and subjected to SDS-PAGE analysis. (**b**) *Lane 1*: PageRuler™ Prestained Protein Ladder, 10–180 kDa molecular weight marker (Thermo Fisher Scientific, Rockford, IL, USA) describe that the gel was stained with Coomassie (R250). *Lane 2*: The upper stained band (low electrophoretic mobility) shows the AcrB protein at an apparent molecular weight of ~100 kDa (calculated weight AcrB with His-tag: ~114.64 kDa) and the lower band (high electrophoretic mobility) shows the DARPin at an apparent molecular weight of ~18 kDa (theoretical DARPin plus His-tag ~18.29 kDa)

the AcrB/DARPin complex solution is concentrated to 200 μL (Fig. 2) (*see* **Note 14**).

7. Determine AcrB/DARPin complex protein concentration by BCA method (*see* **Note 10**). Dilute the AcrB/DARPin complex sample to 13 mg/mL with AcrB GF Buffer.

3.7 Crystallization of AcrB/DARPin Complex Protein by Hanging-Drop Vapor Diffusion

1. Label 24 pieces of 1.5 mL reaction tubes with A1–A6, B1–B6, C1–C6 and D1–D6 as shown in Table 1.
2. Prepare reservoir solution by mixing all the components: 1 M ADA, pH 6.6, 50% glycerol, 50% PEG4000, 2 M ammonium sulfate and MilliQ water in 24 different tubes according to the indicated volume (Table 1). Vortex all the tubes containing crystallization solutions until completely mixed.

Table 1
Reservoir solutions for setting up crystallization plates

1	2	3	4	5	6
50 μL of 1 M ADA, pH 6.6 (50 mM) 100 μL of 50% glycerol (5%) 120 μL of 50% PEG4000 (6%) 55 μL of 2 M $(NH4)_2SO_4$ (110 mM) 675 μL H_2O	50 μL of 1 M ADA, pH 6.6 (50 mM) 100 μL of 50% glycerol (5%) 120 μL of 50% PEG4000 (6%) 60 μL of 2 M $(NH4)_2SO_4$ (120 mM) 670 μL H_2O	50 μL of 1 M ADA, pH 6.6 (50 mM) 100 μL of 50% glycerol (5%) 120 μL of 50% PEG4000 (6%) 65 μL of 2 M $(NH4)_2SO_4$ (130 mM) 665 μL H_2O	50 μL of 1 M ADA, pH 6.6 (50 mM) 100 μL of 50% glycerol (5%) 120 μL of 50% PEG4000 (6%) 70 μL of 2 M $(NH4)_2SO_4$ (140 mM) 660 μL H_2O	50 μL of 1 M ADA, pH 6.6 (50 mM) 100 μL of 50% glycerol (5%) 120 μL of 50% PEG4000 (6%) 75 μL of 2 M $(NH4)_2SO_4$ (150 mM) 655 μL H_2O	50 μL of 1 M ADA, pH 6.6 (50 mM) 100 μL of 50% glycerol (5%) 120 μL of 50% PEG4000 (6%) 80 μL of 2 M $(NH4)_2SO_4$ (160 mM) 650 μL H_2O
50 μL of 1 M ADA, pH 6.6 (50 mM) 100 μL of 50% glycerol (5%) 140 μL of 50% PEG4000 (7%) 55 μL of 2 M $(NH4)_2SO_4$ (110 mM) 655 μL H_2O	50 μL of 1 M ADA, pH 6.6 (50 mM) 100 μL of 50% glycerol (5%) 140 μL of 50% PEG4000 (7%) 60 μL of 2 M $(NH4)_2SO_4$ (120 mM) 650 μL H_2O	50 μL of 1 M ADA, pH 6.6 (50 mM) 100 μL of 50% glycerol (5%) 140 μL of 50% PEG4000 (7%) 65 μL of 2 M $(NH4)_2SO_4$ (130 mM) 645 μL H_2O	50 μL of 1 M ADA, pH 6.6 (50 mM) 100 μL of 50% glycerol (5%) 140 μL of 50% PEG4000 (7%) 70 μL of 2 M $(NH4)_2SO_4$ (140 mM) 640 μL H_2O	50 μL of 1 M ADA, pH 6.6 (50 mM) 100 μL of 50% glycerol (5%) 140 μL of 50% PEG4000 (7%) 75 μL of 2 M $(NH4)_2SO_4$ (150 mM) 635 μL H_2O	50 μL of 1 M ADA, pH 6.6 (50 mM) 100 μL of 50% glycerol (5%) 140 μL of 50% PEG4000 (7%) 80 μL of 2 M $(NH4)_2SO_4$ (160 mM) 630 μL H_2O
50 μL of 1 M ADA, pH 6.6 (50 mM) 100 μL of 50% glycerol (5%) 160 μL of 50% PEG4000 (8%) 55 μL of 2 M $(NH4)_2SO_4$ (110 mM) 635 μL H_2O	50 μL of 1 M ADA, pH 6.6 (50 mM) 100 μL of 50% glycerol (5%) 160 μL of 50% PEG4000 (8%) 60 μL of 2 M $(NH4)_2SO_4$ (120 mM) 630 μL H_2O	50 μL of 1 M ADA, pH 6.6 (50 mM) 100 μL of 50% glycerol (5%) 160 μL of 50% PEG4000 (8%) 65 μL of 2 M $(NH4)_2SO_4$ (130 mM) 625 μL H_2O	50 μL of 1 M ADA, pH 6.6 (50 mM) 100 μL of 50% glycerol (5%) 160 μL of 50% PEG4000 (8%) 70 μL of 2 M $(NH4)_2SO_4$ (140 mM) 620 μL H_2O	50 μL of 1 M ADA, pH 6.6 (50 mM) 100 μL of 50% glycerol (5%) 160 μL of 50% PEG4000 (8%) 75 μL of 2 M $(NH4)_2SO_4$ (150 mM) 615 μL H_2O	50 μL of 1 M ADA, pH 6.6 (50 mM) 100 μL of 50% glycerol (5%) 160 μL of 50% PEG4000 (8%) 80 μL of 2 M $(NH4)_2SO_4$ (160 mM) 610 μL H_2O

(continued)

Table 1
(continued)

1	2	3	4	5	6
50 μL of 1 M ADA, pH 6.6 (50 mM) 100 μL of 50% glycerol (5%) 180 μL of 50% PEG4000 (9%) 55 μL of 2 M $(NH4)_2SO_4$ (110 mM) 615 μL H_2O	50 μL of 1 M ADA, pH 6.6 (50 mM) 100 μL of 50% glycerol (5%) 180 μL of 50% PEG4000 (9%) 60 μL of 2 M $(NH4)_2SO_4$ (120 mM) 610 μL H_2O	50 μL of 1 M ADA, pH 6.6 (50 mM) 100 μL of 50% glycerol (5%) 180 μL of 50% PEG4000 (9%) 65 μL of 2 M $(NH4)_2SO_4$ (130 mM) 605 μL H_2O	50 μL of 1 M ADA, pH 6.6 (50 mM) 100 μL of 50% glycerol (5%) 180 μL of 50% PEG4000 (9%) 70 μL of 2 M $(NH4)_2SO_4$ (140 mM) 600 μL H_2O	50 μL of 1 M ADA, pH 6.6 (50 mM) 100 μL of 50% glycerol (5%) 180 μL of 50% PEG4000 (9%) 75 μL of 2 M $(NH4)_2SO_4$ (150 mM) 595 μL H_2O	50 μL of 1 M ADA, pH 6.6 (50 mM) 100 μL of 50% glycerol (5%) 180 μL of 50% PEG4000 (9%) 80 μL of 2 M $(NH4)_2SO_4$ (160 mM) 590 μL H_2O

There are 24 solutions starting from A1–A6 to D1–D6

3. Pipette 1 mL of each crystallizing solution in the indicated wells of the Linbro 24-well plate. This is called reservoir solution (*see* **Note 15**). Prepare one piece of siliconized circular cover slides.
4. Pipette 1 μL of 13 mg/mL AcrB/DARPin complex protein solution on the siliconized circular cover slide. Pipette 1 μL of reservoir solution from the well A1 on the AcrB/DARPin complex protein drop on the same slide. Mixing of the protein and reservoir solution is not required (*see* **Note 16**).
5. Use a forceps to transfer the AcrB/DARPin complex protein and reservoir solution droplet containing siliconized circular cover slide and place it on the well A1 in upside down orientation. Gently press the cover slide on the grease to fix it properly on the well (*see* **Note 17**).
6. Continue with **steps 3–5** until every well contains a cover slide with droplet of AcrB/DARPin complex protein and respective reservoir solution together on it.
7. Observe all droplets under a light microscope. Incubate the crystallization plate at 18 °C for 2–3 weeks. Observation of crystallization droplets in every 3 days of interval is recommended.

3.8 AcrB/DARPin Complex Crystal Harvesting

1. Observe the crystallization plate under a light microscope. Try to look for crystals with three-dimensional shape and sharp edges. Use a forceps to transfer siliconized circular cover slide on the microscope (*see* **Note 18**).

2. Insert a cryo puck into the liquid nitrogen foam dewar and fill it with liquid nitrogen until the puck is completely submerged. Insert an empty cryo vial into the cryo puck in liquid nitrogen.
3. Pipette 1.5 μL of 5% cryo solution on the AcrB/DARPin complex crystal droplet to prevent dehydration or drying up. Pipette 2.0 μL of 15% cryo solution and 28% cryo solution side by side on a new cover slide.
4. Hold a magnetic cryo loop (appropriate diameter in size, depending on crystal size) with a magnetic cryo wand and use it to harvest the crystals located in 5% cryo solution (*see* **Note 19**). Transfer the crystals to 15% cryo solution and leave it for 30 s. After 30 s, harvest the crystals, transfer it to 28% cryo solution, and incubate for another 30 s.
5. Harvest the crystals located in 28% cryo solution, plunge the magnetic cryo loop containing crystal into the liquid nitrogen, and carefully transfer it into the empty cryo vial placed in cryo puck in liquid nitrogen (*see* **Note 20**).
6. Store the crystal in liquid nitrogen storage dewar until data collection at synchrotron.

4 Notes

1. Carefully handle liquid nitrogen. Wear cryogenic gloves and safety goggles while working with it. Any accident can cause irreversible damage to skin or eyes.
2. Our experience shows that it is better to use freshly prepared PEG4000 solution. However, the solution can be stored for maximum 1 month. It is advisable to store it under dark condition.
3. Linbro 24-well plates, pre-greased format can be used as well.
4. Selection of cryo loops with different diameter is dependent on the size of the crystals.
5. Always use freshly transformed cells for growing cell cultures. Using old glycerol stock for pre-culture can affect the quality as well as quantity of the protein produced.
6. *E. coli* XL1 Blue cells grow slowly and it will take 4–5 h to reach $OD_{600} = 0.7$.
7. Make sure the cell pellet is suspended completely to prevent clogging in the cell disruptor, instead the suspension can be filtered by a thin sieve.
8. Refer to manufacturer's protocol for assembly of 150 and 50 mL superloops with the Äkta Prime System.

9. Refer to manufacturer's protocol for setting up the Äkta Prime System and assembly of His-Trap HP affinity column on it. Instead of using superloop and Äkta system for affinity purification, separate peristaltic pump together with His-Trap column can also be used. Fresh Ni^{2+}-NTA beads can also be used for protein binding and purification by gravity flow method.
10. Refer to manufacturer's protocol. Other protein concentration determination methods can be applied.
11. Cell pellet can be flash frozen with liquid nitrogen and stored at −80 °C until further use or can be directly proceed to cell lysis.
12. Gently mix the detergent with the membrane vesicle suspension and rotate it at 4 °C on a rocker at 10 rpm. Vigorous mixing will produce huge amount of foam which will make it difficult to handle.
13. Gently mix the detergent with the membrane vesicle suspension and rotate it at 4 °C on a rocker at 10 rpm. Vigorous mixing will produce huge amount of foam, which will make it difficult to handle.
14. Centrifuge the AcrB/DAPRin complex protein solution at 2800 × *g* for only 15 min every time and use a pipette to mix the concentrated protein mainly at the bottom region of the concentrator to prevent any precipitation.
15. Preparation of crystallization solution and setting up crystallization plates can be done at room temperature. Do not pipette all 24 solutions in the crystallization plate together. It is better to pipette only six solutions at a time.
16. A slight precipitation will form after adding the protein to reservoir solution. However, this does not prevent the AcrB/DARPin complex crystal formation.
17. It is necessary to fix the cover slide on the crystallization well properly. If not, the drop will dry out very fast.
18. Cover slide is fixed with the well via grease. Therefore, gently pick it up with a forceps otherwise it can break easily.
19. Crystal harvesting is done under light microscope. Magnetic loop is inserted in to the drop very slowly in order to minimize the liquid displacement of the drop as well as to avoid any damage to the crystal.
20. It is necessary to transfer the crystals as fast as possible from one drop to another and finally to liquid nitrogen for freezing. Leaving it too long in the air will dry the crystals, which can affect the diffraction data quality.

Acknowledgements

Work in the Pos lab is supported by the German Research Foundation (SFB 807, Transport and Communication across Biological Membranes and FOR2251, Adaptation and persistence of the emerging pathogen *Acinetobacter baumannii*), the DFG-EXC115 (Cluster of Excellence Macromolecular Complexes at the Goethe-University Frankfurt), Innovative Medicines Initiative Joint Undertaking Project Translocation (IMI-Translocation), EU Marie Curie Actions ITN, Human Frontiers Science Program (HFSP) and the German-Israeli Foundation (GIF).

References

1. Tseng TT, Gratwick KS, Kollman J, Park D, Nies DH, Goffeau A, Saier MH (1999) The RND permease superfamily: an ancient, ubiquitous and diverse family that includes human disease and development proteins. J Mol Microbiol Biotechnol 1:107–125
2. Saier MH, Paulsen IT (2001) Phylogeny of multidrug transporters. Semin Cell Dev Biol 12:205–213
3. Yu EW, Aires JR, Nikaido H (2003) AcrB multidrug efflux pump of *Escherichia coli*: composite substrate-binding cavity of exceptional flexibility generates its extremely wide substrate specificity. J Bacteriol 185:5657–5664
4. Ma D, Cook DN, Hearst JE, Nikaido H (1994) Efflux pumps and drug resistance in Gram-negative bacteria. Trends Microbiol 2:489–493
5. Okusu H, Ma D, Nikaido H (1996) AcrAB efflux pump plays a major role in the antibiotic resistance phenotype of *Escherichia coli* multiple-antibiotic-resistance (Mar) mutants. J Bacteriol 178:306–308
6. Murakami S, Nakashima R, Yamashita E, Yamaguchi A (2002) Crystal structure of bacterial multidrug efflux transporter AcrB. Nature 419:587–593
7. Seeger MA, Schiefner A, Eicher T, Verrey F, Diederichs K, Pos KM (2006) Structural asymmetry of AcrB trimer suggests a peristaltic pump mechanism. Science 313:1295–1298
8. Murakami S, Nakashima R, Yamashita E, Matsumoto T, Yamaguchi A (2006) Crystal structures of a multidrug transporter reveal a functionally rotating mechanism. Nature 443:173–179
9. Sennhauser G, Amstutz P, Briand C, Storchenegger O, Grütter MG (2007) Drug export pathway of multidrug exporter AcrB revealed by DARPin inhibitors. PLoS Biol 5:106–113
10. Eicher T, Seeger MA, Anselmi C, Zhou W, Brandstätter L, Verrey F, Diederichs K, Faraldo-Gómez JD, Pos KM (2014) Coupling of remote alternating-access transport mechanisms for protons and substrates in the multidrug efflux pump AcrB. elife 3:e03145
11. Touzé T, Eswaran J, Bokma E, Koronakis E, Hughes C, Koronakis V (2004) Interactions underlying assembly of the *Escherichia coli* AcrAB-TolC multidrug efflux system. Mol Microbiol 53:697–706
12. Tikhonova EB, Zgurskaya HI (2004) AcrA, AcrB, and TolC of *Escherichia coli* form a stable intermembrane multidrug efflux complex. J Biol Chem 279:32116–32124
13. Pos KM (2009) Drug transport mechanism of the AcrB efflux pump. Biochim Biophys Acta 1794:782–793
14. Yamaguchi A, Nakashima R, Sakurai K (2015) Structural basis of RND-type multidrug exporters. Front Microbiol 6:327
15. Ostermeier C, Iwata S, Ludwig B, Michel H (1995) Fv fragment-mediated crystallization of the membrane protein bacterial cytochrome *c* oxidase. Nat Struct Biol 2:842–846
16. Rasmussen SGF, DeVree BT, Zou Y, Kruse AC, Chung KY, Kobilka TS, Thian FS, Chae PS, Pardon E, Calinski D, Mathiesen JM, Shah STA, Lyons JA, Caffrey M, Gellman SH, Steyaert J, Skiniotis G, Weis WI, Sunahara RK, Kobilka BK (2011) Crystal structure of the β_2 adrenergic receptor-Gs protein complex. Nature 477:549–555
17. Bukowska MA, Grütter MG (2013) New concepts and aids to facilitate crystallization. Curr Opin Struct Biol 23:409–416
18. Eicher T, Cha H, Seeger MA, Brandstätter L, Bohnert JA, Kern WV, Verrey F, Grütter MG,

Diederichs K, Pos KM (2012) Transport of drugs by the multidrug transporter AcrB involves an access and a deep binding pocket that are separated by a switch-loop. Proc Natl Acad Sci U S A 109:5687–5692

19. Binz HK, Stumpp MT, Forrer P, Amstutz P, Plückthun A (2003) Designing repeat proteins: well-expressed, soluble and stable proteins from combinatorial libraries of consensus ankyrin repeat proteins. J Mol Biol 332:489–503
20. Hanes J, Plückthun A (1997) In vitro selection and evolution of functional proteins by using ribosome display. Proc Natl Acad Sci U S A 94:4937–4942
21. Smith GP (1985) Filamentous fusion phage: novel expression vectors that display cloned antigens on the virion surface. Science 228:1315–1317
22. Binz HK, Amstutz P, Kohl A, Stumpp MT, Briand C, Forrer P, Grütter MG, Plückthun A (2004) High-affinity binders selected from designed ankyrin repeat protein libraries. Nat Biotechnol 22:575–582
23. Sennhauser G, Grütter MG (2008) Chaperone-assisted crystallography with DARPins. Structure 16:1443–1453
24. Brandstätter L, Sokolova L, Eicher T, Seeger MA, Briand C, Cha H, Cernescu M, Bohnert J, Kern WV, Brutschy B, Pos KM (2011) Analysis of AcrB and AcrB/DARPin ligand complexes by LILBID MS. Biochim Biophys Acta 1808:2189–2196
25. Du D, Wang Z, James NR, Voss JE, Klimont E, Ohene-Agyei T, Venter H, Chiu W, Luisi BF (2014) Structure of the AcrAB-TolC multidrug efflux pump. Nature 509:512–515

Chapter 2

Crystallographic Analysis of Drug and Inhibitor-Binding Structure of RND-Type Multidrug Exporter AcrB in Physiologically Relevant Asymmetric Crystals

Ryosuke Nakashima, Keisuke Sakurai, and Akihito Yamaguchi

Abstract

Xenobiotic extruding pumps have recently been known to be widely distributed in living organisms from mammalian to bacteria as a host-defense mechanism in cellular level. These pumps not only confer multidrug resistance of cancer cells and pathogenic bacteria but also cause hereditary diseases through the mutation. Our purposes are to elucidate the molecular structures and mechanisms of these xenobiotic exporters.

We had succeeded to determine the crystal structure of bacterial major multidrug exporter AcrB at 3.5 Å resolution (Murakami et al., Nature 419:587–593, 2002) and elucidated the structural bases of substrate recognition that the pump recognize the places and thus act as a "membrane vacuum cleaner." After that we also determined the crystal structure of the drug-binding form of AcrB in space group *C*2 in which asymmetric unit contains structurally asymmetric homo-trimer of AcrB (Murakami et al., Nature 443:173–179, 2006; Nakashima et al., Nature 480:565–569, 2011; Nakashima et al., Nature 500:120–126, 2013). Analyses revealed the existence of a specific mechanism to recognize numerous substrates that the multisite binding is the base of multidrug recognition rather than induced-fit, and functional-rotation mechanism in which three monomers undergo a strictly coordinated sequential conformational change cycle of access, binding, and extrusion. Determination of physiological asymmetric AcrB structure was crucially important to understand these transport mechanisms.

Key words Multidrug exporter AcrB, RND family, Crystallization, X-ray, Crystal structure, Functional-rotation mechanism, Peristaltic pump mechanism

1 Introduction

AcrB and its homologues are the major multidrug transporters in Gram-negative bacteria and play important role in drug resistance. AcrB, the resistance-nodulation-cell division (RND) transporter, is a homotrimer that acts as a tripartite complex with the multifunctional outer membrane channel, TolC and the membrane fusion protein (MFP), AcrA. This complex is constitutively expressed in

Akihito Yamaguchi and Kunihiko Nishino (eds.), *Bacterial Multidrug Exporters: Methods and Protocols*, Methods in Molecular Biology, vol. 1700, https://doi.org/10.1007/978-1-4939-7454-2_2, © Springer Science+Business Media, LLC 2018

bacterial cells and is responsible for the natural resistance against toxic compounds like dyes, detergents, and antibiotics.

In spite of the relative abundance of AcrB in nature, it is still difficult to get the protein in high amount and in crystallization compatible quality from a natural source. Therefor we used bacterial recombinant expression system. It must contribute to reducing future labor to optimize protein expression condition in advance of purification/crystallization trials. We inspected combination of some types of plasmid, host strains, broth, induction (timing, concentration of the inducer, incubation time) and found that, in the case of AcrB, when pACBH and JM109 was used as expression vector and host, respectively, protein expression level became maximum. In this chapter, we introduce the actual situation from bacterial culture to crystallographic analysis of AcrB.

2 Materials

All solutions were prepared with ultrapure water (Milli-Q Integral 3) and guaranteed or higher-grade reagent.

2.1 Expression

1. His-tagged and tag-free AcrB was inserted in pUC118 vector, and *E. coli* bacteria strain JM109 was transformed with these AcrB-containing plasmids.
2. Medium A [1]: 17.5 g K_2HPO_4, 7.5 g KH_2PO_4, 1.25 g Na_3 citrate·$3H_2O$, 0.25 g $MgSO_4 \cdot 7H_2O$, 2.5 g $(NH_4)_2SO_4$ in a final volume of 2.5 L.
3. 20% (w/v) glucose and 10% (w/v) casamino acids were sterilized separately with Medium A to avoid Maillard reaction.
4. Bacterium cultivation was performed with Bioshaker BR-3000LF (TAITEC Corp., Japan) at 37 °C and 180 rpm. It has the ability to load eight flasks of 5 L.

2.2 Purification

1. CR22 high-speed refrigerated centrifuge with R18C continuous flow rotor (Hitachi Koki., Ltd.) was used to harvest bacteria.
2. Buffer A: 50 mM Tris–HCl pH 7.0, 10% (v/v) glycerol, 1 mM $MgCl_2$, 0.5 mM EDTA.
3. Buffer B: 5 mM Tris–HCl pH 7.0, 0.5 mM EDTA.
4. Buffer C: 50 mM Tris–HCl pH 7.0, 10% (v/v) glycerol.
5. Bacterial lysis was performed with laboratory homogenizer APV 1000 (SPX Corporation) at 4 °C and 950 bar.
6. Sucrose Monododecanoate (Dodecanoyl sucrose/DDS) was purchased from Affymetrix (former Anatrace).

7. Iminodiacetic acid sepharose beads (IDA) for immobilized metal ion affinity chromatography were purchased from GE healthcare (Chelating Sepharose™ Fast Flow). It can be immobilized various metal ions such as Cu^{2+}, Zn^{2+}, Ni^{2+}, and Fe^{3+}. Among them, Ni^{2+} ion was immobilized.
8. Column chromatography was performed with 12 ml Econo-column (Bio-Rad).
9. Buffer D: 20 mM Tris–HCl pH 7.5, 10% (v/v) glycerol, 300 mM NaCl, 0.1% (w/v) DDS.
10. Buffer E: 20 mM Tris–HCl pH 7.5, 10% (v/v) glycerol, 300 mM NaCl, 0.1% (w/v) DDS, 25 mM Imidazole-HCl pH 7.5.
11. Buffer F: 20 mM Tris–HCl pH 7.5, 10% (v/v) glycerol, 300 mM NaCl, 0.1% (w/v) DDS, 100 mM Imidazole-HCl pH 7.5.
12. Buffer G: 20 mM Tris–HCl pH 7.5, 10% (v/v) glycerol, 300 mM NaCl, 0.1% (w/v) DDS, 300 mM Imidazole-HCl pH 7.5.
13. Amicon-stirred cell (Model 8010, Merck Millipore) and Biomax Ultrafiltration membrane (NMWL: 100 kDa, Merck Millipore) was used to concentrate protein.
14. Buffer H: 20 mM Sodium phosphate pH 6.2, 10% (v/v) glycerol, 0.1% (w/v) DDS.
15. BCA assay was used to determine protein concentration. BSA was used as a protein standard.
16. Buffer I: 20 mM Tris–HCl pH 7.5, 10% (v/v) glycerol, 0.1% (w/v) DDS.
17. Buffer J: 20 mM Tris–HCl pH 7.5, 10% (v/v) glycerol, 0.1% (w/v) DDS, 300 mM NaCl.
18. n-Dodecyl β-D-maltoside was from GLYCON Biochemicals (Germany).
19. Q sepharose HP was from GE healthcare.

2.3 Crystallization

1. Polyethylene Glycol 4000 and 400 were purchased from Hampton Research.
2. CrystalClear D Strips was from Hampton Research.
3. Buffer K: 100 mM Sodium phosphate pH 6.2, 100 mM NaCl, 12% (w/v) Polyethylene Glycol 4000.
4. Buffer L: 100 mM Sodium phosphate pH 6.2, 100 mM NaCl, 14% (w/v) Polyethylene Glycol 4000.
5. Buffer M: 60 mM Sodium phosphate pH 6.2, 50 mM NaCl, 5% (w/v) glycerol, 0.1% (w/v) DDS, 10% (w/v) PEG4000.

6. Buffer N: 60 mM Sodium phosphate pH 6.2, 50 mM NaCl, 30% (w/v) glycerol, 0.1% (w/v) DDS, 11% (w/v) PEG4000.
7. LithoLoops (Protein Wave Corp., Japan) was used for crystal mounting.

3 Methods

Carry out all procedures at 4 °C (in cold chamber or ice bath) unless otherwise specified.

3.1 Bacterial Membrane Preparation

1. The plasmid encoding his-tagged AcrB is previously constructed [2]. The cloned chromosomal *acr* locus from *E. coli* W3104 was inserted into pUC118 vector (pAC8), followed by addition of a $(His)_4$ sequence at the 3′ end of the *acrB* gene (pACBH). Since AcrB has intrinsic His-His sequence at its C-terminal, the resulting AcrB protein has six histidine tag. This plasmid contains *acrR*, *acrA*, *acrB*, and their promoter region, and hence it expresses encoding proteins without induction.
2. *E. coli* strain JM109 harboring pACBH was cultured in 2xYT broth containing 100 μg/ml ampicillin at 37 °C until OD_{600} reaches 0.7–1.0, and then cell cultures were diluted into 100-fold volume of minimal medium (Medium A) (*see* **Note 1**) supplemented with 0.2% (w/v) glucose and 0.1% (w/v) casamino acids and cultured more 7 h at 37 °C (*see* **Note 2**). Incubation was performed in 5 L Erlenmeyer flask in bioshaker with rotating mode. Cultivation of 30 L in total was carried out with two shakers and 12 flasks (*see* **Note 3**).
3. The cells were collected by centrifugation at 24,000 × *g* with R18C continuous flow rotor at 4 °C. 80 g wet cells are usually obtained from 30 L culture.
4. The cells were suspended in buffer A at 5 ml/g wet cell and disrupted with laboratory homogenizer APV 1000 at 950 bar.
5. After elimination of debris by several high-speed centrifugation, membrane fractions were collected by ultracentrifugation at 158,000 × *g* for 90 min. The resulting precipitate was suspended with buffer B to wash out membrane-associated proteins and then ultracentrifuged again. The purified membrane fractions were suspended in buffer C and divided into approximately 200 mg protein/tube. And then they were frozen in liquid nitrogen and stored at −80 °C until further protein purification.

3.2 Protein Purification (Six Histidine-Tagged AcrB)

1. The protocol used to purify AcrB is previously described in [3]. The membrane fraction was thawed and adjusted to 10 mg/ml protein, then it was solubilized by adding 20% DDS solution quickly to become 1.5% (w/v) of final concentration while mixing it.

2. After 20 min gently stirring, the solubilized membrane fraction was centrifuged for 1 h at 172,000 × *g*, and the supernatant was applied to a 5 ml Ni Sepharose resin pre-equilibrated with buffer D with batch mode.
3. After incubation for 1 h, resin was collected into a column, and proteins were purified by column chromatography. The resin was washed with 10 column volume each of buffer E and F.
4. The proteins were eluted with 10 column volume of buffer G.
5. Buffer composition of eluate was exchanged into buffer H by three successive concentration–dilution steps using an Amicon-stirred cell with 100 kDa cutoff ultrafiltration membrane, which was pressurized with five nines nitrogen gas, and proteins were finally concentrated to approximately 35 mg/ml (*see* **Notes 4** and **5**). About 12 mg highly purified proteins as shown in Fig. 1 are usually obtained at once. Purified AcrB can be cryopreserved but preferable are immediately submitted to crystallization set up.
6. Protein concentration was determined using BCA assay.

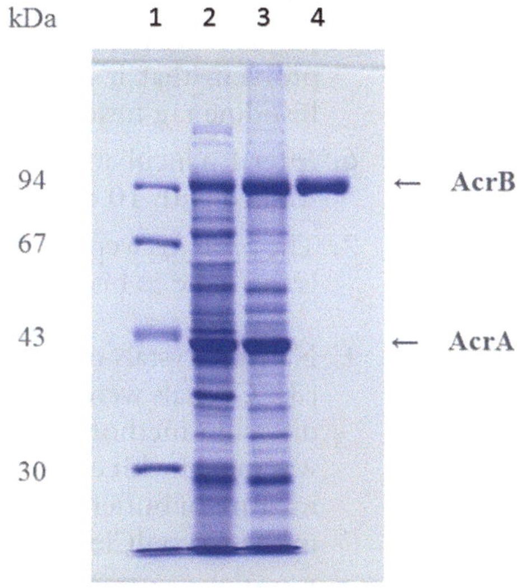

Fig. 1 Analysis of protein purification by SDS-PAGE. Proteins were separated by 10% PAGE and stained by Coomassie Brilliant Blue. *Lane 1*: Molecular maker (LMW, GE Healthcare), *lane 2*: whole cell, *lane 3*: membrane fraction, *lane 4*: purified AcrB. 30% (w/v) Acrylamide/Bis Mixed Solution (37.5:1) was used to prepare a 1 mm thick, 10% gel based on Laemli method (8 cm × 19 cm × 0.1 cm wide gel, Seema-biotech Co., Ltd., Japan). Sample buffer in reductive condition was used to prepare loading samples. Loading samples were not boiled to avoid AcrB aggregation. Electrophoresis was performed at 45 mA constant current. The gels were stained with 0.1% (w/v) CBB R-250/40% ethanol/10% acetic acid for 30 min and destained until a background disappeared in 40% ethanol/10% acetic acid

3.3 Protein Purification (Tagless AcrB)

1. The method of cultivation and membrane preparation are same as that of His-tagged AcrB except for using bacteria harboring pAC8.
2. Membrane fraction was solubilized by 2.0% (w/v) DDS. After incubation for 20 min, it was centrifuged for 1 h at 172,000 × *g*, and then the resulting supernatant was applied to a 5 ml Q sepharose HP resin pre-equilibrated with buffer I with batch mode.
3. After 1 h incubation, resin was collected into a column, and proteins were purified afterwards by column chromatography. The column was washed with 10 column volume of equilibration buffer I.
4. The proteins were eluted with a linear gradient (100 ml total volume) of buffer I and J.
5. Fractions containing AcrB judged from electrophoresis were pooled and was applied directly (without buffer exchange) to a 5 ml Ni Sepharose resin pre-equilibrated with buffer K with batch mode again. Since AcrB has intrinsic His-His sequence at its C-terminal as described above, native AcrB trimer has six histidine tags in total. For this reason, Ni affinity chromatography is available. But this feature of AcrB often causes the problem that it is easy to contaminate in the preparation of histidine tag fusion proteins [4].
6. After 1 h incubation, resin was collected into a column and was washed with 10 column volume of buffer E.
7. The proteins were eluted with 10 column volume of buffer F. It is the same as histidine-tagged AcrB afterward.

3.4 Crystallization of Native AcrB

1. Native Crystals of AcrB were prepared as previously described [3]. Crystals were grown at 25 °C using the sitting drop vapor diffusion method. Protein solution of 28 mg/ml in buffer H was mixed with equal volume of reservoir solution which was a mixture of buffer K and L. 4 μl crystallization drops were setup in the CrystalClear D Strips and sealed with microplate sealer.
2. For the crystallization of space group C2 in which the asymmetric unit contains AcrB trimer, the microseeding method was used to induce crystallization (*see* **Note 6**). A tungsten needle was used to dislodge seeds from a crystal. The needle was touched to all crystallization drop surface to promote crystal growth from the drop surface (Fig. 2a) before strips were sealed. The crystals came to be visible within few hours. The crystals were harvested after growing up 1 week (*see* **Note 7**).

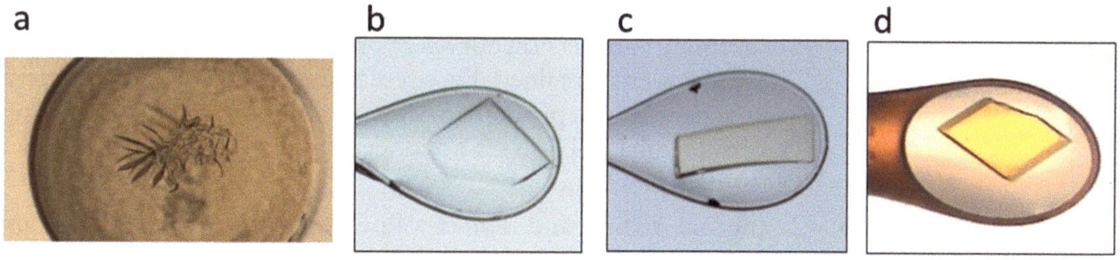

Fig. 2 Photographs of AcrB *C2* crystal. (**a**) A photograph of sitting drop, (**b**) native (drug-free), (**c**) minocycline complex, (**d**) rifampicin complex

3.5 Crystallization of AcrB-Drug (Inhibitor) Complex

3.5.1 In Case of Hydrophilic Compounds

1. The crystallization condition of drug complex was the same as native crystal except for containing drug at molar ratio of 3–50. The drugs were dissolved in buffer H necessary to dilute protein solution to 28 mg/ml. To ensure AcrB–drug complex formation, substrate was added to the purified protein and incubated for a few hours at 4 °C prior to crystallization set up.
2. Microseeding was also performed in drug complex crystallization same as native crystallization in which native crystal was used as seed.

3.5.2 In Case of Hydrophobic Compounds

1. Before buffer exchange of protein fraction using amicon, substrates that were dissolved in ethanol or DMSO were added to eluate from a Ni-column at approximately tenfold of molar ratio with the protein. After incubation of several hours, the solvent was washed out with buffer exchange by three concentration–dilution steps (*see* **Note 8**).
2. The substrate concentration in the final protein sample can be determined by either spectroscopic measurement or by LC–MS/MS for quantitative analysis [5].
3. Microseeding was also performed in drug complex crystallization same as native crystallization.

3.6 Cryo Protection

1. Cryoprotection was achieved by raising the glycerol concentration stepwise to 30% (v/v) (buffer N) in 5% increments, and each step took 10 min (*see* **Notes 9** and **10**). Crystals were picked using LithoLoops for flash cooling in a cold nitrogen gas stream (100 K) from a cryostat (*see* **Note 11**). In most cases of AcrB, it would function as a crystal harvesting buffer if the concentration of precipitant is raised a little in comparison to initial condition when crystallization drop was set up (buffer M). For drug complex crystal, drug was also contained in crystal harvesting buffer.

3.7 Characterization: SAXS (Small Angle X-ray Scattering)

1. Protein sample for crystallization was used in the SAXS measurement. Dispersion curves of protein in each protein concentration were obtained by circular averaging of measurement images with the program SAXSGui and background reduction with Primus (ATSAS). The concentration-dependent multimer formation was not observed.
2. Inertial radius and Dmax value calculated by the program Gnom (ATSAS) using the data in the range up to $q = 0.67$ Å^{-1} were 79.46 and 208.0 Å, respectively.
3. Dummy Residues Model (DRM) was built using Gasbor (ATSAS). The solution structure was obtained by averaging four results of Gasbor (Fig. 3). This result revealed that the protein forms a dimer (MexB trimer × 2) in solution. Although this is the result measured with MexB sample not AcrB sample, we think it is almost identical to that of AcrB because the AcrB/DDS sample includes great amount of an ingredient to elute in a void volume of size exclusion chromatography using Superdex 200 10/300 GL (*see* **Note 12**).

3.8 Crystallographic Analysis

1. Each data set was collected using the BL44XU beamline at SPring-8 with an MX225-HE charge-coupled-device detector (Rayonix) at 100 K.
2. The diffraction data were processed and scaled using the HKL2000 [6] package or MosFlm in the CCP4 program suite [7].
3. The initial phases were first determined by molecular replacement with MOLREP [8] using the atomic coordinates of AcrB (Protein Data Bank ID, 2DHH) as a search model.

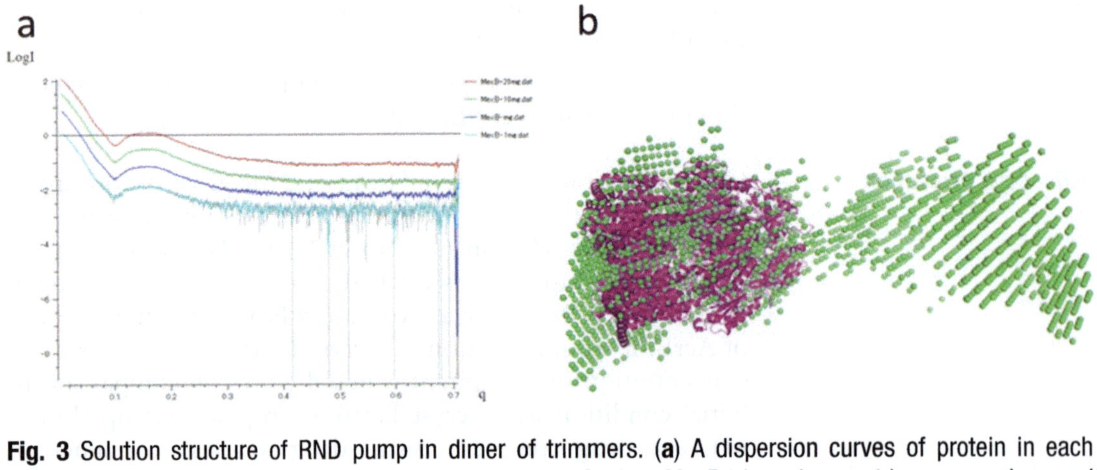

Fig. 3 Solution structure of RND pump in dimer of trimmers. (**a**) A dispersion curves of protein in each concentration, (**b**) the solution structure calculated using Gasbor. MexB trimer (*magenta*) was superimposed onto solution structure (*green*). Measuring conditions were as follows: X-ray Generator/Camera: FR-E+/BioSAXS-1000 (Rigaku Corp.), Camera length: 500 mm, Sample condition: MexB, range of 20–0.1 mg/ml (10 mM Tris–HCl pH 7.5, 50 mM NaCl, 0.05% DDM), sample volume: 30 μl, Room temperature

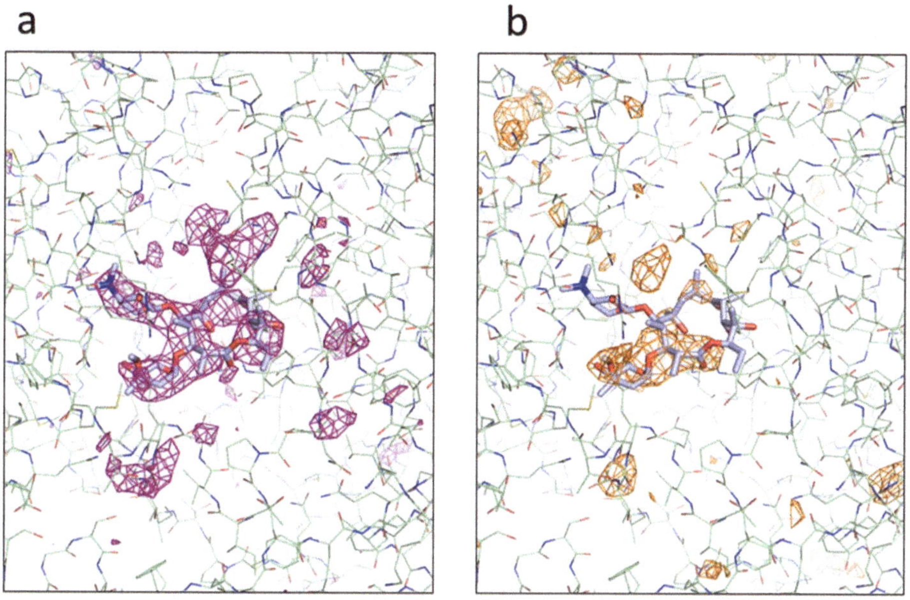

Fig. 4 Comparison of initial $F_{O(liganded)} - F_{O(unliganded)}$ and $F_o - F_c$ electron density maps for the region of erythromycin binding site. Protein moiety and erythromycin are drawn in wire model and gray stick model, respectively. (**a**) $F_{O(liganded)} - F_{O(unliganded)}$ map contoured at 3σ is shown in *magenta*, (b) $F_o - F_c$ map contoured at 3σ is shown in *orange*. We calculated maps after molecular replacement (Molrep) and refinement (restrained refinement with Refmac5). Suitable solution was provided when the initial map was calculated with the differences between the observed structure factors as coefficient. Because $F_o - F_o$ map requires that the isomorphism being kept between the crystals and substrate-free crystal exist, this method is not applicable to the proteins of which substrate-free crystal is not available like MexB

4. Model rebuilding was performed using COOT [9], and model refinement was performed using CNS [10] and REFMAC [11].
5. To identify drug molecules in the AcrB–drug complexed crystals, different Fourier electron density maps were calculated with coefficients of $(|F_{O(liganded)}| - |F_{O(unliganded)}|)$ exp. $(i\alpha_{unliganded})$, where $F_{O(liganded)}$ is the structure factor in the presence of a bound drug, $F_{O(unliganded)}$ is the structure factor of the drug-free crystal, and $\alpha_{unliganded}$ is the phase of the drug-free crystal (Fig. 4). Refinement was performed with constrained structures for the drug molecules.

4 Notes

1. Because the RND family has the tendency that protein expression level was low in rich broth like 2xYT, the minimal medium was used for protein expression.

2. Bacterial cells were harvested within 7 h from a main culture start because bacteriolysis was observed in the stationary phase of pACBH/JM109 cultivation.
3. For unknown reasons, protein expression level was also low in large scale cultivation in 50 L or larger Jar fermentor. Then we limited culture scale up to 2.5 L in flask.
4. The protein sample changes its solubility after size exclusion chromatography, probably because the number of phospholipid remaining in final protein sample was reduced. However, it was not adopted as protein purification step because it didn't improve the quality of the crystal.
5. Although detergent was concentrated with increasing protein concentration, excessive detergent composition pass into the filtrate because micelle size of DDS was below the membrane molecular weight cut-off. For this reason, there is no necessary to treat concentrated protein solution by Bio-Beads to remove excess detergent.
6. Because crystallization condition of the space group *C*2 produced many protein precipitate, seeding that induced crystal growth from drop surface (Fig. 2a) was necessary to obtain good crystals. When the concentration of precipitant and/or protein were reduced, it resulted in a small crystal.
7. Although the crystal grew up for one month and reached the maximum size of up to 1 mm, there was considerable decrease in crystal quality after growing. In consideration of size and quality, crystals were harvested within 7–10 days.
8. The method of Subheading 3.5.2 has the problem that it is difficult to control drug concentration. However, it is necessary to remove the organic solvent because it usually has bad effect on crystallization.
9. At first, the cryoprotectant should be chosen among the components of mother liquor. In case of AcrB *C*2 crystal, glycerol was chosen [2] and in case of MexB, PEG400 was raised as protectant [5]. If these reagents that work as cryoprotectants are not included in mother liquor, I will test glycerol, low molecular weight PEG like PEG200 and 400, ethylene glycol, and so on.
10. A crystal cracks when the cryo-protection process is hurried, and a crystal changes in quality into sponge state without appearances change when it is soaked against the solution of high concentration glycerol for more than a day. Therefor the processing by dialysis to be seen elsewhere is not applicable to AcrB *C*2 crystal. It seems to be sufficient spending 10 min on each step.

11. In case of AcrB *C2* crystal, cryostream was used to freeze crystal instead of liquid nitrogen. When crystal was soaked into liquid nitrogen directly, the isomorphism (it is important for analysis of substrate-binding structure) was lost, and ability to diffract X-ray was also decreased. On the other hand in case of a crystal of other homolog, we know that the direct method has better results. It is not decided which method is superior, therefore it is necessary to test both stream and direct method at first.
12. There are many reports on crystal structure of RND-type transporter including symmetric {AcrB [3, 12–16], AcrB/YajC [17], AcrB/AcrZ [18], CusA [19], CusAB [20], MtrD [21]}, and asymmetric {AcrB [22, 23, 24], AcrB/DARPin [25], MexB [5, 26], ZneA [27]} trimer. The symmetric structure is regarded as resting form, while the asymmetric structure is regarded as physiologically relevant structure and is essential to elucidate the mechanism of RND transporter. Because RND-type transporters have a pseud threefold symmetry axis at the center of the trimer, it might have a tendency to crystallize easily into the space group having crystallographic threefold symmetry. Therefore we think that using the protein binder such as DARPin and the dimer of trimer formation are effective to obtain the physiologically relevant asymmetric crystals.

Acknowledgements

This work was supported by CREST, JST.

References

1. McMurry L, Petrucci RE Jr, Levy SB (1980) Active efflux of tetracycline encoded by four genetically different tetracycline resistance determinants in *Escherichia coli*. Proc Natl Acad Sci U S A 77:3974–3977
2. Fujihira E, Tamura N, Yamaguchi A (2002) Membrane topology of a multidrug efflux transporter, AcrB, in *Escherichia coli*. J Biochem (Tokyo) 131:145–151
3. Murakami S, Nakashima R, Yamashita E, Yamaguchi A (2002) Crystal structure of bacterial multidrug efflux transporter AcrB. Nature 419:587–593
4. Veesler D, Blangy S, Cambillau C, Sciara G (2008) There is a baby in the bath water: AcrB contamination is a major problem in membrane-protein crystallization. Acta Crystallogr Sect F Struct Biol Cryst Commun 64:880–885
5. Nakashima R, Sakurai K, Yamasaki S, Hayashi K, Nagata C, Hoshino K, Onodera Y, Nishino K, Yamaguchi A (2013) Structural basis for the inhibition of bacterial multidrug exporters. Nature 500:120–126
6. Otwinowski Z, Minor W (1997) Processing of X-ray diffraction data collected in oscillation mode. Methods Enzymol 276:307–326
7. Collaborative Computational Project (1994) Number 4. The CCP4 suite: programs for protein crystallography. Acta Crystallogr D 50:760–763
8. Vagin A, Teplyakov A (1997) MOLREP: an automated program for molecular replacement. J Appl Crystallogr 30:1022–1025

9. Emsley P, Cowtan K (2004) Coot: model-building tools for molecular graphics. Acta Crystallogr D 60:2126–2132
10. Brunger AT (2007) Version 1.2 of the crystallography and NMR system. Nat Protoc 2:2728–2733
11. Murshudov GN, Vagin AA, Dodson EJ (1997) Refinement of macromolecular structures by the maximum-likelihood method. Acta Crystallogr D 53:240–255
12. Yu EW, McDermott G, Zgurskaya HI, Nikaido H, Koshland DE Jr (2003) Structural basis of multiple drug binding capacity of the AcrB multidrug efflux pump. Science 300:976–980
13. Pos KM, Schiefner A, Seeger MA, Diederichs K (2004) Crystallographic analysis of AcrB. FEBS Lett 564:333–339
14. Das D, Xu QS, Lee JY, Ankoudinova I, Huang C, Lou Y, DeGiovanni A, Kim R, Kim SH (2007) Crystal structure of the multidrug efflux transporter AcrB at 3.1 Å resolution reveals the N-terminal region with conserved amino acids. J Struct Biol 158:494–502
15. Drew D, Klepsch MM, Newstead S, Flaig R, De Gier JW, Iwata S, Beis K (2008) The structure of the efflux pump AcrB in complex with bile acid. Mol Membr Biol 25:677–682
16. Hung LW, Kim HB, Murakami S, Gupta G, Kim CY, Terwilliger TC (2013) Crystal structure of AcrB complexed with linezolid at 3.5 Å resolution. J Struct Funct Genom 14:71–75
17. Törnroth-Horsefield S, Gourdon P, Horsefield R, Brive L, Yamamoto N, Mori H, Snijder A, Neutze R (2007) Crystal structure of AcrB in complex with a single transmembrane subunit reveals another twist. Structure 15:1663–1673
18. Du D, Wang Z, James NR, Voss JE, Klimont E, Ohene-Agyei T, Venter H, Chiu W, Luisi BF (2014) Structure of the AcrAB-TolC multidrug efflux pump. Nature 509:512–515
19. Long F, Su CC, Zimmermann MT, Boyken SE, Rajashankar KR, Jernigan RL, Yu EW (2010) Crystal structures of the CusA efflux pump suggest methionine-mediated metal transport. Nature 467:484–488
20. Su CC, Long F, Zimmermann MT, Rajashankar KR, Jernigan RL, Yu EW (2011) Crystal structure of the CusBA heavy-metal efflux complex of *Escherichia coli*. Nature 470:558–562
21. Lei HT, Chou TH, Su CC, Bolla JR, Kumar N, Radhakrishnan A, Long F, Delmar JA, Do SV, Rajashankar KR, Shafer WM, Yu EW (2015) Crystal structure of the *Neisseria gonorrhoeae* MtrD inner membrane multidrug efflux pump. PLoS One 6:e97475
22. Murakami S, Nakashima R, Yamashita E, Matsumoto T, Yamaguchi A (2006) Crystal structures of a multidrug transporter reveal a functionally rotating mechanism. Nature 443:173–179
23. Seeger MA, Schiefner A, Eicher T, Verrey F, Diederichs K, Pos KM (2006) Structural asymmetry of AcrB trimer suggests a peristaltic pump mechanism. Science 313:1295–1298
24. Sennhauser G, Amstutz P, Briand C, Storchenegger O, Grütter MG (2006) Drug export pathway of multidrug exporter AcrB revealed by Darpin inhibitors. PLoS Biol 5:e7
25. Sennhauser G, Bukowska MA, Briand C, Grütter MG (2009) Crystal structure of the multidrug exporter MexB from *Pseudomonas aeruginosa*. J Mol Biol 389:134–145
26. Pak JE, Ekendé EN, Kifle EG, O'Connell JD 3rd, De Angelis F, Tessema MB, Derfoufi KM, Robles-Colmenares Y, Robbins RA, Goormaghtigh E, Vandenbussche G, Stroud RM (2013) Structures of intermediate transport states of ZneA, a Zn(II)/proton antiporter. Proc Natl Acad Sci U S A 110:18484–18489
27. Nakashima R, Sakurai K, Yamasaki S, Nishino K, Yamaguchi A (2011) Structures of the multidrug exporter AcrB reveal a proximal multisite drug-binding pocket. Nature 480:565–569

Chapter 3

Crystallographic Analysis of MATE-Type Multidrug Exporter with Its Inhibitors

Tsukasa Kusakizako, Yoshiki Tanaka, Christopher J. Hipolito, Hiroaki Suga, and Osamu Nureki

Abstract

Multidrug exporters expressed in pathogens efflux substrate drugs such as antibiotics, and thus, the development of inhibitors against them has eagerly been anticipated. Furthermore, the crystal structures of multidrug exporters with their inhibitors provide novel insights into the inhibitory mechanism and the development of more specific and effective inhibitors. We previously reported the complex structures of the Multidrug And Toxic compound Extrusion (MATE)-type multidrug exporter with the macrocyclic peptides, which inhibit the efflux of substrates by the MATE-type multidrug exporter (Tanaka et al., Nature 496:247–251, 2013). In this chapter, we describe methodologies of the screening and synthesis of macrocyclic peptides as inhibitors, as well as the purification, crystallization, and structure determination of the complexes of the MATE-type multidrug exporter with its inhibitors.

Key words Multidrug resistance, Multidrug exporter, Transporter, Membrane proteins, X-ray crystallography, Structural analysis, Macrocyclic peptide, Inhibitors, RaPID system

1 Introduction

Multidrug exporters function to maintain a cellular life by exporting toxic compounds [1, 2]. A Multidrug And Toxic compound Extrusion (MATE) transporter is one of multidrug exporter families, which exports the invading xenobiotics in cells using the electrochemical energy of a H^+ or Na^+ gradient across the membrane [3, 4]. Since MATE transporters expressed in pathogens contribute to the multidrug resistance of pathogens [5, 6], they are clinically important drug targets and the development of inhibitors against them has eagerly been anticipated.

The crystal structure of the complex of a multidrug exporter and its inhibitors provides useful information for the further design of effective inhibitors (e.g., Structure-Based Drug Design) as well as the structure of the complex of a multidrug exporter with substrate drugs. To date the several crystal structures of multidrug

Akihito Yamaguchi and Kunihiko Nishino (eds.), *Bacterial Multidrug Exporters: Methods and Protocols*, Methods in Molecular Biology, vol. 1700, https://doi.org/10.1007/978-1-4939-7454-2_3, © Springer Science+Business Media, LLC 2018

exporter family transporters have been reported, however the complex structures with inhibitors are relatively unknown. AcrB and its homolog MexB, both belonging to Resistance-Nodulation-Division (RND) family, are two of the few multidrug exporters whose complex structures with their inhibitors have been determined [7]. The complex structures revealed that the pyridopyrimidine derivatives, the AcrB- and MexB-specific inhibitors, bind to a phenylalanine cluster in the drug-binding pocket and sterically prevent AcrB and MexB from functionally rotating to export substrates. This study provides structural insights for the further development of inhibitors.

Although several crystal structures of MATE transporters have been available [8–12], the complex structures of MATE transporters with their inhibitors had not been reported due to the lagging of the development of inhibitors. We reported the first inhibitor-bound structures of a MATE transporter in 2013 [13]. In this chapter, we describe the methodology of crystallographic analysis of MATE transporter with its inhibitors. We discuss the screening and synthesis of macrocyclic peptides as inhibitors, as well as the purification, crystallization and structure determination of the complexes of MATE transporter with its inhibitors.

2 Materials

2.1 Purification of *Pyrococcus furiosus* MATE Transporter (PfMATE)

1. *E. coli* C41(DE3) ΔacrB strain, which can be requested from Dr. Koichi Ito (The University of Tokyo).
2. Luria Bertani (LB) media containing 100 μg/mL of ampicillin.
3. *E. coli* C41(DE3) Met minus strain.
4. L-Selenomethionine.
5. Se-Met core media (WAKO).
6. Isopropyl-β-D-thiogalactopyranoside (IPTG).
7. n-Dodecyl-β-D-maltoside (DDM; Calbiochem).
8. 6-Cyclohexyl-1-hexyl-β-D-maltoside (Cymal-6; Anatrace).
9. Phenylmethylsulfonyl fluoride (PMSF).
10. Lysis buffer (50 mM HEPES-NaOH, pH 7.0, 150 mM NaCl, 1 mM PMSF).
11. Nickel-nitrilotriacetic acid resin (Ni-NTA; Qiagen).
12. Buffer A: 20 mM HEPES-NaOH, pH 7.0, 150 mM NaCl, 0.05% DDM, 20 mM imidazole.
13. Buffer B: 20 mM HEPES-NaOH, pH 7.0, 150 mM NaCl, 0.05% DDM, 50 mM imidazole.
14. Buffer C: 20 mM HEPES-NaOH, pH 7.0, 150 mM NaCl, 0.05% DDM, 250 mM imidazole.

15. Buffer D: 20 mM HEPES-NaOH, pH 7.0, 150 mM NaCl, 0.02% DDM.
16. Buffer E: 20 mM HEPES-NaOH, pH 7.0, 20 mM NaCl, 0.06% Cymal-6.
17. Amicon Ultra centrifugal filter 50,000 MWCO (Millipore).
18. Superdex200 10/300 GL size-exclusion column (GE Healthcare).

2.2 Random Nonstandard Peptide Integrated Discovery (RaPID) System

2.2.1 N-Chloroacetyl-$^{L/D}$Phe-tRNA$^{fMet}{}_{CAU}$

1. 250 μM tRNA$^{fMet}{}_{CAU}$.
2. 3.0 M Magnesium chloride ($MgCl_2$).
3. 25 mM *N*-(2-chloroacetyl)-L-phenylalanine cyanomethyl ester.
4. 25 mM *N*-(2-chloroacetyl)-D-phenylalanine cyanomethyl ester.
5. 250 μM Flexizyme eFx.
6. 0.3 M Sodium acetate (AcONa).
7. 0.1 M AcONa in 70% ethanol.

2.2.2 Library Construction

1. 100 μM Reverse primer NNK$_n$ (*see* Table 1).
2. 100 μM Forward primer T7g10M.F48 (*see* Table 1).
3. 100 μM Reverse primer CGS3an13.R39 (*see* Table 1).
4. 3.0 M Sodium chloride (NaCl).
5. 25:24:1 Phenol:chloroform:isoamyl alcohol.
6. 24:1 Chloroform:isoamyl alcohol.
7. DNase I.
8. 100 mM Ethylenediaminetetracetic acid (EDTA).
9. 40% (19:1) Acrylamide:bis-acrylamide.

2.2.3 Puromycin Linker Ligation

1. Dimethyl sulfoxide (DMSO).
2. 1× T4 RNA Ligase buffer.

Table 1
Oligonucleotides used for construction of the mRNA library, reverse transcription, recovery of cDNA from the beads, and PCR amplifications

Oligonucleotides used for selection	Sequence (5′–3′)
T7g10M.F48	TAATACGACTCACTATAGGGTTAA*CTTTAAGAAGGAGATATACATATG*
NNK$_n$	**GCTGCCGCTGCCGCTGCCGCA** (MNN)$_n$ *CATATGTATATCTCCTTCTTAAAG*
CGS3an13.R39	TTTCCGCCCCCCGTCCTA**GCTGCCGCTGCCGCTGCCGCA**
Puromycin linker	d(pCTCCCGCCCCCCGTCC)-(SPC18)5-d(CC)-puromycin

3. 7.5 μM Puromycin linker.
4. T4 Ligase.

2.2.4 In Vitro Translation and Reverse Transcription

1. In vitro transcription kit (New England Biolabs).
2. 5× MMLV RT buffer.
3. 25 mM dNTP.
4. RNasin ribonuclease inhibitor (Promega, Madison, WI, USA).
5. M-MLV reverse transcriptase (Promega).

2.2.5 Desalting the Peptides

1. G-25 Sephadex.
2. 20 mM HEPES, pH 7.0, 150 mM NaCl, 0.1% (w/v) Cymal-6.
3. 2× Blocking solution (0.2% (w/v) acetylated bovine serum albumin, 20 mM HEPES, pH 7.0, 150 mM NaCl, 0.1% (w/v) Cymal-6).

2.2.6 Negative Selection

1. Magnetic beads (Dynabeads® His-Tag Isolation and Pulldown, Life Technologies).

2.2.7 Immobilization of PfMATE on Magnetic Beads

1. Purified His-tagged PfMATE.

2.2.8 Recovery and Quantification of cDNA from the Magnetic Beads

1. 1× Taq PCR Buffer.

2.3 Peptide Syntheses

1. NovaPEG resin (EMD Millipore).
2. Chloroacetyl *N*-hydroxysuccinimide.
3. Trifluoroacetic acid (TFA).
4. Diethyl ether.
5. 0.1% (v/v) Trifluoroacetic acid in 1:1 acetonitrile:water.
6. Triethylamine.

2.4 Crystallization

1. Purified protein sample.
2. Cyclic peptides as inhibitors.
3. Dimethyl sulfoxide (DMSO).
4. 1-Oleoyl-rac-glycerol (monoolein, Nu-Chek Prep).
5. Buffer E: 20 mM HEPES-NaOH, pH 7.0, 20 mM NaCl, 0.06% Cymal-6 (same as buffer used at the final step of the purification).
6. Harvest solution (derived from crystallization conditions).
7. Crystal screening kit (e.g., MemMeso™, Molecular Dimensions).

8. Gas-tight syringes (100 μL) × 2 (Hamilton Research).
9. LCP Syringe Couplers (Hamilton Research).
10. 96-Well plastic plate (SwissCi).

3 Methods

3.1 Purification of *Pyrococcus furiosus* MATE Transporter (PfMATE)

Although the expression level of membrane proteins is generally low, we should prepare ~mg of highly pure samples to determine crystal structures of membrane proteins. We screened MATE transporters suitable for structural analysis among ~20 organisms based on the expression level by western blotting (*see* **Note 1**). MATE transporter genes were cloned from ~20 organisms by a PCR method and inserted into the modified pET11a expression vector (Novagen, *see* **Note 2**). The pET11a-MATE plasmid was introduced into the *E. coli* C41(DE3) ΔacrB strain and the cells were grown in 3 mL of LB media containing 100 μg/mL at 37 °C. When an absorbance of 600 nm (A_{600}) reaches 0.4–0.6, the expression was induced with 1 mM IPTG and the expression condition was tested by for 3 h at 37 °C for 20 h at 37 °C and for 20 h at 20 °C. The cultured cells were collected by centrifugation at 4000 × *g* for 15 min. The resulting pellets were resuspended in 200 μL of the lysis buffer (50 mM HEPES-NaOH, pH 7.0, 150 mM NaCl, 1 mM PMSF) and lysed by sonication at 4 °C. The lysate was centrifuged at 15,000 × *g* for 15 min and the supernatant was subjected to SDS-PAGE. The expression level were confirmed by western blotting using the His-probe (H-15; Santa Cruz Biotechnology) as a first antibody and the mouse anti-rabbit IgG-HRP (Santa Cruz Biotechnology) as a second antibody. Chemi-Lumi One (Nacalai tesque) was used for the chemiluminescence reagent, and the chemiluminescence was detected by LAS-3000 (FUJIFILM). As a result, a MATE transporter from hyperthermophilic archaeal *Pyrococcus furiosus* (PfMATE) was selected as the best candidate for the structural analysis of a MATE transporter. The method below describes the purification from the large-scale culture enough to carry out the crystallization screening. All purification procedures are carried out at 4 °C unless described.

1. Inoculate a single colony of *E. coli* cells containing pET11a-PfMATE plasmids in 50 mL of LB media containing 50 μg/mL ampicillin and grow the cells for more than 8 h at 37 °C (*see* **Note 3**).
2. Inoculate the pre-culture in 5 L of LB media containing 50 μg/mL ampicillin and grow the cells for about 2.5 h at 37 °C.
3. When A_{600} reaches 0.4–0.6, the culture is chilled and the expression is induced with 0.5 mM IPTG for 20 h at 20 °C.
4. The cells are collected by centrifugation at 5000 × *g*.

5. Resuspend the cells in the lysis buffer and dilute cells to ~100 mL.
6. Centrifuge the lysate at 10,000 × *g* and discard the pellets.
7. Ultracentrifuge the supernatant at 106,000 × *g* for 1 h and resuspend the pellets in the lysis buffer. You can freeze the membrane fractions in liquid nitrogen and store them at −80 °C at this point.
8. Solubilize the membrane fractions in Buffer A with 2% DDM by stirring for 1 h at 4 °C.
9. Ultracentrifuge at 125,000 × *g* and discard the resulting pellets.
10. Mix the supernatant and 5 mL of Ni-NTA resin equilibrium with Buffer A and stir the mixture for 1 h at 4 °C.
11. Load the mixture on the econo-column and discard the flow-through fraction.
12. Wash the resin with tenfold column volumes of buffer B.
13. Elute the sample-containing fractions with two- to threefold column volumes of buffer C.
14. Add trypsin (Invitrogen) to a final concentration of 1 μg/mL into samples to cleave His_6-tag (*see* **Note 4**).
15. Dialyze trypsin-containing samples in Buffer D at 4 °C overnight.
16. The dialyzed samples are loaded on the Ni-NTA column equilibrated with Buffer D again to remove the cleaved His_6-tag.
17. Concentrate the flow-through fraction of the re-chromatogram in an Amicon Ultra-15 50,000 Da MWCO to less than 500 μL.
18. Concentrated samples are further purified on Superdex 200 10/300 GL (GE Healthcare) equilibrated with Buffer E.
19. The purity of the peak fractions can be confirmed by SDS-PAGE.
20. Pool the appropriate fractions and concentrate to ~8 mg/mL by an Amicon Ultra-4 50,000 Da MWCO (Millipore).
21. Concentrated samples are divided by appropriate volume (~8 μL) and can be flash frozen in liquid nitrogen and stored at −80 °C until use.

The selenomethionine-derivatized PfMATE (SeMet PfMATE) was prepared for the phase determination. The pET11-PfMATE plasmid was introduced into the *E. coli* C41(DE3) Met minus strain cells and the cells were grown in core media containing L-selenomethionine. The following purification procedures are performed in the same manner as the above native protein.

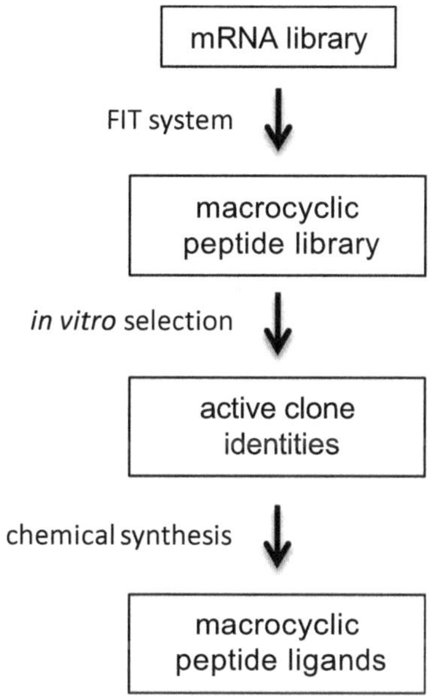

Fig. 1 General scheme of an in vitro selection for and production of macrocyclic peptide ligands

3.2 Random Peptide Integrated Discovery (RaPID) System

To discover potent molecules for inhibiting the transport activity of PfMATE, specific binding ligands for PfMATE were identified using the Random nonstandard Peptide Integrated Discovery (RaPID) system, an in vitro selection technique combining mRNA-display and the Flexible In vitro Translation (FIT) system [14] (Fig. 1, *see* **Note 5**).

3.2.1 N-Chloroacetyl-$^{L/D}$Phe-$tRNA^{fMet}_{CAU}$

In vitro translation is commonly initiated with a nonstandard *N*-chloroacetylated amino acid. Introduction of the nonstandard *N*-chloroacetyl-$^{L/D}$Phe into the translation system is accomplished by first charging initiator $tRNA^{fMet}_{CAU}$ with the nonstandard amino acid by using flexizyme eFx and chemically activated *N*-chloroacetyl-$^{L/D}$Phenylalanine cyanomethyl ester (Fig. 2).

1. Charge $tRNA^{fMet}_{CAU}$ (Round 1, 5.25 nmol; Round 2^+, 175 pmol) using eFx (Round 1, 5.25 nmol; Round 2^+, 175 pmol), magnesium (600 mM) and *N*-(2-chloroacetyl)-L-phenylalanine cyanomethyl ester or *N*-(2-chloroacetyl)-D-phenylalanine cyanomethyl ester (Round 1, 1.05 μmol; Round 2^+, 35 nmol) in a total volume of 210 μL for Round 1 or 7.0 μL for Round 2^+. Incubate reactions for 2 h on ice.
2. Precipitate the *N*-(2-chloroacetyl)-aminoacyl-$tRNA^{fMet}_{CAU}$ by adding 4 equivalent volumes of 0.3 M AcONa and 10 equivalent volumes of ethanol.
3. Centrifuge the mixture at 15,200 × *g* for 15 min and discard the supernatant.

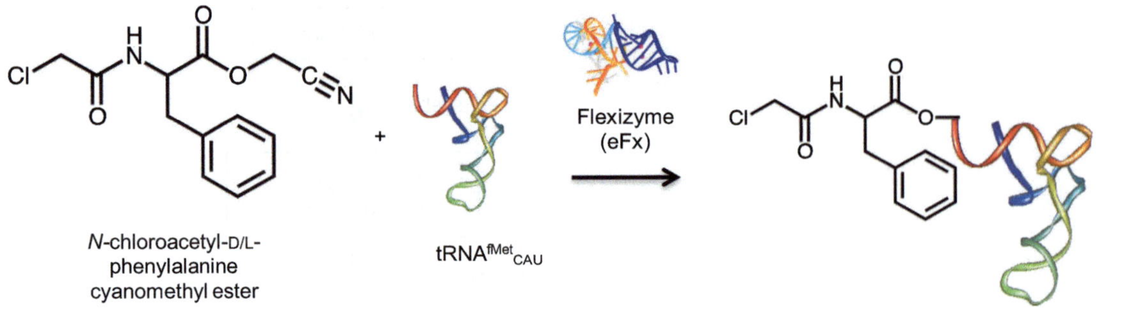

Fig. 2 Charging of $tRNA^{fMet}{}_{CAU}$ with a nonstandard amino acid

4. Wash the pellet with 5 equivalent volumes of 0.1 M AcONa in 70% ethanol with vigorous agitation.
5. Centrifuge the suspension at 15,200 × *g* for 15 min.
6. Repeat **steps 4** and **5**.
7. Wash the pellet once with 5 equivalent volumes of 70% ethanol.
8. Air-dry the pellet.
9. Store pellet at −80 °C for up to 1 month.

3.2.2 Library Construction

The nucleic acid library is designed to have a T7 promoter region, a defined initiation codon, a region of variable length of degenerate residue positions, a defined cysteine codon, a peptide linker region composed of codons coding for Gly-Ser-Gly-Ser-Gly-Ser and an amber stop codon (UAG). The library can be constructed with a random region coding for 7–15 degenerate amino acid positions using the codon NNK (*see* **Note 6**), where N represents any of the four 2′-deoxynucleosides and K represents either thymidine or 2′-deoxyguanosine. For convenience, the libraries will be referred to according to their random-region length, e.g., the library containing seven degenerate codons is called "NNK7."

1. Perform primer extension by adding equimolar amounts of forward primer T7g10M.F48 and one of the NNK_n reverse primers (Table 1). Thermocycle five times (94 °C/20 s; 55 °C/20 s; 72 °C/20 s) (*see* **Note 7**).
2. Add in 5 stoichiometric equivalents of forward primer T7g10M.F48 and reverse primer CGS3an13.R39 and thermocycle five times (94 °C/20 s; 55 °C/20 s; 72 °C/20 s).
3. Add 0.1 equivalent volume of 3 M NaCl and mix.
4. Extract proteins using 1 equivalent volume of 25:24:1 phenol:chloroform:isoamyl alcohol.
5. Recover the aqueous phase and extract with 24:1 chloroform:isoamyl alcohol.
6. Recover the aqueous phase.

7. Precipitate the amplicon by using 2 equivalents of ethanol.
8. Centrifuge at 15,200 × *g* for 15 min.
9. Discard the supernatant.
10. Dissolve the amplicon and add transcription reaction mixture.
11. Incubate the transcription reaction for at least 2 h at 37 °C.
12. Treat with DNase I and incubate for another 30 min at 37 °C.
13. Add EDTA to the reaction and precipitate the RNA with 0.8 equivalents of isopropanol.
14. Purify mRNA using an 8% denaturing polyacrylamide gel.
15. Standardize mRNA library concentrations to 20 μM.

3.2.3 Puromycin Linker Ligation

The puromycin moiety is a critical feature of mRNA-display [15, 16], which allows for the covalent linkage between nascent peptide and the mRNA that codes for it.

1. Mix mRNA libraries in a ratio of 160:40:10:2.5:0.625:0.156: 0.391:0.00977:0.00244 pmol for library templates NNK15: NNK14:NNK13:NNK12:NNK11:NNK10:NNK9:NNK8: NNK7, respectively.
2. Ligate mRNA with a puromycin linker (Table 1) using the following reaction: 20% DMSO, 1× T4 RNA ligase buffer, 1.5 μM puromycin linker (Round 1, 300 pmol; Round 2^+, 60 pmol), 1 μM mRNA (Round 1, 200 pmol; Round 2^+, 40 pmol), and T4 ligase (Round 1, 236 pmol; Round 2^+, 47.2 pmol) in a total volume of 200 μL for Round 1 and 40 μL for Round 2^+. Incubate the reaction mixture for 30 min at room temperature.
3. For Round 2^+ only, add 40 μL water.
4. Add a stop solution (0.6 M NaCl, 10 mM EDTA, pH 7.5; Round 1, 200 μL; Round 2^+, 80 μL).
5. Extract unwanted proteins with 1 equivalent volume of 25:24:1 phenol:chloroform:isoamyl alcohol. Discard the organic phase.
6. Extract a second time with 1 equivalent volume of 24:1 chloroform:isoamyl alcohol. Discard the organic phase.
7. Precipitate the mRNA-puromycin molecules with the addition of 2 equivalent volume of ethanol, mix, and centrifuge at 15,200 × *g*. Discard supernatant and wash pellet with 70% ethanol.
8. Dissolve mRNA-puromycin molecules in water and dilute to a final concentration of 5 μM.

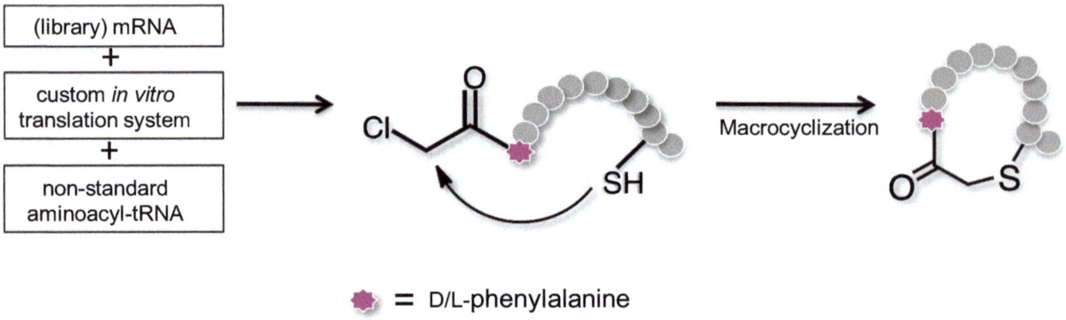

Fig. 3 Production of macrocyclic peptides using the FIT system

3.2.4 In Vitro Translation and Reverse Transcription

The puromycin-linked mRNA library constructed in Subheading 3.2.3 can be applied to a reconstituted in vitro translation system [17, 18] lacking methionine and release factor 1 (Fig. 3, *see* **Note 8**). With the exception of Round 1, the reverse transcription is often performed immediately after the in vitro translation for the production of mRNA-cDNA hybrids, and before the selection step, to avoid the isolation of RNA aptamers (*see* **Note 9**).

1. Add the mRNA-puromycin (Round 1, 200 pmol; Round 2^+, 7.5 pmol) to an in vitro translation mixture lacking methionine and release factor 1 (*see* **Note 10**). Add *N*-(2-chloroacetyl)-L-phenylalanine-$tRNA^{fMet}{}_{CAU}$ or *N*-(2-chloroacetyl)-D-phenylalanine-$tRNA^{fMet}{}_{CAU}$ (Round 1, 5.25 nmol; Round 2^+, 175 pmol) to the in vitro translation mixture and incubate reaction for 30 min at 37 °C in a total volume of 150 μL for Round 1 and 5 μL for Round 2^+. Incubate for 30 min at 37 °C.
2. After 30 min of incubation at 37 °C, place the reaction at room temperature and let stand for 12 min.
3. After 12 min at room temperature, add an EDTA solution (Round 1, 200 mM, 15 μL; Round 2^+, 100 mM, 1 μL), and incubate for 30 min at 37 °C.
4. For Round 1, add a reverse transcription mixture (1.25× MMLV RT buffer, 0.625 mM dNTP, 3.125 μM primer CGS3an13.R39, 8 U RNasin ribonuclease inhibitor (Promega, Madison, WI, USA), 300 U M-MLV reverse transcriptase (Promega), 40 μL) to the solution containing macrocyclic peptide-mRNA hybrids and incubate for 1 h at 42 °C. For Round 2^+, add a reverse transcription mixture (73.41 mM Tris–HCl, pH 8.3, 43.73 mM $Mg(OAc)_2$, 7.34 μM dNTPs, 4.93 μM CGS3an13.R39, 50 U M-MLV reverse transcriptase (−H) (Promega), 3.43 μL).
5. Add selection buffer to the transcription reaction (Round 1, 165 μL; Round 2^+, 21 μL).

3.2.5 Desalting the Peptides

The processes of in vitro translation followed by reverse transcription involve the addition of EDTA to sequester those divalent metals. Desalting the library helps remove potentially harmful EDTA.

1. Prepare a column of G-25 sephadex in 1 mL disposable syringes using media pre-swollen in selection buffer 20 mM HEPES, pH 7.0, 150 mM NaCl, 0.1% (w/v) Cymal-6 (Round 1, 2× 1 mL; Round 2^+, 0.7 mL).
2. Place each column in a 15 mL tube, and spin the columns gently at 800 × *g* for 1 min to remove excess buffer.
3. Place columns in a clean 15 mL tube for collection of the macrocyclic peptide library.
4. Add macrocyclic peptide library to the column and gently spin the columns at 800 × *g* for 1 min (*see* **Note 11**).
5. Add 1 equivalent volume of 2× blocking solution (0.2% (w/v) acetylated bovine serum albumin, 20 mM HEPES, pH 7.0, 150 mM NaCl, 0.1% (w/v) Cymal-6) to the macrocyclic peptide library (Round 1, 165 μL; Round 2^+, 30 μL).

3.2.6 Negative Selection

Prior to the selection step, the macrocyclic peptide library is "precleared" of unwanted nonspecific bead binders and translation components by incubating the library with magnetic beads.

1. Wash magnetic beads (Round 1, 150 μL worth, 6 mg, Round 2^+, 1 μL worth, 40 μg) with selection buffer (20 mM HEPES, pH 7.0, 150 mM NaCl, 0.1% (w/v) Cymal-6).
2. Incubate peptide library with the magnetic beads for 30 min at 4 °C with rotation. Separate the magnetic beads from the supernatant. Keep the supernatant and discard the beads. For Round 2^+, this process should be repeated at least three times; five times is recommended.
3. Separate and save 0.5 μL of "precleared" solution for qPCR analysis.

3.2.7 Immobilization of PfMATE on Magnetic Beads

His_6-tagged PfMATE is bound to the magnetic beads, and all unbound protein is washed away from the beads. Any remaining supernatant is separated from the beads prior to the application of the macrocyclic peptide (*see* **Note 12**).

1. Determine amount of protein loaded per μg of magnetic beads (*see* **Note 13**).
2. Wash magnetic beads (Round 1, 600 μg; Round 2^+, 40 μg) with selection buffer and incubate His_6-tagged PfMATE (Round 1, 751 pmol; Round 2^+, 1 μL, 50 pmol) with the magnetic beads in total volume of 150 μL of selection buffer for Round 1 and 10 μL for Round 2^+.

3. Incubate the beads and PfMATE for 20 min at room temperature with rotation.
4. After the 20 min at room temperature, thoroughly wash the beads 3–5 times with selection buffer. On the last wash, transfer the beads to a new tube.

3.2.8 Selection for Macrocyclic Peptides that Bind to MATE-Type Transporters

The selection step involves incubating the macrocyclic peptide library with the magnetic-bead-bound PfMATE transporter followed by removal of nonbinding macrocyclic peptides by washing. After incubation with the target, discard supernatant and thoroughly wash the beads to remove nonbinding peptides (*see* **Note 14**).

1. Incubate the "precleared" macrocyclic peptide library prepared in Subheading 3.2.6 and with the bead-bound PfMATE prepared in Subheading 3.2.7 for 1 h at 4°C with rotation.
2. After incubation, wash the beads at least three times with selection buffer to remove weakly binding and non-binding macrocyclic peptides.

3.2.9 Recovery and Quantification of cDNA from the Magnetic Beads

cDNA is recovered by resuspending the beads in a PCR reaction mixture lacking polymerase and heating the suspension to separate cDNA from mRNA.

1. Resuspend the beads in a PCR reaction mixture containing no polymerase, only 1 × Taq PCR Buffer, 2.5 mM $MgCl_2$, 0.5 mM dNTP, 0.25 μM primer T7g10M.F48, 0.25 μM primer CGS3an13.R39 (Round 1, 200 μL; Round 2^+, 100 μL).
2. Incubate the resuspended magnetic beads for 5 min at 95 °C.
3. After heating, collect the supernatant.
4. Separate 1 μL of this supernatant for quantification of the recovered cDNA by qPCR.
5. To the remaining supernatant, add Taq or another appropriate polymerase for PCR amplification (*see* **Note 15**).
6. Isolate and transcribe amplicon DNA into RNA for the next round of selection (if needed).

The procedure involving ligation of puromycin, reverse transcription, negative selection, positive selection, PCR, and transcription constitutes a round of selection (Fig. 4). After approximate six rounds of selection, cDNA can be submitted to sequencing, and the macrocyclic peptide identities are decoded (Fig. 5).

3.3 Peptide Syntheses

Peptide synthesis is carried out using standard Fmoc solid phase peptide synthesis and NovaPEG rink amide resin for the production of a C-terminal carboxamide.

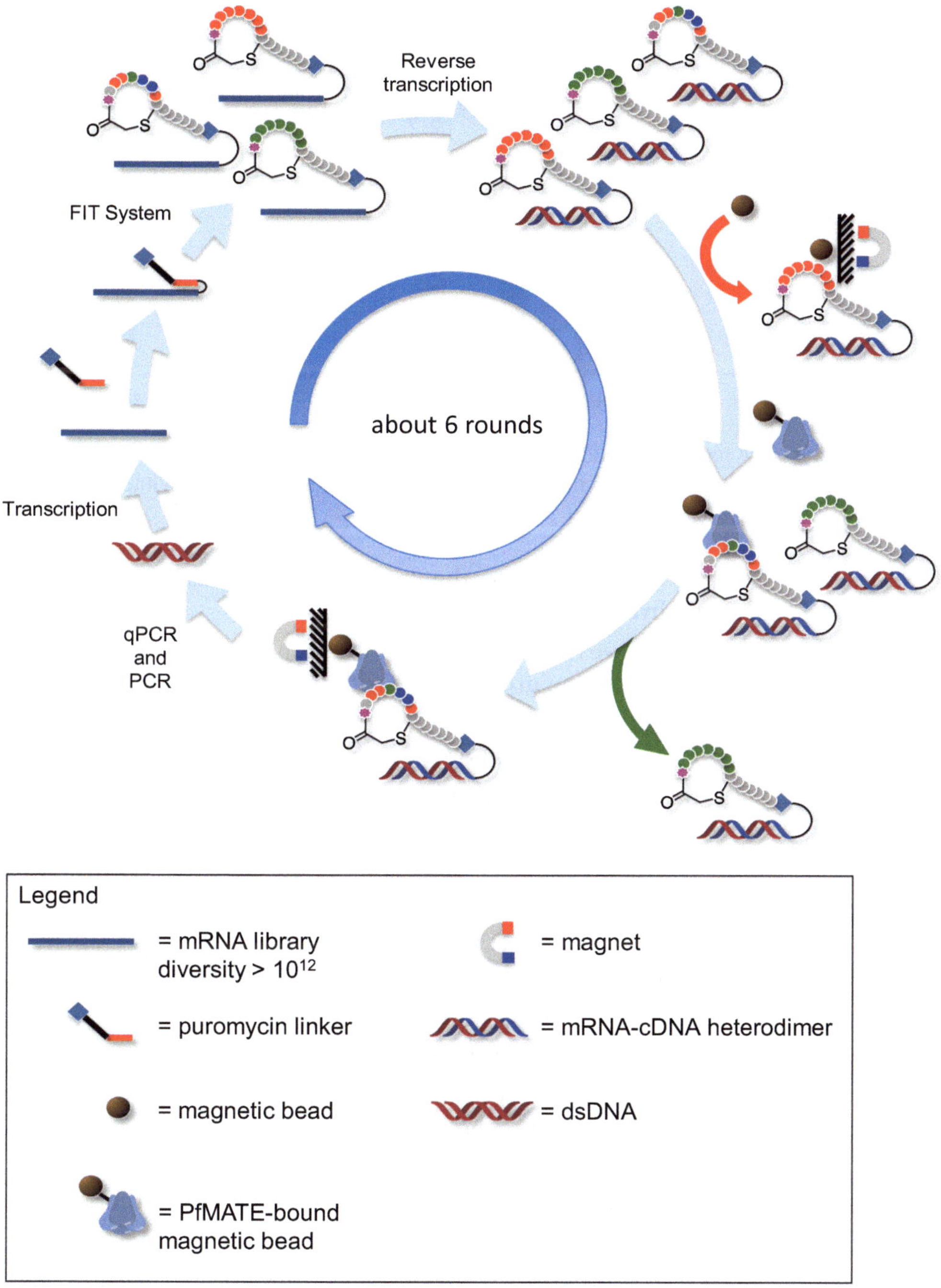

Fig. 4 Scheme of a selection for macrocyclic peptide ligands that bind to PfMATE

1. Perform standard Fmoc-based solid phase peptide synthesis to produce the linear peptide backbone.
2. Deprotect the N-terminal Fmoc protecting group.

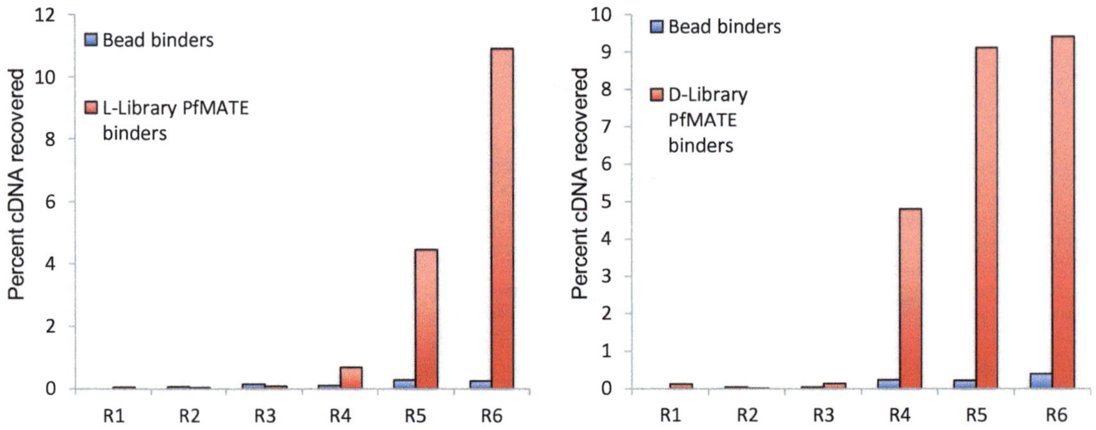

Fig. 5 Percent of the cDNA recovered during the various rounds of selection. Percent of the DNA recovered is calculated by dividing the number of cDNA molecules recovered by the total cDNA input into the selection step times 100%

3. Chloroacetylate the free N-terminus using chloroacetyl *N*-hydroxysuccinimide.
4. Cleave the peptides off of the resin and deprotect the acid-labile protecting groups using trifluoroacetic acid (1 mL per 50 μmol of peptide).
5. Collect the trifluoroacetic acid containing the cleaved peptide.
6. Add 10 mL of diethyl ether to precipitate the peptide.
7. Collect the precipitated peptide as a pellet by using a hand crank centrifuge. Do not use an electrical centrifuge with a diethyl ether solvent.
8. Wash the pellet five times with 10 mL of diethyl ether.
9. Dissolve the pellet in 50 mL 0.1% (v/v) trifluoroacetic acid in 1:1 acetonitrile:water.
10. Raise pH above 9 using about 30 μL of triethylamine.
11. Incubate for 1 h at room temperature with rotation. Periodically, monitor the progress of the macrocyclization and check the pH (*see* **Note 16**).
12. After macrocyclization is confirmed, remove the solvent using a rotary evaporator.
13. Dissolve the peptide in DMSO and purify using HPLC.

The obtained peptides were confirmed to have the inhibitory activities of the substrate transport by PfMATE, using the ethidium bromide (EtBr) accumulation assay [13, 14].

3.4 Crystallization

In general, a crystal of membrane proteins obtained by a usual vapor diffusion method is difficult to diffract a X-ray beam to high resolution since the hydrophobic regions of membrane proteins solubilized in micelles are covered by detergent molecules and consequently the contact regions of membrane proteins are reduced and the crystal packing tend to be poor. To overcome this situation, a lipidic cubic phase (LCP) crystallization method was developed by Dr. Landau and Dr. Rosenbusch instead of a vapor diffusion method [19]. In the LCP method, membrane proteins are reconstituted into the three-dimensional lipid bilayer, which is a more native membrane-like environment, and the tighter crystal packing is anticipated. In fact, many crystal structures of membrane proteins have been determined using the LCP crystallization method. Therefore, the LCP crystallization method can be applied to various membrane proteins more than expected at first. To determine high-resolution complex structures of PfMATE with macrocyclic peptides, the LCP crystallization method was carried out as described previously (*see* [20] and **Note 17**).

1. Macrocyclic peptides as inhibitors are dissolved in dimethyl sulfoxide (DMSO) to a 20 mM concentration because of their hydrophobicity.
2. Macrocyclic peptides dissolving in DMSO are added to the purified protein sample in a 20:1 protein to peptide ratio (v/v) and incubate for 1 h at 4 °C (*see* **Note 18**).
3. Monoolein stored at −20 °C is thawed in a block incubator for more than 15 min at 42 °C (*see* **Note 19**).
4. Attach a coupling part to one gas-tight syringe (Hamilton Research) and weigh the syringe-coupling part.
5. Load liquefied monoolein into the syringe-coupling part and weigh the monoolein.
6. Wash the internal of the other gas-tight syringe attached to a needle with Buffer E several times.
7. The incubated sample is sucked into the washed syringe in a 3:2 lipid to protein ratio (w/w).
8. Combine the two syringes by a coupler part and mix the monoolein and the protein well (*see* **Note 20**).
9. Using a crystallization robot, Mosquito LCP (TTP LabTech), 100 nL of the LCP sample is spotted on the 96-well plates and 1 μL the reservoir solution (e.g., MemMeso™ crystallization screening kit, Molecular Dimensions) is overlaid (*see* **Note 21**). The 96-well plates are sealed with a plastic film and crystallization drops are sandwiched (*see* **Notes 22** and **23**).
10. Crystals are grown to full size for 5–7 days at 20 °C.

For the phase determination, the SeMet PfMATE sample was also crystallized in the same manner as described above. The cocrystallization of SeMet PfMATE with the macrocyclic peptide MaL6 improved the quality and reproducibility of the crystals, and thus facilitate the structure determination. The reservoir solution in which crystallization was performed is as below:

For the SeMet PfMATE with the macrocyclic peptide MaL6; 30–32% PEG400, 100 mM MES-NaOH, pH 6.5, 100 mM magnesium acetate.

For the native PfMATE; 28–30% PEG400, 50 mM MES-NaOH, pH 6.0–8.0, 20 mM $CaCl_2$, 100 mM NaSCN.

For the native PfMATE with macrocyclic peptide; 26–28% PEG550MME, 100 mM Tris-HCl, pH 8.0, 100 mM Li_2SO_4.

3.5 Data Collection and Structure Determination

Although the obtained crystals using the LCP crystallization method were so tiny, the X-ray diffraction data sets could be collected from a single crystal using a micro-focus beam with 1 μm width and 5 μm height on BL32XU at SPring-8 [21]. The crystal structure of PfMATE was determined from the data set using the SeMet PfMATE-MaL6 complex crystal by the single anomalous diffraction (SAD) phasing method. In fact, although the only SeMet PfMATE crystal was poor diffraction-quality, the SeMet PfMATE-MaL6 complex crystal showed the high reproducibility and improved diffraction-quality, facilitating the structure determination. The data sets were collected by the helical data collection method [22] and processed using the HKL2000 program (HKL Research). The heavy atom sites were identified using the program SnB [23] and the experimental phases were calculated with the program autoSHARP [24]. The initial homology model was built from NorM-VC (PDB ID: 3MKU) with the program MODELLER [25]. The model was manually rebuilt using the program Coot [26] and refined with the programs PHENIX [27] and AutoBUSTER [28]. The crystal structures of the native PfMATE and the complexes with macrocyclic peptides MaL6, MaD5, and MaD3S were determined at 2.4–3.0 Å resolutions by molecular replacement with the structure of SeMet PfMATE with MaL6 using the program Phaser [29] (Fig. 6).

4 Notes

1. For a pre-crystallization screening, the Fluorescence-Detection Size-Exclusion (FSEC) is a high throughput and effective method [30, 31]. Recently, many membrane proteins whose crystal structures have been reported were screened using the FSEC method. In this typical method, a target gene fused a green fluorescent protein (GFP) is expressed first. Cells expressing GFP-fused proteins are solubilized in detergents and then applied into a gel filtration column without purified. The potential for the crystallization of membrane proteins is evaluated by the expression level and the monodispersity of candidates based on the chromatography profile.

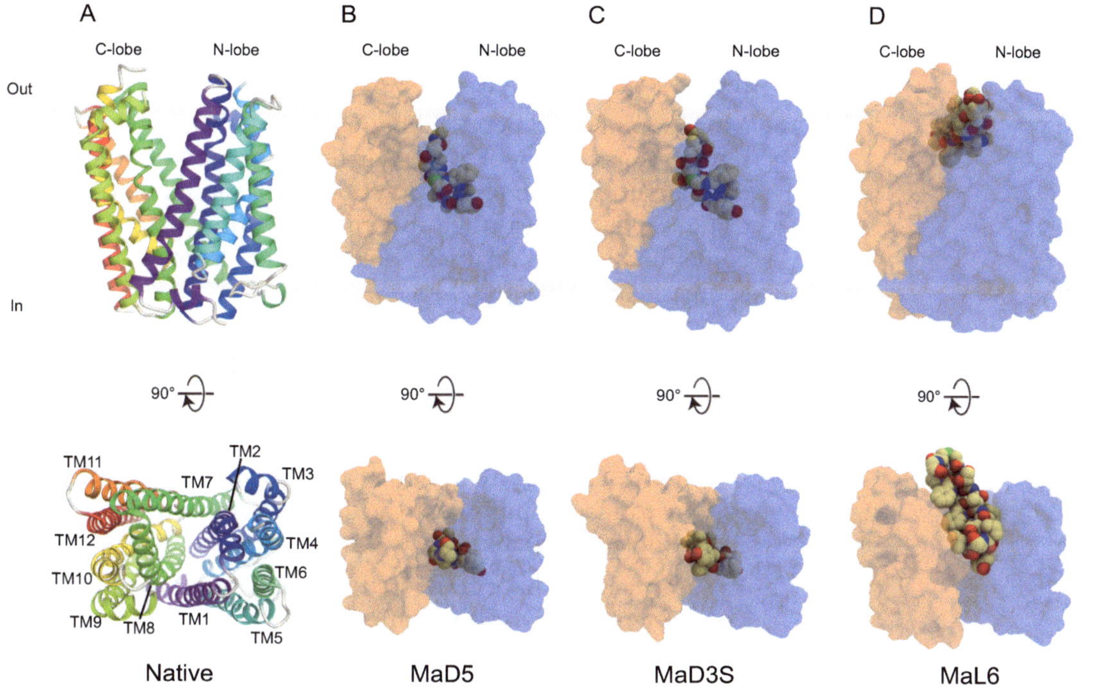

Fig. 6 (**a**) Ribbon diagram of the crystal structure of the native PfMATE (PDB ID: 3VVN). The 12 transmembrane helices (TM) are divided into the N-lobe (TM1–6) and the C-lobe (TM7–12). (**b–d**) Surface models of complex structures of PfMATE with macrocyclic peptides MaD5, MaD3S and MaL6, respectively (PDB ID: 3WBN, 3VVR and 3VVS, respectively). The peptides bind to the central cleft, and inhibit the EtBr efflux activity of PfMATE [13, 14]. The molecular graphics were created with the program CueMol (http:/www.cuemol.org/)

2. In the modified pET11a vector, which has the T7 expression system, the hexa-histidine tag (His_6-tag) was introduced into the C-terminus of the inserted gene construct to detect by anti-His-tag antibodies.
3. It is useful to make a glycerol stock from the *E. coli* C41(DE3) ΔacrB strain transformed by the pET11a-PfMATE plasmid for further studies.
4. We performed the limited digestion of PfMATE by trypsin. As a result, a single band was observed by SDS-PAGE. The resulting sequence was confirmed to be from Ser2 to Lys452 by N-terminal sequencing and mass spectroscopy in collaborated with Dr. Domae (RIKEN Advanced Science Institute).
5. For a detailed description of a fully customizable in vitro translation system called the Flexible In vitro Translation (FIT) system [32].
6. Use of NNK reduces the appearance of stop codons, but still allows for the appearance of the codon AUG in the elongation step, which is typically misread and leads to misincorporation of leucine, isoleucine, or valine.

7. The number of cycles during the construction is kept low to minimize the copy number for any one sequence and maximize the diversity in a given sample size.
8. More complicated peptide libraries, such as those with heavily reprogrammed genetic codes for the introduction of nonstandard amino acids in elongation positions, should be translated using a fully reconstituted in vitro translation system to avoid contamination by natural amino acid from endogenous sources, e.g., ARSs [32].
9. Due to the larger scale of Round 1, reverse transcription is performed after the selection step using a less-expensive wild-type reverse transcriptase.
10. The original selection was performed using an in-house-produced translation system, but an alternative in vitro translation system can be composed from commercial sources, such as New England Biolabs.
11. Ensure that you have recovered the same volume of liquid as you have put in. Discard used columns.
12. After the transporter is applied to the beads, the beads may aggregate if left too long as a pellet. Resuspend the beads as quickly as possible to avoid aggregation.
13. The author recommends adding enough protein during the immobilization to bind 50% of the protein to the magnetic beads and leave 50% remaining in the supernatant.
14. Beads may still aggregate if left as a pellet for too long. Resuspend as soon as possible.
15. Use the crosspoint observed during the qPCR to determine the number of cycles to use in the amplification of the isolated cDNA. Typically, the crosspoint +2–4 cycles is enough to provide an observable amplicon in agarose gel analysis.
16. Since one of the by-products is hydrochloric acid, the pH will decrease as the macrocyclization reaction proceeds. Add additional triethylamine if the reaction has not reached completion.
17. We have tried the vapor diffusion crystallization method prior to the LCP method. As a result of the initial crystallization screening, preliminary crystals were obtained under several crystallization and detergent conditions. However, the crystals diffracted X-rays to about 5 Å at a maximum in spite of the large size of crystals (~300 μm). In contrast, crystals obtained from the initial crystallization screening using the LCP method diffracted X-rays to 2.5 Å. Since then, we optimized the conditions using the LCP method.
18. Be sure not to exceed the volume DMSO dissolving macrocyclic peptides more than 1/20th of the total volume because

DMSO may interfere the phase transition of lipids. Fortunately, it is not critical to contain 5% (v/v) of DMSO in the protein-peptide-monoolein mixtures.

19. Thawed monoolein is often solidified at room temperature especially in the winter season. After weighing the monoolein loaded in the syringe attached to the coupler part, it is good that they are incubated at 42 °C during the preparation of the other protein sample-loaded syringe.
20. Mix the monoolein and protein sample well by shuttling both plungers about 100–500 times until the protein and lipid mixture looks transparent. When the mixture remains a white turbidity, mixing it on ice is effective.
21. The conditions of usual crystallization screening kits are necessarily suitable for the LCP method because they are optimized for the vapor diffusion method. We made the screening kit optimized for the LCP conditions, MemMeso™ screening kit (Molecular Dimensions). In our laboratory, the initial crystallization screening of membrane proteins in the LCP has been performed using MemMeso™ screening kit.
22. The crystallization trial in the LCP method can be manually performed using a Repeating Dispenser (Hamilton).
23. When the crystallization condition is optimized, the sitting drop or the hanging drop method should be tried instead of the sandwich method. The crystals of the sitting drop or the hanging drop plate would be easy to pick up compared to that of the sandwich plate.

References

1. Dawson RJP, Locher KP (2006) Structure of a bacterial multidrug ABC transporter. Nature 443:180–185. https://doi.org/10.1038/nature05155
2. Murakami S, Nakashima R, Yamashita E et al (2006) Crystal structures of a multidrug transporter reveal a functionally rotating mechanism. Nature 443:173–179. https://doi.org/10.2142/biophys.47.309
3. Brown MH, Paulsen IT, Skurray RA (1999) The multidrug efflux protein NorM is a prototype of a new family of transporters. Mol Microbiol 31:394–395. https://doi.org/10.1046/j.1365-2958.1999.01162.x
4. He G, Kuroda T, Mima T et al (2004) An H+-coupled multidrug efflux pump, PmpM, a member of the MATE family of transporters, from Pseudomonas aeruginosa. J Bacteriol 186:262–265. https://doi.org/10.1128/JB.186.1.262
5. Kaatz GW, Mcaleese F, Seo SM (2005) Multidrug resistance in Staphylococcus aureus due to overexpression of a novel multidrug and toxin extrusion (MATE) transport protein. Antimicrob Agents Chmother 49:1857–1864. https://doi.org/10.1128/AAC.49.5.1857
6. McAleese F, Petersen P, Ruzin A et al (2005) A novel MATE family efflux pump contributes to the reduced susceptibility of laboratory-derived Staphylococcus aureus mutants to tigecycline. Antimicrob Agents Chemother 49:1865–1871. https://doi.org/10.1128/AAC.49.5.1865-1871.2005
7. Nakashima R, Sakurai K, Yamasaki S et al (2013) Structural basis for the inhibition of bacterial multidrug exporters. Nature 500:102–106. https://doi.org/10.1038/nature12300
8. He X, Szewczyk P, Karyakin A et al (2010) Structure of a cation-bound multidrug and toxic compound extrusion transporter. Nature

467:991–994. https://doi.org/10.1038/nature09408

9. Lu M, Symersky J, Radchenko M et al (2013) Structures of a Na+-coupled, substrate-bound MATE multidrug transporter. Proc Natl Acad Sci U S A 110:2099–2104. https://doi.org/10.1073/pnas.1219901110
10. Lu M, Radchenko M, Symersky J et al (2013) Structural insights into H+-coupled multidrug extrusion by a MATE transporter. Nat Struct Mol Biol 20:1310–1317. https://doi.org/10.1038/nsmb.2687
11. Radchenko M, Symersky J, Nie R, Lu M (2015) Structural basis for the blockade of MATE multidrug efflux pumps. Nat Commun 6:7995. https://doi.org/10.1038/ncomms8995
12. Mousa JJ, Yang Y, Tomkovich S et al (2016) MATE transport of the E. coli-derived genotoxin colibactin. Nat Microbiol 1:15009. https://doi.org/10.1038/nmicrobiol.2015.9
13. Tanaka Y, Hipolito CJ, Maturana AD et al (2013) Structural basis for the drug extrusion mechanism by a MATE multidrug transporter. Nature 496:247–251. https://doi.org/10.1038/nature12014
14. Hipolito CJ, Tanaka Y, Katoh T et al (2013) A macrocyclic peptide that serves as a cocrystallization ligand and inhibits the function of a MATE family transporter. Molecules 18:10514–10530. https://doi.org/10.3390/molecules180910514
15. Roberts RW, Szostak JW (1997) RNA-peptide fusions for the in vitro selection of peptides and proteins. Proc Natl Acad Sci U S A 94:12297–12302. https://doi.org/10.1073/pnas.94.23.12297
16. Nemoto N, Miyamoto-Sato E, Husimi Y, Yanagawa H (1997) In vitro virus: bonding of mRNA bearing puromycin at the 3′-terminal end to the C-terminal end of its encoded protein on the ribosome in vitro. FEBS Lett 414:405–408. https://doi.org/10.1016/S0014-5793(97)01026-0
17. Shimizu Y, Inoue A, Tomari Y et al (2001) Cell-free translation reconstituted with purified components. Nat Biotechnol 19:751–755. https://doi.org/10.1038/90802
18. Shimizu Y, Kanamori T, Ueda T (2005) Protein synthesis by pure translation systems. Methods 36:299–304. https://doi.org/10.1016/j.ymeth.2005.04.006
19. Landau EM, Rosenbusch JP (1996) Lipidic cubic phases: a novel concept for the crystallization of membrane proteins. Proc Natl Acad Sci U S A 93:14532–14535. https://doi.org/10.1073/pnas.93.25.14532
20. Caffrey M, Cherezov V (2009) Crystallizing membrane proteins for structure-function studies using lipidic mesophases. Nat Protoc 4:706–731. https://doi.org/10.1007/978-94-007-6232-9-4
21. Hirata K, Kawano Y, Ueno G et al (2013) Achievement of protein micro-crystallography at SPring-8 beamline BL32XU. J Phys Conf Ser 425:012002. https://doi.org/10.1088/1742-6596/425/1/012002
22. Flot D, Mairs T, Giraud T et al (2010) The ID23-2 structural biology microfocus beamline at the ESRF. J Synchrotron Radiat 17:107–118. https://doi.org/10.1107/S0909049509041168
23. Xu H, Smith AB, Sahinidis NV, Weeks CM (2008) SnB version 2.3: triplet sieve phasing for centrosymmetric structures. J Appl Crystallogr 41:644–646. https://doi.org/10.1107/S0021889808007966
24. Vonrhein C, Blanc E, Roversi P, Bricogne G (2007) Automated structure solution with autoSHARP. Methods Mol Biol 364:215–230. https://doi.org/10.1385/1-59745-266-1:215
25. Eswar N, Webb B, Marti-Renom MA et al (2007) Comparative protein structure modeling using MODELLER. Curr Protoc Protein Sci. https://doi.org/10.1002/0471140864.ps0209s50
26. Emsley P, Lohkamp B, Scott WG, Cowtan K (2010) Features and development of Coot. Acta Crystallogr Sect D Biol Crystallogr 66:486–501. https://doi.org/10.1107/S0907444910007493
27. Adams PD, Afonine PV, Bunkóczi G et al (2010) PHENIX: a comprehensive python-based system for macromolecular structure solution. Acta Crystallogr Sect D Biol Crystallogr 66:213–221. https://doi.org/10.1107/S0907444909052925
28. Smart OS, TO W, Flensburg C et al (2012) Exploiting structure similarity in refinement: automated NCS and target-structure restraints in BUSTER. Acta Crystallogr Sect D Biol Crystallogr 68:368–380. https://doi.org/10.1107/S0907444911056058
29. McCoy AJ, Grosse-Kunstleve RW, Adams PD et al (2007) Phaser crystallographic software. J Appl Crystallogr 40:658–674. https://doi.org/10.1107/S0021889807021206
30. Kawate T, Gouaux E (2006) Fluorescence-detection size-exclusion chromatography for precrystallization screening of integral membrane proteins. Structure 14:673–681. https://doi.org/10.1016/j.str.2006.01.013

31. Hattori M, Hibbs RE, Gouaux E (2012) A fluorescence-detection size-exclusion chromatography-based thermostability assay for membrane protein precrystallization screening. Structure 20:1293–1299. https://doi.org/10.1016/j.str.2012.06.009

32. Goto Y, Katoh T, Suga H (2011) Flexizymes for genetic code reprogramming. Nat Protoc 6:779–790. https://doi.org/10.1038/nprot.2011.331

Chapter 4

Crystallographic Analysis of the CusBA Heavy-Metal Efflux Complex of *Escherichia coli*

Jared A. Delmar and Edward W. Yu

Abstract

Crystallization is one of the most successful techniques used to determine protein structure, especially for membrane proteins. However, the application of this technique is not straightforward and often hampered by the difficulties associated with expression, purification, and crystallization. Here we present our protocol and methodology for crystallizing the CusBA adaptor–transporter complex of *Escherichia coli*. Using these procedures, we were able to produce the first co-crystal structure of a resistance-nodulation-cell division (RND) transporter in complex with its associated membrane fusion protein.

Key words Membrane protein, X-ray crystallography, Vapor diffusion, Heavy-metal efflux, Antimicrobial resistance

1 Introduction

Approximately 90% of all protein structures have been solved by X-ray crystallography, which is by far the most successful technique used to study structures of bio-macromolecules. However, only approximately 1% of all structures appearing in the Protein Data Bank (PDB) are membrane proteins [1]. Understanding the structures and action mechanisms of these important membrane proteins is crucial for the development of novel therapeutic strategies to combat diseases and often hinges on the results of difficult crystallization experiments.

In *E.* coli, the CusA efflux pump is the only transporter that belongs to the heavy-metal efflux RND (HME-RND) family. This inner membrane efflux pump (or transporter) forms a complex with the CusB membrane fusion protein (or adaptor), CusC outer membrane channel, and CusF periplasmic metallochaperone to specifically recognize and confer resistance to Cu^+ and Ag^+ ions. Our laboratory has determined crystal structures of the CusA inner membrane transporter [2] and CusB periplasmic adaptor [3]. We have also crystallized two mutants of the CusC outer membrane

Akihito Yamaguchi and Kunihiko Nishino (eds.), *Bacterial Multidrug Exporters: Methods and Protocols*, Methods in Molecular Biology, vol. 1700, https://doi.org/10.1007/978-1-4939-7454-2_4, © Springer Science+Business Media, LLC 2018

channel and improved the resolution of the crystal structure of wild-type CusC [4], which has been determined previously by Kulathila et al. [5]. In addition, we have resolved the first crystal structure of the CusBA adaptor–transporter complex [6, 7]. This invaluable structural information has allowed us to elucidate how these proteins assemble to form a functional complex and understand how they cooperate to expel toxic Cu^{+} and Ag^{+} ions from the bacterial cell. It is our hope that by presenting our available methodologies, a rational approach can eventually be made to the crystallization of these important membrane proteins.

2 Materials

Prepare all solutions using ultrapure water (prepared by purifying deionized water to attain a sensitivity of 18 MΩ cm at 25 °C) and analytical-grade reagents. Diligently follow all waste disposal regulations when disposing waste materials. We do not add sodium azide to the reagents.

2.1 Transformation of cusA

1. Plasmid: 100 ng/μL: Amplify the open reading frame of *cusA* from *Escherichia coli* K12 genomic DNA by PCR using the primers 5′ AAACATATGATTGAATGGATTATTCGTCGCTCGGTGG 3′ and 5′ AAACTCGAGTTATTTCCGTACCCGATGTCGGTGCAGC 3′. Purify the 3146 basepair PCR fragments using a gel extraction kit. Digest using NdeI and XhoI restriction enzymes. Ligate the digested products and the pET15b expression vector with T4 DNA ligase to generate pET15bΩ*cusA*. Transform the recombinant plasmid into DH5α cells and select on LB agar plates. Purify the plasmid with a final concentration of 100 ng/μL using a Spin Miniprep Kit. Verify the construction using Sanger sequencing. Store plasmid stock at −20 °C Dilute to 10 ng/μ L with autoclaved water before use.
2. Competent cells: Prepare competent cells from BL21(DE3)Δ *acrB* cells using the $CaCl_2$ method [8] (*see* **Note 1**).
3. Petri dishes: Round.
4. LB agar plates: 25 mg/mL Luria Broth (LB), 15 mg/mL Bacto agar, 100 μg/mL ampicillin. Add 300 mL of water to a 500 mL glass Erlenmeyer flask. Weigh 7.5 g of LB and 4.5 g Bacto agar and add to the flask. Cover and autoclave at 121 °C for 20 min. Incubate in a 65 °C water bath until the solution cools to 65 °C (approximately 1 h). Add 300 μL ampicillin solution and stir until completely dissolved. Pour immediately into cell-culture dishes, approximately 20 mL per plate. Avoid bubble formation. Allow gel to cool before use (approximately 3 h).

2.2 Bacterial Growth and Expression of CusA

1. Ampicillin solution: 100 mg/mL ampicillin (*see* **Note 2**).
2. 60 mL LB media: 25 mg/mL LB. Add 60 mL of water to a 250 mL Erlenmeyer flask. Weigh 1.5 g of LB and add to the flask. Cover and autoclave at 121 °C for 20 min. Cool to 25 °C before use.
3. 1 L LB media: 25 mg/mL LB. Add 1 L of water to a 1 L Erlenmeyer flask. Weigh 25 g of LB and add to the flask. Cover and autoclave at 121 °C for 20 min. Cool to 25 °C before use.
4. IPTG solution: 1 M isopropyl β-D-1-thiogalactopyranoside (IPTG). Weigh 285 mg of IPTG and add to a 1.7 mL microtube. Add autoclaved water to a final volume of 1.2 mL and vortex until completely dissolved (*see* **Note 3**).

2.3 Purification of CusA

1. Tissue grinder.
2. PMSF solution: 200 mM phenylmethanesulfonylfluoride (PMSF). Weigh 1.742 g of PMSF and add to a new 50 mL falcon tube. Add 100% ethanol to a final volume of 50 mL and vortex until dissolved. Store at −20 °C.
3. Low-salt buffer: 500 mL, 100 mM sodium phosphate (pH 7.4), 10% glycerol, 5 mM ethylenediaminetetraacetic acid (EDTA) and 1 mM PMSF. Add 300 mL water to a clean beaker. Weigh 0.422 g NaH_2PO_4, 6.61 g Na_2HPO_4 and 0.9305 g EDTA and add slowly while stirring until dissolved. Add 50 mL 100% glycerol solution and stir until dissolved. Add water to a final volume of 500 mL and store in a closed container at 4 °C (*see* **Note 4**).
4. High-salt buffer: 500 mL, 20 mM sodium phosphate (pH 7.4), 2 M KCl, 5 mM EDTA, and 1 mM PMSF. Add 300 mL water to a clean beaker. Weigh 0.0844 g NaH_2PO_4, 1.322 g Na_2HPO_4, 74.56 g KCl, and 0.9305 g EDTA and add slowly while stirring until dissolved. Add water to a final volume of 500 mL and store in a closed container at 4 °C. Before use, add 500 μL PMSF to 100 mL of high salt buffer (*see* **Note 4**).
5. Final buffer: 20 mM 4-(2-Hydroxyethyl)piperazine-1-ethanesulfonic acid sodium salt (Na-HEPES) (pH 7.5).
6. Ni-NTA purification system.
7. 3500 MWCO dialysis cassette.
8. 50,000 MWCO centrifugal filter.
9. Pierce BCA protein assay kit.

2.4 Transformation of cusB

1. Plasmid: 100 ng/μL. Amplify the open reading frame of *cusB* from *Escherichia coli* K12 genomic DNA by PCR using the primers 5′ AAACCATGGGCAAAAAAATCGCGCTTATTATCGGC 3′ and 5′ AAAGGATCCTCAATGGTGATGGTGATGATGATGCGCATGGGTAGCACTTTCAG 3′. Purify

the 1224 basepair PCR fragments using a gel extraction kit. Digest using NcoI and BamHI restriction enzymes. Ligate the digested products and the pET15b expression vector with T4 DNA ligase to generate pET15bΩ*cusB*. Transform the recombinant plasmid into DH5α cells and select on LB agar plates. Purify the plasmid with a final concentration of 100 ng/μL using a Spin Miniprep Kit. Verify the construction using Sanger sequencing. Store plasmid stock at −20 °C Dilute to 10 ng/μL with autoclaved water before use.

2. Competent cells: Prepare competent cells from BL21(DE3) host cells using the $CaCl_2$ method [8].
3. Petri dishes: Round.
4. LB agar plates: 25 mg/mL LB, 15 mg/mL Bacto agar, 100 μg/mL ampicillin (*see* Subheading 2.1, **item 4**).

2.5 Bacterial Growth and Expression of CusB

1. Ampicillin solution: 100 mg/mL ampicillin (*see* **Note 2**).
2. 60 mL LB media: 25 mg/mL LB (*see* Subheading 2.2, **item 2**).
3. 1 L LB media: 25 mg/mL LB. (*see* Subheading 2.2, **item 3**).
4. IPTG solution: 1 M IPTG (*see* Subheading 2.2, **item 4**).

2.6 Purification of CusB

1. Wash buffer: 20 mM Na-HEPES (pH 7.5) and 150 mM NaCl.
2. Final buffer: 20 mM Na-HEPES (pH 7.5).
3. Ni-NTA purification system.
4. 3500 MWCO dialysis cassette.
5. 10,000 MWCO centrifugal filter.
6. Pierce BCA protein assay kit.

2.7 Crystallization of the CusBA Complex

1. 24 well crystallization plates.
2. Glass cover slides.
3. High vacuum grease.
4. 50% polyethylene glycol (PEG) 6000 (*see* **Note 5**).
5. 1 M Na-HEPES (pH 7.5).
6. 1 M ammonium acetate.
7. 50% glycerol.
8. Cryoloop.

3 Methods

3.1 Transformation of cusA

1. Mix 1 μL of pET15bΩ*cusA* plasmid (10 ng/μL) with 50 μL of competent cells. Incubate on ice for 15 min. Stir every 5 min.

2. Heat shock cells by incubating in a 42 °C water bath for 45 s. Return to ice immediately.
3. Plate cells evenly to an LB agar plate containing 100 μg/mL ampicillin to select for the transformed cells. Incubate the plate for 14–16 h at 37 °C. Small distinct colonies should be observed which correspond to transformed BL21(DE3)Δ *acrB* cells containing the plasmid pET15bΩ*cusA* (*see* **Note 6**).

3.2 Bacterial Growth and Expression of CusA

1. Add 1 μL of ampicillin solution into 1 mL LB media. Inoculate a single colony of the transformed BL21(DE3)Δ*acrB*/pET15bΩ*cusA* cells into the LB media.
2. Shake culture overnight at 37 °C.
3. Add the overnight culture to 60 mL LB media. Add 60 μL of ampicillin solution. Shake the culture at 37 °C until the optical density at $\lambda = 600$ nm ($OD_{600\ nm}$) reaches 1.5 (approximately 2 h).
4. Add 10 mL of the small culture to 1 L of LB media. Add 1 mL of ampicillin solution. Repeat for each of six large cultures.
5. Shake large cultures at 37 °C until $OD_{600\ nm}$ reaches 0.4–0.5 (approximately 2.5 h).
6. Add 200 μL of IPTG solution to each culture.
7. Continue to shake each culture at 37 °C for 3 h.
8. Harvest cells via centrifuge at 4000 × *g* for 10 min.

3.3 Purification of CusA

1. Resuspend the harvested cells in 100 mL chilled low-salt buffer (*see* **Note** 7).
2. Disrupt cell membranes using a French pressure cell.
3. Collect membrane fractions via ultracentrifugation at 150,000 × *g* for 45 min.
4. Discard supernatant. Homogenize pellet in 100 mL chilled high-salt buffer.
5. Collect membrane fractions via ultracentrifugation at 150,000 × *g* for 45 min.
6. Repeat **steps 4** and **5**.
7. Discard supernatant. Homogenize pellet in 50 mL final buffer.
8. Add 1 g of 6-cyclohexyl-1-hexyl-β-D-maltoside (CYMAL-6) (2% w/v final concentration) and stir the membrane suspension for 3 h using a magnetic stirrer.
9. Remove insoluble material via ultracentrifugation at 150,000 × *g* for 45 min.
10. Pre-equilibrate Ni-NTA affinity column by washing with three column volumes of final buffer, supplemented with 0.05% (w/v) CYMAL-6.

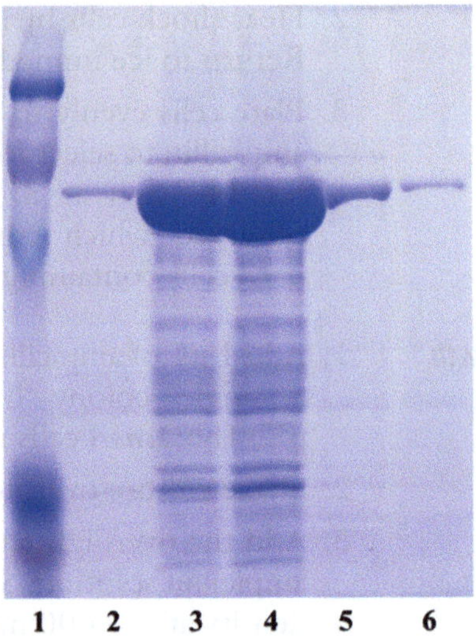

Fig. 1 10% SDS-PAGE of the purified CusA protein. Each lane corresponds to 10 μL of each corresponding 2 mL elution fraction. Fractions were collected which contained >90% pure CusA protein (Marker, *lane 1*; Purified CusA protein, *lanes 2–6*)

11. Collect raw protein solution from **step 9** and load into pre-equilibrated Ni-NTA affinity column (*see* **Note 8**).
12. Wash loaded affinity column with five column volumes of final buffer supplemented with 20 mM imidazole and 0.05% CYMAL-6, followed by three column volumes supplemented with 50 mM imidazole and 0.05% CYMAL-6.
13. Elute protein with three column volumes of final buffer supplemented with 300 mM imidazole and 0.05% CYMAL-6. Collect eluted protein solution in fractions of approximately 2 mL each.
14. Judge purity of fractions by 10% SDS-PAGE (Fig. 1). Collect only fractions with >90% purity.
15. Dialyze collected fractions against 1 L of final buffer for approximately 6–8 h. Repeat twice with freshly prepared buffer.
16. Judge concentration of collected fractions by BCA protein assay. Concentrate protein to 20 mg/mL using a 50 kDa cutoff centrifugal concentrator.

3.4 Transformation of cusB

1. Mix 1 μL of pET15bΩ*cusB* plasmid (10 ng/μL) with 50 μL of BL21(DE3) competent cells. Incubate on ice for 15 min. Stir every 5 min.
2. Heat shock cells by incubating in a 42 °C water bath for 45 s. Return to ice immediately.

3. Plate cells evenly to an LB agar plate containing 100 μg/mL ampicillin to select for the transformed cells. Incubate the plate for 14–16 h at 37 °C. Small distinct colonies should be observed which correspond to transformed BL21(DE3) cells containing the plasmid pET15bΩ*cusB* (*see* **Note 6**).

3.5 Bacterial Growth and Expression of CusB

1. Add 1 μL of ampicillin solution into 1 mL LB media. Inoculate a single colony of the transformed BL21(DE3)/pET15bΩ*cusB* cells into the LB media.
2. Shake culture overnight at 37 °C.
3. Add the overnight culture to 60 mL LB media. Add 60 μL of ampicillin solution. Shake the culture at 37 °C until $OD_{600\ nm}$ reaches 1.5 (approximately 2 h).
4. Add 10 mL of the small culture to 1 L of LB media. Add 1 mL of ampicillin solution. Repeat for each of six large cultures.
5. Shake large cultures at 37 °C until $OD_{600\ nm}$ reaches 0.4–0.5 (approximately 2.5 h).
6. Add 200 μL of IPTG solution to each culture.
7. Continue to shake each culture at 37 °C for 3 h.
8. Harvest cells via centrifuge at 4000 × *g* for 10 min.

3.6 Purification of CusB

1. Resuspend collected cells from growth **step** 7 in 100 mL wash buffer (*see* **Note** 7).
2. Disrupt cell membranes using a French pressure cell.
3. Separate insoluble membrane fractions via ultracentrifugation at 150,000 × *g* for 45 min.
4. Pre-equilibrate Ni-NTA affinity column by washing with two column volumes of wash buffer.
5. Collect raw protein solution from **step 4** and load into pre-equilibrated Ni-NTA affinity column (*see* **Note 8**).
6. Wash loaded affinity column with five column volumes of wash buffer supplemented with 20 mM imidazole, followed by three column volumes supplemented with 50 mM imidazole.
7. Elute protein with three column volumes of final buffer supplemented with 300 mM imidazole. Collect eluted protein solution in fractions of approximately 2 mL each.
8. Judge purity of fractions by 12.5% SDS-PAGE (Fig. 2). Collect only fractions with >90% purity.
9. Dialyze collected fractions against 1 L of final buffer for approximately 6–8 h. Repeat twice with freshly prepared buffer.
10. Judge concentration of collected fractions by BCA protein assay. Concentrate protein to 20 mg/mL using a 10 kDa cutoff centrifugal concentrator.

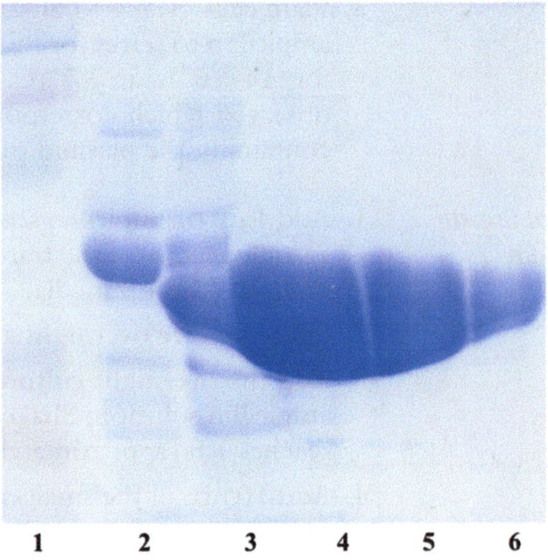

Fig. 2 12.5% SDS-PAGE of the purified CusB protein. Each lane corresponds to 10 μL of each corresponding 2 mL elution fraction. Fractions were collected which contained >90% pure CusB protein (Marker, *lane 1*; Purified CusB protein, *lanes 2–6*)

3.7 Crystallization of the CusBA Complex

1. Mix 36.3 μL purified CusA solution with 13.7 μL purified CusB solution (CusA:CusB molar ratio of 1:1). Adjust the final concentration of CYMAL-6 to 0.05% (w/v) by adding additional CYMAL-6 detergent solution (10% stock) into the protein mixture. Incubate at 4 °C for 2 h.
2. Coat the rims of a 24-well plate with high vacuum grease.
3. In each well, combine 100 μL PEG 6000, 50 μL Na-HEPES (pH 7.5), 50 μL ammonium acetate, 200 μL glycerol, and 100 μL water. Mix thoroughly to form the well solution.
4. Pipette 2 μL of the concentrated CusBA protein complex solution and drop it to the center of a new cover slide. Pipette 2 μL of well solution from the well and place it to the top of the protein drop.
5. Invert the cover slide over the well and center it. Press down gently on the outer edge of the slide to seal the well.
6. Repeat crystallization **steps 3–5** for each well of the crystallization plate.
7. Incubate tray at room temperature until crystals grow to full size (Fig. 3) (approximately 2 months) (*see* **Note 9**).
8. In a new well, combine 100 μL PEG 6000, 50 μL Na-HEPES (pH 7.5), 50 μL ammonium acetate, 250 μL glycerol, and 50 μL water. Mix thoroughly to form a cryoprotectant solution.
9. To the center of a new cover slide, pipette 2 μL of the well solution from the previous step.

Fig. 3 Crystal of the CusBA adaptor–transporter complex. Crystals of CusBA were obtained using vapor diffusion at room temperature. These crystals grew to full size in drops within 2 months

10. Using a nylon CryoLoop, carefully transfer a fully grown crystal of CusBA to the new drop.
11. Invert the cover slide over the new well and center it. Press down gently on the outer edge of the slide to seal the well.
12. Incubate the crystal at room temperature for 15 min.
13. In a new well, combine 100 μL PEG 6000, 50 μL Na-HEPES (pH 7.5), 50 μL ammonium acetate and 300 μL glycerol. Mix thoroughly to form a cryoprotectant solution.
14. Repeat **steps 9–12**.
15. Invert the cover slide and place it on a microscope.
16. Using a nylon CryoLoop, carefully pick up the incubated CusBA crystal. Freeze the crystal in liquid nitrogen.

3.8 Structural Determination and Refinement of the CusBA Complex

1. Collect diffraction data for both the CusA (native)-CusB (native) complex and the CusA (native)-CusB (selenomethionine substitute) complex at 100K using the ADSC Quantum 315 CCD_based detector in beamline 24ID-C at the Advanced Photon Source (Fig. 4).
2. With the program Phaser [9], use single-wavelength anomalous dispersion phasing data for the CusA (native)-CusB (selenomethionine substitute) cocrystal to obtain experimental phases, in addition to the phases from the structural model of the apo-form of CusA.
3. Subject phases to density modification and phase extension to 2.90 Å resolution using the program RESOLVE [10]. The full-length CusB protein contains nine methionine residues, and six selenium sites per CusB molecule (12 selenium sites per asymmetrical unit) can be identified (Fig. 5).

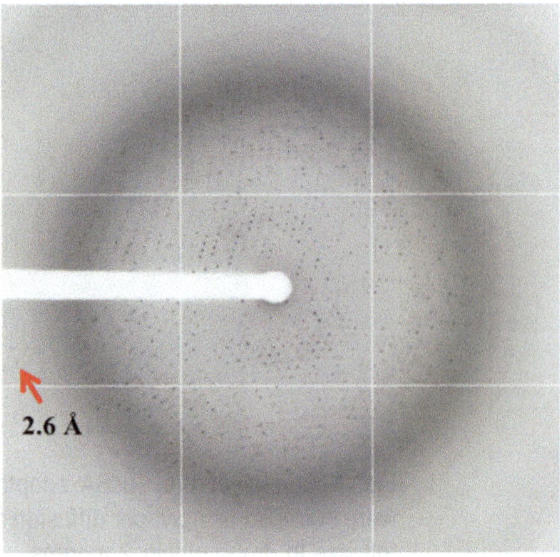

Fig. 4. X-ray diffraction pattern of the native CusBA crystal. The crystal diffracted X-rays beyond a resolution of 2.6 Å

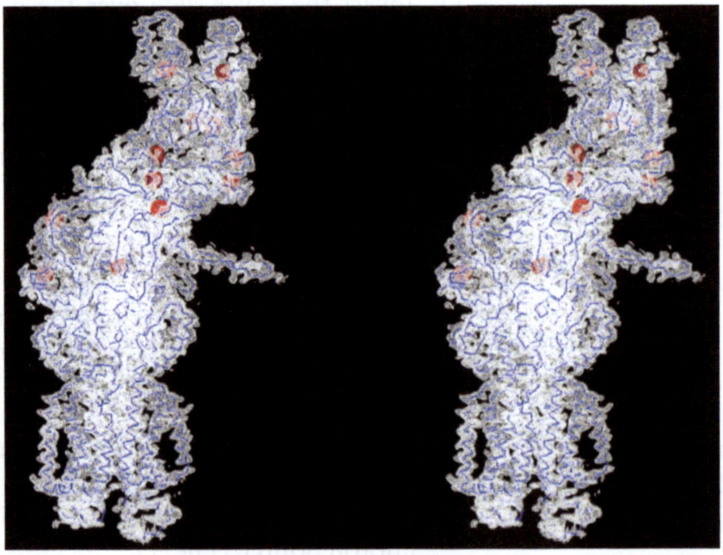

Fig. 5 Stereo view of the experimental electron density map at a resolution of 2.9 Å. The electron density map contoured at 1.2σ is in *white*. Each subunit of CusB consists of nine methionines. Twelve selenium sites, six from each CusB molecule, are found within the asymmetric unit of the CusA (native)-CusB (SeMet) cocrystal. The anomalous maps of the selenium sites contoured at 4σ are in *red*. The Cα traces of the CusA and two CusB molecules are in *blue*

4. After tracing the initial model manually using the program Coot [11], refine the model against the native data at 2.90 Å resolution using translation, libration and screw rotation (TLS) refinement, adopting a single TLS body as implemented in the

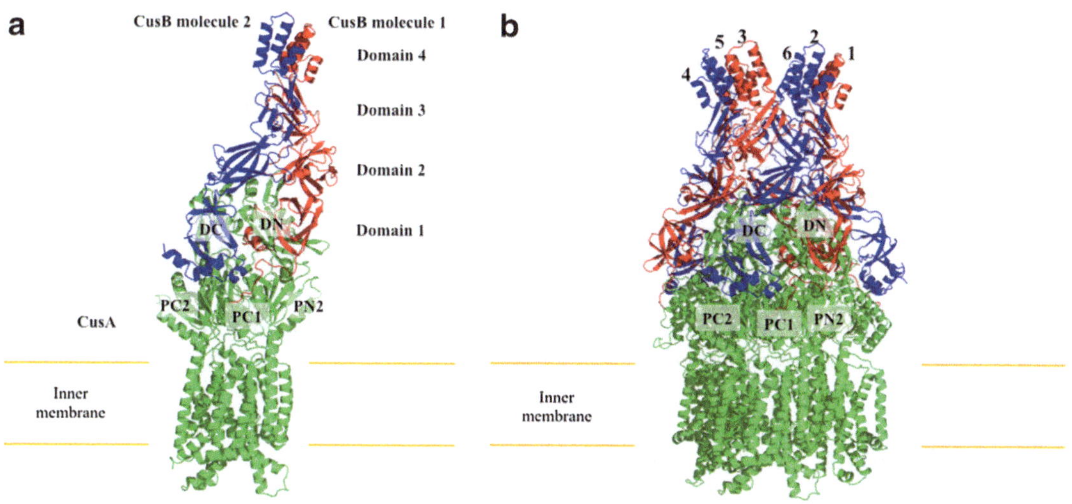

Fig. 6 Crystal structure of the CusBA efflux complex. (**a**) Ribbon diagram of the structures of one CusA protomer (*green*) and two CusB protomers (*red* and *blue*) in the asymmetric unit of the crystal lattice. (**b**) Side view of the CusBA efflux complex. Each subunit of CusA is colored green. Molecules 1, 3, and 5 of CusB are colored *red*. Molecule 2, 4, and 6 of CusB are in *blue*

program PHENIX [12], leaving 5% of the reflections in the free R set. Iterations of refinement using PHENIX [12] and model building in Coot [11] led to the current model, which consists of 1686 amino acid residues (residues 4–1043 of CusA, residues 79–400 of molecule 1 of CusB and residues 79–402 of molecule 2 of CusB) (Fig. 6).

4 Notes

1. During preliminary crystallization trials, we found that the protein AcrB co-purified with CusA and contaminated our sample. Mass spectrometry confirmed our initial crystals were of AcrB instead of CusA. Thus, we made an *E. coli* knockout strain BL21(DE3)Δ*acrB*, which harbors a deletion in the chromosomal *acrB* gene.
2. Due to its instability in solution, ampicillin should be stored dry at 4 °C and solutions prepared immediately before use. Use autoclaved water only. Filter solution using a 0.2 μm sterile syringe filter.
3. Similarly, IPTG should be stored dry at −20 °C and solutions prepared immediately before use. Use autoclaved water only. Filter solution using a 0.2 μm sterile syringe filter.
4. Low-salt and high-salt buffer should be stored without PMSF. Add PMSF as needed immediately before use.

5. All crystallization solutions filtered using a 0.2 μm sterile syringe filter.
6. Satellite colonies must be avoided, which may develop for longer incubation times.
7. All purification steps at 4 °C unless noted. Chill all buffers before use.
8. All cell lysate and wash buffers should pass through nickel column at approximately 5 μL/s.
9. Crystallization tray should remain undisturbed as vibrations adversely affect crystal growth.

Acknowledgements

This work was supported by an NIH grant R01AI114629 to E.W.Y.

References

1. White SH (2015) Membrane proteins of known 3D structure. http://blanco.biomol.uci.edu/mpstruc/
2. Long F, Su CC, Zimmermann MT, Boyken SE, Rajashankar KR, Jernigan RL, Yu EW (2010) Crystal structures of the CusA efflux pump suggest methionine-mediated metal transport. Nature 467:484–488
3. Su CC, Yang F, Long F, Reyon D, Routh MD, Kuo DW, Mokhtari AK, Van Ornam JD, Rabe KL, Hoy JA, Lee YJ, Rajashankar KR, Yu EW (2009) Crystal structure of the membrane fusion protein CusB from *Escherichia coli*. J Mol Biol 393:342–355
4. Lei HT, Bolla JR, Bishop NR, Su CC, Yu EW (2014) Crystal structures of CusC review conformational changes accompanying folding and transmembrane channel formation. J Mol Biol 426:403–411
5. Kulathila R, Kulathila R, Indic M, van den Berg B (2011) Crystal structure of *Escherichia coli* CusC, the outer membrane component of a heavy metal efflux pump. PLoS One 6:e15610
6. Su CC, Long F, Zimmermann MT, Rajashankar KR, Jernigan RL, Yu EW (2011) Crystal structure of the CusBA heavy-metal efflux complex of *Escherichia coli*. Nature 470:558–563
7. Su CC, Long F, Lei HT, Bolla JR, Do SV, Rajashankar KR, Yu EW (2012) Charged amino acids (R83, E567, D617, E625, R669, and K678) of CusA are required for metal ion transport in the Cus efflux system. J Mol Biol 422:429–441
8. Pope B, Kent HM (1996) High efficiency 5 min transformation of *Escherichia coli*. Nucleic Acids Res 43:536–537
9. McCoy AJ, Grosse-Kunstleve RW, Adams PD, Winn MD, Storoni LC, Read RJ (2007) *Phaser* crystallographic software. J Appl Crystallogr 40:658–674
10. Terwilliger TC (2001) Maximum-likelihood density modification using pattern recognition of structural motifs. Acta Crystallogr D 57:1755–1762
11. Emsley P, Cowtan K (2004) Coot: model-building tools for molecular graphics. Acta Crystallogr D 60:2126–2132
12. Adams PD, Grosse-Kunstleve RW, Hung LW, Ioerger TR, McCoy AJ, Moriarty NW, Read RJ, Sacchettini JC, Sauter NK, Terwilliger TC (2002) PHENIX: building new software for automated crystallographic structure determination. Acta Crystallogr 58:1948–1954

Chapter 5

Purification of AcrAB-TolC Multidrug Efflux Pump for Cryo-EM Analysis

Dijun Du, Zhao Wang, Wah Chiu, and Ben F. Luisi

Abstract

The cell envelope of Gram-negative bacteria comprises an outer membrane, a cytoplasmic inner membrane, and an interstitial space. The tripartite multidrug transporter AcrAB-TolC, which uses proton electrochemical gradients to vectorially drive the efflux of drugs from the cell, spans this envelope. We describe here details of the methods used to prepare the recombinant tripartite assembly for high-resolution structure determination by cryo-EM.

Key words Transmembrane protein assembly, Outer membrane protein, Inner membrane, Cryo-EM, Three-dimensional structure, Transport mechanism, RND family

1 Introduction

The method of electron cryo-microscopy (cryo-EM) has become well established as the state-of-the-art tool to elucidate the three-dimensional structures of biologically important macromolecules [1]. The wide success of cryo-EM to determine structures has resulted from the development of high-quality microscopes with optics that deliver more coherent electron beams, and more recently to the advent of sensitive recording devices that have high spatial resolution and detective quantum efficiency. These key hardware developments have gone hand in hand with powerful algorithms such as maximum likelihood principles to classify noisy images of particles into groups of two-dimensional projections for three-dimensional reconstruction.

The specimens to be visualized by cryo-EM are embedded in a thin layer of vitreous ice in small holes within a carbon layer on an EM grid and are routinely prepared by blotting off excess solution followed by rapid cooling by plunging into liquid ethane. The molecular weight limit for the method is steadily dropping, as is the requirement for conformational heterogeneity. However, cryo-EM structure determination is critically dependent on finding

Akihito Yamaguchi and Kunihiko Nishino (eds.), *Bacterial Multidrug Exporters: Methods and Protocols*, Methods in Molecular Biology, vol. 1700, https://doi.org/10.1007/978-1-4939-7454-2_5, © Springer Science+Business Media, LLC 2018

conditions to prepare specimens with good number of particles that are intact and have withstood the forces of blotting, surface tension and the rapid cooling process. Another criterion is that the particles should also be well separated and not aggregated into incommensurate bodies.

Studying membrane proteins by cryo-EM has the additional consideration of using detergents that are needed to mask the hydrophobic, transmembrane portion of the protein. Several successful studies have identified suitable detergents or mixtures of detergents and lipids that sustain the stability and structural homogeneity. The optimal choice of detergent will differ for individual proteins, but popular detergents such as beta-dodecylmaltoside and amphipathic polymers have proven very effective for several EM studies with small particles [2–6].

When the membrane protein of interest is involved in large complexes that span more than one membrane, as in the case of tripartite assemblies like AcrAB-TolC that span the cell envelope, the additional problem arises as to how to maintain two different membrane proteins stably in solution, and how to mimic the periplasm organization and keep the components together. Simply mixing the detergent-solubilized components together has limited success, due to the small fraction of full-assembled pumps that are generated spontaneously. We have developed a procedure of fusing the components together to increase their association through chelate cooperative effects. Here, we describe the details of the experimental approach to prepare the specimens for structural and biophysical analysis.

2 Materials

Prepare all solutions using double-deionized water ($_{dd}H_2O$). Prepare and store all reagents and solutions at 4 °C unless otherwise indicated.

2.1 Reagents for Molecular Cloning

1. Primers for PCR amplification and site-directed mutagenesis (*see* Table 1).
2. Phire Hot Start II DNA Polymerase (Thermo Scientific).
3. *PfuTurbo* DNA Polymerase (2.5 U/μl) (Agilent Technologies).
4. In-Fusion® HD Cloning Kit (Clontech).
5. 10 mM each of dATP, dTTP, dGTP, dCTP.
6. Restriction Enzymes: *NcoI*, *SalI*, *NdeI*, *XhoI*, *BamHI*, and 10xCutSmart® Buffer (New England Biolabs).
7. QIAquick Gel Extraction Kit (Qiagen).
8. LigaFast™ Rapid DNA Ligation System (Promega).

Table 1
Primers for the generation of constructs

$AcrB_{NdeI}$_F	5′-ATGCATATGCCTAATTTCTTTATCGATCGCCCG-3′
$AcrB_{XhoI}$_R	5′-GAGCTCGAGATGATGATCGACAGTATGGCTGTGCTC-3′
$AcrB_{C}$_F	5′-TGAAGATATCGAGCACAGCCATACTGTCGATTGAGAT CCGGCTGCTAACAAAGCCC-3′
$AcrB_{C}$_R	5′-GGGCTTTGTTAGCAGCCGGATCTCAATC GACAGTATGGCTGTGCTCGATATCTTCA-3′
$AcrZ_{NcoI}$_F	5′-TGGCCATGGGCTTAGAGTTATTAAAAAG TCTGGTATTCGCCGTAATCATGG-3′
$AcrZ_{His5\,SalI}$_R	5′-GACGTCGACTCAGTGGTGGTGGTGGTGA TGATTTTGTCCGGGCTGGTCTTTTTTACC-3′
$AcrA_{NcoI}$_F	5′-TGGCCATGGGCAACAAAAACAGAGGGTT TACGCCTCTGGCG-3′
$AcrA_{GSx3}$_R1	5′-TGGCCATGGATCCGCCGCCACCAGAGCCA CCACCGCCGCTCCCACCGCCACCAGACTTGGACTGT TCAGGCTGAGCACC-3′
$TolC_{inf}$_F	5′-AAGGAGATATACATATGAAGAAATTGCTCCC CATTCTTATCGGCC-3′
$TolC_{1392inf}$_R	5′-TTGAGATCTGCCATATGTCAATCAGCAATAG CATTCTGTTCCGGCGT-3′
$AcrB_{D328}$_F	5′-CTGAAAATTGTTTACCCATAGACGGATCCACC ACGCCGTTCGTGAAA-3′
$AcrB_{D328}$_R	5′-TTTCACGAACGGCGTGGTGGATCCGTCGTAT GGGTAAACAATTTTCAG-3′
$AcrA_{25GS}$_F	5′-GGATCCGGTGGGAGCGGTGGCGGCGGTAGTGGCGGTGG TGGCTCTTGTGACGACAAA CAGGCCCAACAAGG-3′

(continued)

Table 1 (continued)

$AcrA_{GSx3}$_R2	5′-GGATCCGCCGCCACCAGAGCCACCACCGCC GCTCCCACCGCCACCAGACTTGGACTGTTCAGGCT GAGCACC-3′
$AcrA_{inf}$_F	5′-CCCATACGACGGATCCGGTGGGAGCGGTGG CGGCGGTA-3′
$AcrA_{inf}$_R	5′-ACGGCGTGGTGGATCCGCCGCCACCAGAGCC ACCACC-3′

9. 10×Tris-Borate-EDTA (TBE) buffer (Sigma).
10. Agarose, molecular biology grade for 1% agarose gels (Melford).
11. Ethidium bromide: 10 mg/ml stock solution.
12. GeneJET Plasmid Miniprep Kit (Thermo Scientific).

2.2 Protein Expression and Purification

1. Isopropyl-β-D-galactopyranoside (IPTG): Stock solution of 1 M (store at −20 °C).
2. Carbenicillin and kanamycin: Stock solution of 100 and 50 mg/ml, respectively (store at −20 °C).
3. Lysozyme.
4. 2×YT Broth: 16 g Bacto Tryptone, 10 g Bacto Yeast Extract, and 5 g NaCl for 1 L medium.
5. *n*-Dodecyl-β-D-Maltopyranoside (DDM), Anagrade (Anatrace).
6. Amphipol A8-35: Stock solution of 100 mg/ml (store at −20 °C).
7. Imidazole: Stock solution of 5 M, pH: 7.5.
8. Deoxyribonuclease I (DNase I): Stock solution of 1000 U/μl (store at −20 °C).
9. EDTA-free protease inhibitor cocktail tablets (cOmplete Tablets, Roche).
10. EmulsiFlex-C5 high-pressure homogenizer.
11. HiTrap™ Chelating HP 1 ml column (GE Healthcare Life Sciences).
12. Lysis buffer: 20 mM Tris pH: 8.0 and 400 mM NaCl.
13. IMAC Wash buffer-I: 20 mM Tris pH: 8.0, 400 mM NaCl, 50 mM Imidazole and 0.03% DDM.
14. IMAC Wash buffer-II: 20 mM Tris pH: 8.0, 400 mM NaCl, 70 mM Imidazole and 0.03% DDM.
15. IMAC elution buffer: 20 mM Tris pH: 8.0, 400 mM NaCl, 500 mM Imidazole and 0.03% DDM.
16. GF buffer: 20 mM Tris pH: 8.0, 400 mM NaCl and 0.03% DDM.
17. Superose™ 6 size-exclusion column (GE Healthcare Life Sciences).
18. SDS-PAGE: Apparatus, accessories, and gels.
19. Bio-Beads™ SM-2 Adsorbent (Bio-Rad).
20. Vivaspin concentrator, MWCO: 100 kDa (Sartorius).

3 Methods

3.1 Construction of Vectors for Overexpression of AcrABZ–TolC Complex

The plasmids pET-*acrAB* and pRSF-$acrAZ_{His5}$-$tolC_{1392}$ are used to coexpress AcrAB, AcrAZ fusion proteins and C-terminal-truncated TolC to reconstitute the AcrABZ-TolC tripartite efflux pump in vivo.

3.1.1 Generation of Construct pET-acrAB

1. For the amplification of *acrB* gene from genomic DNA of *E. coli* W3110 strain, use primers $AcrB_{NdeI}$_F and $AcrB_{XhoI}$_R to introduce the *NdeI* and *XhoI* restriction sites to the 5′- and 3′-end of *acrB*. Insert the fragment of *acrB* bounded by *NdeI* and *XhoI* sites into the multiple cloning site of expression vector pET21a to construct the pET21a-$acrB_{His6}$ vector.
2. To perform site-directed mutagenesis of *acrB*, use pET-$acrB_{His6}$ as a template and primers $AcrB_C$_F and $AcrB_C$_R to remove two C-terminus histidine residues and a 6×His-tag to generate construct pET-$acrB_{\Delta His}$ (*see* **Note 1**). Then use pET-$acrB_{\Delta His}$ as a template and primers $AcrB_{D328}$_F and $AcrB_{D328}$_R to insert a *BamHI* site between Asp-328 and Thr-329 of AcrB to construct the vector pET-$acrB_{\Delta His}BamHI_{328}$.
3. For the amplification of *acrA* gene, use primers $AcrA_{25GS}$_F and $AcrA_{GSx3}$_R2 to add poly-GlySer linkers to both ends of AcrA. Then amplify the PCR product again using primers $AcrA_{inf}$_F and $AcrA_{inf}$_R. Insert the DNA fragment into the *BamHI* site of pET-$acrB_{\Delta His}BamHI_{328}$ using the In-Fusion cloning method to generate the construct pET-*acrAB*.

3.1.2 Generation of Construct pRSF-$acrAZ_{His5}$-$tolC_{1392}$

1. For the amplification of *acrZ* gene from genomic DNA of *E. coli* W3110 strain, use primers $AcrZ_{NcoI}$_F and $AcrZ_{His5SalI}$_R to introduce the *NcoI* and *SalI* restriction sites to the 5′- and 3′-end of *acrZ* and a 5×His-tag at the 3′-end. Insert the fragment of $acrZ_{His5}$, bounded by *NcoI* and *SalI* sites, into the multiple cloning site of expression vector pRSFDuet-1 to generate the pRSF-$acrZ_{His5}$ vector.
2. For the amplification of *acrA* gene, use primers $AcrA_{NcoI}$_F and $AcrA_{GSx3}$_R1 to add a poly-GlySer linker to the C-terminus and an *NcoI* site at both ends. Insert the *NcoI*-bounded *acrA-polyGlySer* into the *NcoI* site of pRSF-$acrZ_{His5}$ to generate the construct pRSF-$acrAZ_{His5}$.
3. For the amplification of *tolC* gene, use primers $TolC_{inf}$_F and $TolC_{1392inf}$_R to remove 29 unstructured residues at the C-terminus. Insert the PCR product into the *NdeI* site of pRSF-$acrAZ_{His5}$ using the In-Fusion cloning method to generate the construct pRSF-$acrAZ_{His5}$-$tolC_{1392}$.

3.1.3 Procedures for Molecular Cloning

1. To amplify *acrA*, *AcrB*, *acrZ*, and *tolC* genes, set up the PCR reactions as follows:

   10 μl 5×Phire Reaction Buffer, 1 μl 10 mM dNTPs, 0.5 μl forward primer (100 pmol/μl), 0.5 μl reverse primer (100 pmol/μl), 1 μl Phire Hot Start II DNA Polymerase, 0.5 μl template DNA (200 ng/μl), 36.5 μl deionized H_2O.
   Run the PCRs with the following program:
   30 s 98 °C; five cycles: 5 s 98 °C, 5 s 70 °C, 20 s per 1 kb of template DNA 72 °C; 30 cycles: 5 s 98 °C, 20 s per 1 kb of template DNA 72 °C; 5 min 72 °C.
2. To perform site-directed mutagenesis of *acrB* using two-stage PCR, set up two PCR reactions as follows:

   Reaction-I: 5 μl 10×Cloned *pfu* DNA polymerase Reaction Buffer, 1 μl 10 mM dNTPs, 0.5 μl forward primer (100 pmol/μl), 1 μl *PfuTurbo* DNA Polymerase, 0.5 μl template plasmid DNA (200 ng/μl), 42 μl deionized H_2O.
   Reaction-II: 5 μl 10×Cloned *pfu* DNA polymerase Reaction Buffer, 1 μl 10 mM dNTPs, 0.5 μl reverse primer (100 pmol/μl), 1 μl *PfuTurbo* DNA Polymerase, 0.5 μl template plasmid DNA (200 ng/μl), 42 μl deionized H_2O.
   Run the PCR with the following program:
   2 min 95 °C; five cycles: 30 s 95 °C, 30 s 55 °C, 8 min 72 °C; 10 min 72 °C.
   Mix 25 μl of Reaction-I with equal volume of Reaction-II and 1 μl *PfuTurbo* DNA Polymerase, run PCR with the above program for 16 cycles. Add 20 U *DpnI* to the reaction and incubate at 37 °C for 2 h to digest template plasmid, then place on ice. Continue to the transformation procedure.
3. Digest the PCR fragments and plasmids with restriction enzymes at 37 °C overnight as follows:

   x μl DNA (2 μg), 20 μl 10× CutSmart® Buffer, *x* μl Restriction enzymes (60 U for each enzyme), add deionized H_2O to 200 μl.
4. Analyze the PCR products and enzyme-digested plasmids on an agarose gel. Run the gel for 60 min with 100 V in 0.5× TBE buffer (*see* **Note 2**). Then extract the DNA fragments with the right size from the gel using the QIAquick Gel Extraction Kit.
5. Set up the In-Fusion cloning reaction for *acrA* and *tolC* as follows:

   2 μl 5× In-Fusion HD Enzyme Premix, 1 μl Linearized Vector (200 ng/μl), 1 μl Purified PCR Fragment (100 ng/μl), 7 μl deionized H_2O.
   Incubate the reaction for 15 min at 50 °C, then place on ice. Continue to the transformation procedure.
6. Insert the digested fragments into the prepared vector by LigaFast™ Rapid DNA Ligation System as follows:

x μl Insert DNA, x μl Vector DNA (molar ratio between Insert DNA and Vector DNA is about 3 to 1), 5 μl 2× Rapid Ligation Buffer, 1.5 μl T4 DNA Ligase (3 U/μl), add deionized H_2O to 10 μl. Incubate at room temperature for 5 min, and then place on ice. Continue to the transformation procedure.

7. Transformation of plasmids into chemical competent *E. coli* strain XL1-Blue. Prepare the plasmid DNA using GeneJET Plasmid Miniprep Kit. Sequence the plasmid to identify plasmid with correct insertion.

3.2 Overexpression and Purification of AcrABZ-TolC Pump

Solutions should be degased, sterile-filtered (0.22 μm), and prechilled at 4 °C, before use. It is also recommended to prechill tubes and rotors for centrifugation.

3.2.1 Expression of AcrABZ-TolC

1. Pick a fresh single colony of *E. coli* C43 (DE3)*ΔacrAB* transformed with plasmids pET21a-*acrAB* and pRSF-$acrAZ_{His5}$-$tolC_{1392}$ and inoculated 20 ml LB broth containing carbenicillin (100 μg/ml) and kanamycin (50 μg/ml) in a 50 ml tube. Incubate shaking at 37 °C, 220 rpm for 4 h.
2. Add 8 ml of the culture to 1 L of 2×YT medium (with carbenicillin and kanamycin) in a 2 L baffled flask. Incubate shaking at 37 °C, 220 rpm while monitoring the OD_{600}.
3. When an OD_{600} of 0.5–0.6 is reached, add IPTG to a final concentration of 0.15 mM to induce expression of the AcrABZ-TolC pump. Drop the temperature to 20 °C and shake at 220 rpm overnight.
4. Transfer the culture to centrifuge tubes and pellet cells by centrifugation at 3635 average g for 25 min at 4 °C. Carefully discard the supernatant and resuspend the pellet from 2-L culture in 50 ml lysis buffer.

3.2.2 Protein Purification by IMAC

1. Add EDTA-free protease inhibitor cocktail tablets to the cell suspension at one tablet/50 ml and DNase I to a final concentration of 5 U/ml. Add lysozyme to a final concentration of 5 mg/ml and incubate at 4 °C with gently stirring for 1 h to improve the efficiency of cell wall fractionation (*see* **Note 3**). Lyse the cells by eight passages through an Emulsiflex-C5 high-pressure homogenizer at 15 Kpsi at 4 °C.
2. Centrifuge the cell lysate at 10,000 × g for 30 min at 4 °C to remove the cell debris and unbroken cells. Collect the supernatant carefully so as not to disturb the pelleted material. Ultracentrifuge the supernatant at 125755 average g (rotor SW32 Ti) for 4 h at 4 °C to spin down the cellular membrane.
3. Resuspend the cellular membrane pellets from 2 L culture in 50 ml lysis buffer (*see* **Note 4**). Add DDM powder to the membrane suspension to a final concentration of 1.5% and

EDTA-free protease inhibitor cocktail tablets at one tablet/ 50 ml. Stir the mixture gently at 4 °C for 3 h (*see* **Note 5**). Ultra-centrifuge the membrane solution at 125755 average *g* (rotor SW32 Ti) for 30 min at 4 °C to spin down the insoluble material. Decant the supernatant liquid carefully so as not to disturb the pelleted material.

4. Immobilize HiTrap Chelating 1 ml column with $Ni^{2+}$and equilibrate the column with lysis buffer containing 0.03% DDM and 20 mM imidazole. Add imidazole to the membrane solution to a final concentration of 15 mM. Load the mixture onto HiTrap Chelating column by periplastic pump at 0.8 ml/ min.
5. Wash column with 40 ml of IMAC wash buffer-I and 20 ml of IMAC wash buffer-II.
6. Elute bound protein with 10 ml of IMAC elution buffer and collect 0.5 ml fractions.
7. Identify positive fractions by SDS-PAGE gel and combine fractions showing a similar protein content.

3.2.3 Protein Purification by Gel Filtration Chromatography

1. Load the IMAC-purified AcrABZ-TolC complex onto a Vivaspin concentrator (MWCO: 100 kDa) and centrifuge at 2400 × *g*, 4 °C to concentrate the protein to 0.5 ml.
2. Equilibrate a Superose™ 6 column with two CV of GF buffer.
3. Load 0.5 ml of IMAC-purified AcrABZ-TolC complex onto the Superose™ 6 column. Elute the column with GF buffer at 0.2 ml/min and collect 0.4 ml fractions.
4. Identify positive fractions by SDS-PAGE gel and combine fractions showing a similar protein content.

3.2.4 Detergent-Exchange to Amphipol A8-35

1. Load the size exclusion-purified AcrABZ-TolC complex onto a Vivaspin concentrator (MWCO: 100 kDa) and centrifuge at 2000 × *g*, 4 °C to concentrate the protein to 0.1 mg/ml (*see* **Note 6**).
2. Add amphipol A8-35 (100 mg/ml) to the protein solution with a mass ratio of amphipol A8-35 to protein of 4:1. Incubated the mixture on ice for 3 h.
3. Add Polystyrene beads (Bio-Beads™ SM-2) to the protein/ DDM/amphipol A8-35 mixture with a mass ratio of Bio-Beads SM-2 to DDM of 10:1. Rotated the mixture gently at 4 °C overnight to remove DDM.
4. Load the detergent-exchanged AcrABZ-TolC complex onto a mini chromatography column to remove the Polystyrene beads.

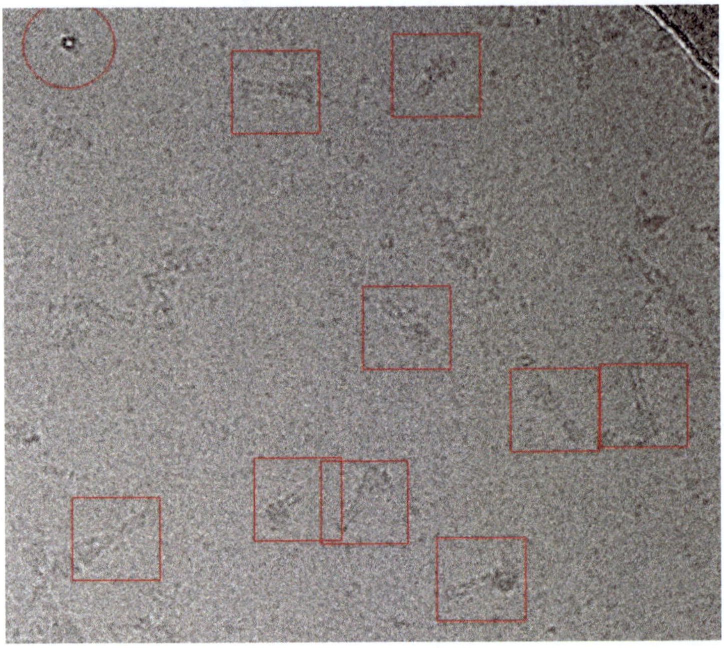

Fig. 1 A representative cryo-electron microscopy image of ice-embedded AcrABZ-TolC pump recorded using a K2 Summit detector. *Red circles* indicate particles with long axis almost normal to the viewing plane; *red squares* show particles with the long axis parallel to the viewing plane

5. Apply the detergent-exchanged AcrABZ-TolC complex onto a Vivaspin concentrator (MWCO: 100 kDa) and centrifuge at 2000 × *g*, 4 °C to concentrate the protein to 2 mg/ml. A representative cryo-electron microscopy image of ice-embedded AcrABZ-TolC pump is shown in Fig. 1.

4 Notes

1. Direct cloning of *NdeI* and *XhoI*-bounded *acrB*, with a stop code and truncation of two histidines at C-terminus, into the *NdeI* and *XhoI* site of pET21a vector can generate the construct pET-$acrB_{\Delta His}$ in one step.
2. Agarose gel: Dissolve 1 g agarose in 100 ml 0.5×TBE buffer and heat the solution in microwave until the agarose is completely solved. Add 10 μl ethidium bromide stock solution, pour it into a gel rack with a comb of desired slot size. When the agarose has solidified, put the gel rack into an electrophoresis chamber and fill the chamber with 0.5×TBE buffer until the gel is completely covered. Mix the DNA sample with 6×DNA Loading Dye (Thermo Scientific). Pipette 10 μl of 1 kb DNA ladder (Hyper Ladder I, Bioline) in the first slot and

the stained DNA in the other slots. Run the gel for 60 min with 100 V. Analyze the gel under UV light (make sure to be quick to avoid UV-induced DNA damage).

3. Resuspend desired amount of lysozyme in lysis buffer to a final concentration of 30–40 mg/ml, centrifuge at 6000 × g for 10 min at 4 °C to remove foam, and then mix it with cell suspension.
4. The cellular membrane from 2-L culture may vary from batch to batch. It is applicable to resuspend 3.5 g cellular membrane in 50 ml lysis buffer.
5. Stir gently after adding DDM powder into the cellular membrane suspension to avoid forming of foam. It is applicable to solve desired amount of DDM to lysis buffer to a final concentration of 20%, centrifuge at 6000 × g for 10 min at 4 °C to remove foam, and then mix it with cellular membrane suspension, add lysis buffer to reach the final volume of 50 ml.
6. For the OD_{280} measurement, use the NanoDrop® Spectrophotometer ND-1000 (Thermo Scientific). For the calculation of protein concentration, use the formula $M = OD_{280}/\varepsilon$ (M = Molarity, ε = extinction coefficient of the protein).

Acknowledgments

Wellcome Trust and Human Frontier Science Program provided funding for this work. The work was also supported by NIH grant NIH 5P41GM103832.

References

1. Nogales E, Scheres SHW (2015) Cryo-EM: a unique tool for the visualization of macromolecular complexity. Mol Cell 58:677–689
2. Du D et al (2014) Structure of the AcrAB-TolC multidrug efflux pump. Nature 509:512–515
3. Bai X et al (2015) An atomic structure of human γ-secretase. Nature 525:212–217
4. Tribet C, Audebert R, Popot JL (1996) Amphipols: polymers that keep membrane proteins soluble in aqueous solutions. Proc Natl Acad Sci U S A 93:15047–15050
5. Popot J-L (2010) Amphipols, nanodiscs, and fluorinated surfactants: three nonconventional approaches to studying membrane proteins in aqueous solutions. Annu Rev Biochem 79:737–775
6. Zoonens M, Popot J-L (2014) Amphipols for each season. J Membr Biol 247:759–796

Chapter 6

NMR Spectroscopy Approach to Study the Structure, Orientation, and Mechanism of the Multidrug Exporter EmrE

Maureen Leninger and Nathaniel J. Traaseth

Abstract

Multidrug exporters are a class of membrane proteins that remove antibiotics from the cytoplasm of bacteria and in the process confer multidrug resistance to the organism. This chapter outlines the sample preparation and optimization of oriented solid-state NMR experiments applied to the study of structure and dynamics for the model transporter EmrE from the small multidrug resistance (SMR) family.

Key words Solid-state NMR, PISEMA, NMR pedagogy, Membrane proteins, Multidrug resistance, Bicelles

1 Introduction

Drug efflux by membrane transport proteins is a primary mechanism bacteria use to confer resistance to antiseptics and antibiotics [1–3]. These multidrug transporters bind and efflux lethal compounds from the cytoplasm and in the process reduce the toxicity to the organism. Several transporter structures have been determined using X-ray crystallography, which have provided detailed insight into the transport process (reviewed in [4]). A complementary method for resolving the mechanism of ion-coupled transport is through the use of nuclear magnetic resonance (NMR) spectroscopy. One advantage of this approach is to study efflux pumps under conditions that mimic the native membrane environment. This chapter describes how oriented solid-state NMR (O-SSNMR) spectroscopy is used to directly interrogate the structure of the drug transporter EmrE from the small multidrug resistance (SMR) family. The anisotropic observables offered from these experiments (e.g., anisotropic chemical shifts and dipolar couplings) are used as restraints in structure determination and are excellent reporters of the tilt and rotation angles for transmembrane proteins within the lipid bilayer (Fig. 1) [5–11].

Akihito Yamaguchi and Kunihiko Nishino (eds.), *Bacterial Multidrug Exporters: Methods and Protocols*, Methods in Molecular Biology, vol. 1700, https://doi.org/10.1007/978-1-4939-7454-2_6, © Springer Science+Business Media, LLC 2018

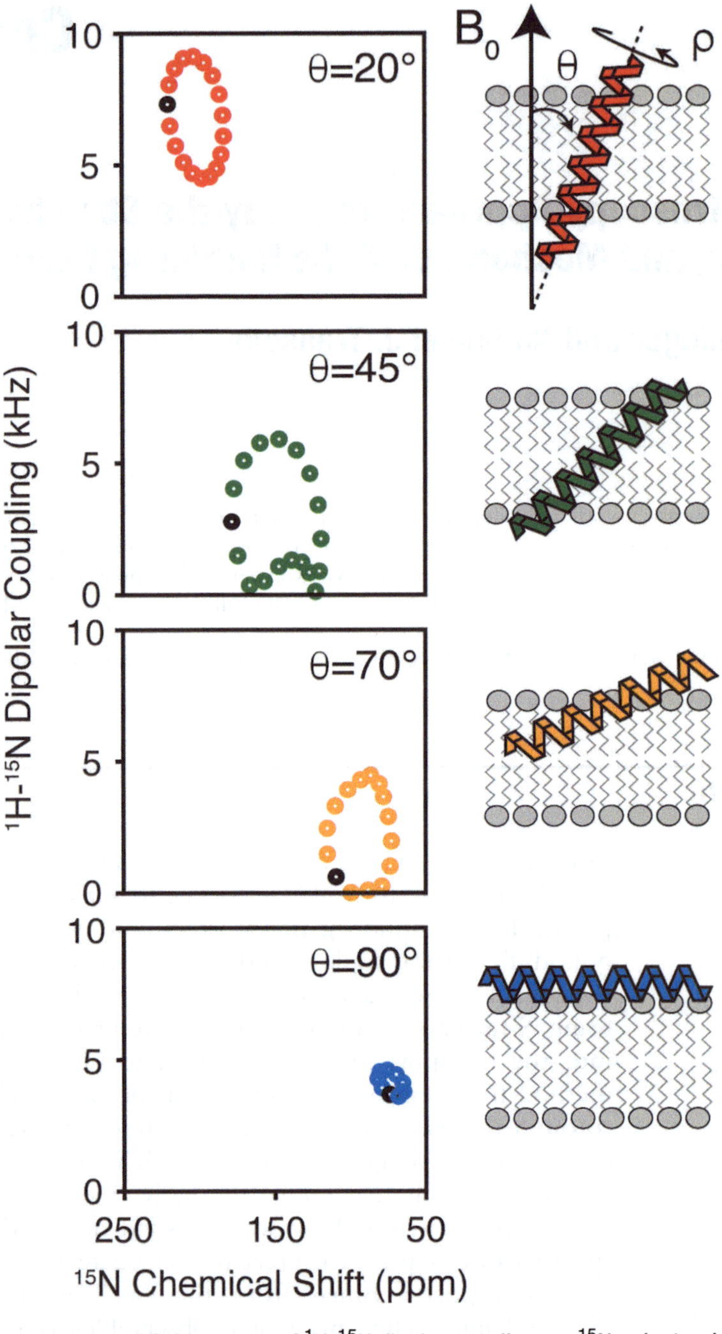

Fig. 1 Calculated separated local field spectra of ^{1}H-^{15}N dipolar coupling vs. ^{15}N anisotropic chemical shift for an ideal helix. The periodic spectral patterns are sensitive to the tilt and rotation angles of the secondary structures with respect to the magnetic field. These patterns have been termed PISA wheels [7, 9]

The prerequisite for using O-SSNMR to characterize membrane protein structures is to align the samples within the magnetic field. This is most commonly accomplished using mechanical alignment of lipid bilayers absorbed onto glass plates [12, 13] or magnetic alignment with lipid bicelles [14]. The former approach has been successfully applied for multiple membrane proteins including M2 from the influenza A virus [15] and the muscle regulatory protein phospholamban [16]. For EmrE, we utilize lipid bicelle technology due to the excellent control of sample hydration and improved alignment with respect to the magnetic field [17]. Specifically, our experiments use a bicelle composition of DMPC (14:0) and DHPC (6:0) at a 3.2/1 molar ratio, which readily forms aligned liquid crystals at 37 °C [18–21]. Replacing the DMPC component with a shorter (12:0) or longer (16:0) chain phosphatidylcholine lipid can be used to lower or raise the alignment temperature, respectively [22–24]. In addition, the usage of POPC in combination with DMPC can also reduce the alignment temperature to 25 °C [17]. One of the most important requirements of bicelle selection is to ensure protein stability in the chosen lipid composition [20]. Finally, since the quality of protein alignment can be protein dependent, the lipid to protein ratio should be optimized in an empirical fashion to ensure bicelle alignment in the magnetic field. For EmrE, we find that a lipid to protein ratio of 150:1 (mol:mol) is effective to achieve good signal to noise without compromising the overall alignment with respect to the magnetic field.

Below we discuss the procedure of bicelle sample preparation containing EmrE and the setup of the polarization inversion spin exchange at the magic angle (PISEMA) experiment [25, 26], which is a sensitive way to probe secondary structural information with respect to the membrane. Using this experiment and others, we have resolved the asymmetric monomer subunits within the EmrE homodimer and quantified conformational exchange dynamics of the transporter that are required for drug efflux [27, 28].

2 Materials

2.1 Sample Preparation

1. Isotopically ^{15}N-enriched protein: modify the expression conditions for your favorite protein with the addition of $^{15}NH_4Cl$ to M9 minimal media to obtain uniformly ^{15}N-enriched proteins or the addition of one or several ^{15}N-labeled amino acids for selectively enriched proteins.
2. *n*-Dodecyl-β-D-maltoside (DDM).
3. *n*-Octyl-β-D-glucoside (OG).
4. Bio-Beads™ SM-2 resin, Bio-Rad Laboratories.
5. Ultracentrifuge: instrument should reach centrifugal forces of at least 150,000 × *g*.
6. 1,2-Dimyristoyl-*sn*-glycero-3-phosphocholine (DMPC), Avanti Polar Lipids, Inc.

7. 1,2-Dihexanoyl-*sn*-glycero-3-phosphocholine (DHPC), Avanti Polar Lipids, Inc.
8. Ytterbium(III) chloride ($YbCl_3$).

2.2 NMR Instrumentation and Data Analysis

1. Bicelle sample holder assembly (design by Peter Gor'kov at NHMFL, Tallahassee, FL):
   (a) 5 mm open-ended glass tube, New Era Enterprises, part No. NE-RG5-P-17-OE-BOE.
   (b) B5 plugs with and without fill hole, Revolution NMR, part No. MP4763-001 and MP4764-001.
2. Solid-state NMR spectrometer: our experiments were carried out using an Agilent DD2 spectrometer operating at a ^{1}H Larmor frequency of 600 MHz.
3. Solid-state NMR probes capable of ^{31}P and ^{15}N detection. Our experiments use the following:
   (a) BioStatic H-X probe with 5 mm bicelle coil tuned to ^{31}P on the X-channel (Agilent).
   (b) *Low-E* ^{1}H-^{15}N probe with 5 mm bicelle coil for ^{15}N detection (probe design by Peter Gor'kov) [29]. This probe technology is commercially available from Revolution NMR, LLC (www.revolutionnmr.com).
4. NMR data processing software: NMRPipe [30] and Sparky v3.113 (T.D. Goddard and D. G. Kneller, SPARKY 3, University of California, San Francisco).

3 Methods

3.1 Sample Preparation

EmrE is expressed from a pMAL™ vector (New England Biolabs) where maltose binding protein (MBP) is positioned on the N-terminal side of the EmrE gene. The expression is performed in *E. coli* BL21(DE3) cells in minimal media using selectively labeled ^{15}N amino acids or $^{15}NH_4Cl$ to uniformly enrich the protein. EmrE is purified with an amylose affinity column, cleaved with TEV protease to remove MBP, and passed over a size exclusion column as previously described [27, 28]. Purified EmrE in DDM detergent micelles is then reconstituted into DMPC lipid bilayers. Below is the reconstitution procedure:

1. 28 mg DMPC is hydrated with 1 mL of 20 mM sodium phosphate and 20 mM NaCl and subjected to multiple cycles of flash freezing.
2. Bath sonicate the sample for 15 min.
3. Add 5.6 mg OG to the mixture and equilibrate for 10 min.

4. Add 7 mL EmrE at a concentration of 0.5 mg/mL in DDM detergent to the sonicated lipids. Additional DDM may be added to ensure the lipids are solubilized, which will be evident when the mixture becomes clear. Typically, a total of ~30 mg DDM is added including the detergent with EmrE.
5. Incubate the sample for 1 h at room temperature.
6. Add 2.25 g of Bio-Beads to give a final ratio of 75 mg per mg of detergent.
7. Gently stir for 12 h at 4 °C. Note that the removal of detergent by the Bio-Beads will instigate formation of proteoliposomes and lead to a slightly cloudy solution.
8. Remove Bio-Beads from the suspension.
9. Centrifuge proteoliposomes in a Beckman Optima MAX_XP TLA 110 rotor at 100,000 × *g* for 1.5 h.
10. Resuspend the liposome pellets in 100 mM HEPES and 20 mM NaCl (pH 7) buffer and freeze thaw 5–10 times to exchange the buffer.
11. Centrifuge the sample in the TLA 110 rotor for 2.5 h at 150,000 × *g*.
12. Add 25 μL of 425 mM DHPC in water to the proteoliposome pellet. This corresponds to a DMPC:DHPC molar ratio ("*q* value") of ~3.2–3.5 (*see* **Note 1**).
13. Mix thoroughly by a combination of vortexing, flash freezing with liquid nitrogen, and heating to 37 °C. After the mixing is complete, the sample should be fluid on ice and a viscous gel at 37 °C (*see* **Note 2**).
14. Centrifuge the sample in a 1.5 mL microcentrifuge tube for 2 min at 500 × *g* using a benchtop centrifuge and place the supernatant in a new microcentrifuge tube.
15. For unflipped bicelle samples (bicelle normal perpendicular to magnetic field), proceed to **step 16**. For flipped samples (bicelle normal parallel to magnetic field), add 6 μL of 100 mM $YbCl_3$ to give a final concentration of 4.5 mM (*see* **Note 3**).
16. Place the sample in the bicelle sample holder for O-SSNMR experiments.

3.2 Check Bicelle Alignment with ^{31}P NMR

The alignment of the bicelle in the magnetic field is checked by recording a one-pulse ^{31}P experiment optimized for a 90° tilt angle followed by acquisition of the free induction decay under ^{1}H decoupling with SPINAL-64 [31] This experiment is a sensitive way to check the sample since ^{31}P is 100% natural abundant, has a relatively high gyromagnetic ratio, and is present in both DMPC and DHPC headgroups. Representative ^{31}P spectra of flipped, isotropic and unflipped bicelles are shown in Fig. 2a–c, respectively (*see* **Note 4**).

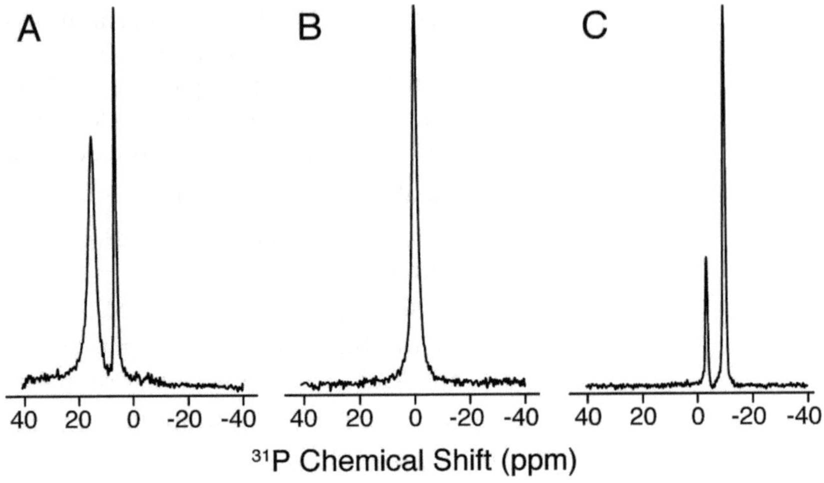

Fig. 2 ^{31}P spectra of lipid bicelle samples. The panels show representative spectra for (**a**) flipped, (**b**) isotropic, and (**c**) unflipped bicelles. The flipped spectra were obtained by the addition of $YbCl_3$

3.3 Optimization of the Polarization Inversion Spin Exchange at the Magic Angle (PISEMA) Experiment

PISEMA is a separated local field experiment to correlate dipolar couplings with anisotropic chemical shifts [25, 26]. The spectral patterns from PISEMA are directly sensitive to the tilt and rotation angles of the protein secondary structures with respect to the magnetic field [7, 9]. PISEMA begins with a cross-polarization (CP) sequence to transfer magnetization from ^1H to ^{15}N spins in order to increase the sensitivity of the experiment. Homonuclear dipolar decoupling in the indirect dimension is achieved by frequency-switched or phase-modulated Lee-Goldburg (FSLG or PMLG) [32–35]. Both of these sequences enable evolution of a scaled heteronuclear dipolar coupling in the absence of ^{15}N chemical shift evolution. After evolving the dipolar coupling in the indirect dimension, the PISEMA experiment detects the free precession of the ^{15}N nucleus under high-power ^1H decoupling. The most common application of PISEMA is to enrich the protein with ^{15}N labeling and correlate the ^1H-^{15}N dipolar coupling with the ^{15}N anisotropic chemical shift (*see* **Note 5**). Advancements in the original PISEMA experiment include developments to reduce the ^1H offset dependence of the indirect dimension [36], a sensitivity-enhanced version to improve signal-to-noise [37, 38] and a constant-time experiment to improve the resolution in the dipolar coupling dimension [39].

Prior to data acquisition on the protein sample, it is recommended to first optimize the pulse sequence parameters with two model compounds: (1) an aqueous buffer composed of 100 mM HEPES and 20 mM NaCl (pH 7.0) and (2) a single crystal of ^{15}N-labeled *N*-acetyl-leucine (NAL).

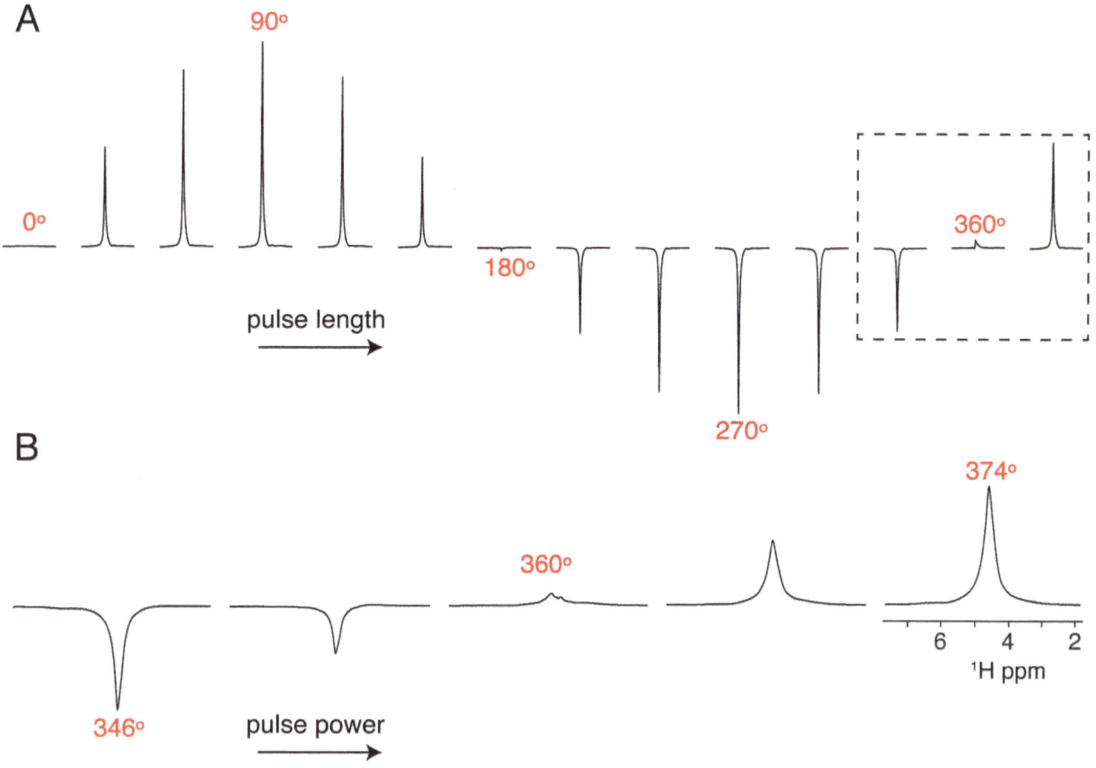

Fig. 3 Optimization of the ^{1}H 90° pulse on the water peak. (**a**) Nutation curve by varying the pulse length at a fixed B_1 amplitude to determine an accurate 360° pulse. (b) Fine tune calibration of the 360° pulse by setting the pulse length to four times the desired 90° pulse and increasing the power level from left to right. The null value corresponds to the 360° pulse

3.3.1 Insert Buffer Sample to Calibrate ^{1}H 90° Pulse Parameters

If the probe requires pulse length calibration for the first time, it is necessary to optimize ^{1}H 90° pulses for a given pulse duration that is within the voltage handling of the probe circuit. This is typically specified by a minimum 90° pulse width that can be applied without damage to the probe. We typically optimize for 5 and 6 μs pulse lengths on the water signal of a sample of 100 mM HEPES and 20 mM NaCl (pH 7.0) by carrying out a full nutation curve where the pulse length is varied at a constant power (Fig. 3a). Next, the pulse power is more finely calibrated to the desired 360° pulse (e.g., 20 or 24 μs, respectively) by adjusting the power level to identify the corresponding null in the spectrum (*see* **Note 6**). An example of the calibration is shown in Fig. 3b.

3.3.2 Insert the N-Acetyl Leucine Single Crystal

1. *Optimize cross-polarization (CP) from ^{1}H to ^{15}N.* There are a few variants of this sequence that provide efficient polarization transfers, including linear and adiabatic ramps as well as CP-MOIST [40, 41] (*see* **Note 7**).

2. *Optimize the 90° pulse for ^{15}N.* The experiment employs a 90° ^{15}N pulse after the CP from ^{1}H to ^{15}N. The signal on ^{15}N is detected under ^{1}H decoupling. The flip back pulse is set to the desired length with the power adjusted in an iterative manner. The null in the spectrum gives the optimal power corresponding to the pulse length.
3. *Calculate the parameters for the indirect dimension of the PISEMA experiment.* For the SEMA portion of PISEMA (i.e., t_1 period), the effective frequency on ^{15}N must match the effective frequency of ^{1}H (i.e., $\omega_{eff,15N} = \omega_{eff,1H}$). The ^{15}N spin-lock during t_1 evolution is on resonance so the effective frequency is equal to applied frequency ($\omega_{1,15N} = \omega_{eff,15N}$) (*see* **Note 8**). Unlike the ^{15}N effective frequency, the ^{1}H effective frequency during FSLG or PMLG is not equal to the applied frequency. For this reason, the frequency offset ($\Delta\omega$) and applied frequency (ω_1) need to be set so that the effective frequency of ^{1}H will match that of ^{15}N ($\omega_{eff,15N} = \omega_{eff,1H}$). A second requirement of PMLG or FSLG is that the effective frequency on ^{1}H is applied at the magic angle in order to average out homonuclear dipolar couplings. Using the trigonometric relationships given in Eqs. 1 and 2 and the desired effective frequencies for ^{15}N and ^{1}H, $\Delta\omega$ and ω_1 can be calculated for ^{1}H (*see* **Note 9**).

$$\tan\theta = \frac{\omega_1}{\Delta\omega} \quad (\theta = 54.7) \tag{1}$$

$$\omega_{eff}^2 = \omega_1^2 + \Delta\omega^2 \tag{2}$$

4. *Collect the 2D PISEMA experiment.* The PISEMA experiment is collected using the optimized pulse sequence parameters and Fourier transformed to give the resulting two-dimensional spectrum (*see* **Note 10**).
5. *Adjust ^{1}H offset to minimize the zero-frequency.* Since the dipolar coupling frequency measured in the indirect dimension is sensitive to the ^{1}H offset, one can adjust the transmitter frequency for ^{1}H to minimize the zero-frequency signals for desired peaks in the single crystal sample. The optimal ^{1}H offset may be different for the protein sample.

3.3.3 Acquire PISEMA Spectrum on Membrane Protein Sample

1. *Optimize the ^{1}H 90° pulse.* Similar to the pulse calibration on the buffer sample, we optimize for 5 and 6 μs 90° pulse lengths on the water signal in the protein/bicelle sample. As described above and shown in Fig. 3b, the pulse length is set to four times the 90° pulse with the power levels adjusted to find the null.

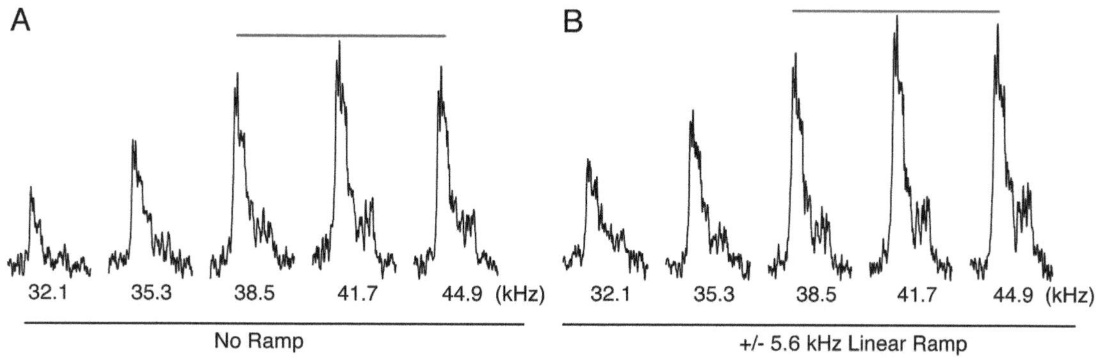

Fig. 4 Optimization of ^1H to ^{15}N cross-polarization as demonstrated for EmrE in lipid bicelles. (**a**) The ^1H power is arrayed around a constant ^{15}N amplitude used for cross-polarization. As seen from the spectra, a linear ramp of 41.7 $\pm$ 5.6 kHz gave the highest signal/noise for all experiments. These data show that a linear ramp improves the signal/noise and emphasizes the importance of optimizing cross-polarization

2. *Optimize the CP condition for protein.* In order to find optimal matching conditions for CP, it is recommended to have a uniformly ^{15}N-labeled sample that gives at least 10:1 signal-to-noise in ~256–512 scans. Similar to that described for the single crystal, the CP values can be optimized directly on the sample by adjusting the ^1H power level (Fig. 4) (*see* **Note 11**).
3. *Optimize the length of the CP contact time.* In addition to the power levels, the contact time of the CP should be iteratively adjusted to maximize signal-to-noise. EmrE in aligned bicelles shows an optimal value of ~0.75 ms.
4. *^1H offset.* The ^1H offset must be set correctly in order to reduce the zero-frequency peaks. For transmembrane helical proteins in flipped bicelles, a ^1H offset in the range of 4.7–7 ppm will minimize the zero-frequency signals [37].
5. *Find optimal recycle delay.* The recycle delay should be set to 3–5 times the ^1H T_1, which can be measured using an inversion recovery experiment on the protein. For EmrE, we normally use a delay of 3 or 4 s.
6. *Acquisition of the PISEMA experiment.* Using the calibrated ^{15}N pulse powers from the single crystal and the ^1H pulse power from the protein sample, one can calculate the applied frequency for ^1H during FSLG or PMLG to match the desired effective frequency. Since the maximum ^1H-^{15}N dipolar couplings are ~10 kHz, we typically employ a t_1 dwell time of 48 μs for protein samples in flipped bicelles, which corresponds to an effective frequency $\omega_{eff}/2\pi = 41.7$ kHz. This spectral width is sufficient to cover the entire breadth of dipolar couplings in the indirect dimension. An example PISEMA spectrum for selectively labeled EmrE with ^{15}N methionine is shown in Fig. 5.

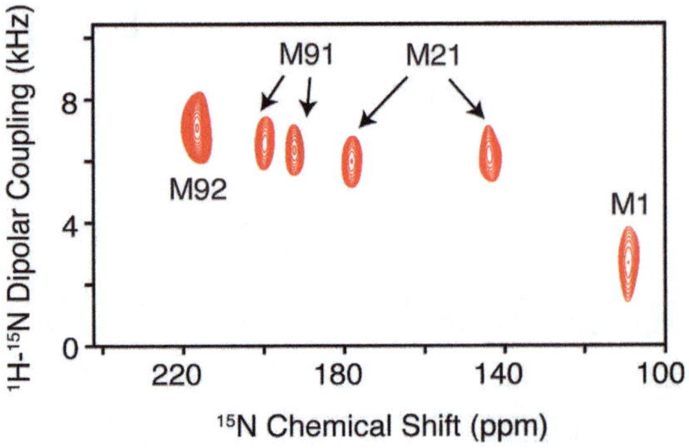

Fig. 5 PISEMA of ^{15}N methionine-labeled EmrE in DMPC/DHPC aligned bicelles. Note that M21 and M91 show clear doubling in the spectrum indicative of an asymmetric dimer. (Reproduced with permission from Ref. 27. © WILEY-VCH Verlag GmbH & Co. KgaA, Weinheim, 2013)

7. *Assignment of PISEMA spectra.* The acquired spectrum needs to be assigned in order to obtain site-resolved structural information. There are several viable approaches including the usage of site-directed mutagenesis [27], selectively labeled samples in combination with periodic wheel approximations [7, 9, 42, 43], and spectroscopic methods that make use of magnetization transfers [44, 45]. Once the assignments are obtained, these can be used as restraints in structural refinement calculations.

We have successfully employed PISEMA to show that both monomers in the functional EmrE dimer give a separate set of peaks (*see* Fig. 5). These results validated the asymmetric dimer structural models available from X-ray crystallography and cryoelectron microscopy [27, 46, 47]. Furthermore, we recently applied the pure exchange (PUREX) method [48] to investigate the conformational dynamics of EmrE in aligned bicelles that can be applied to other membrane protein systems [28]. Notably, we discovered that the change in protonation state of a conserved glutamate residue alters the conformational exchange rate of EmrE that plays a direct role in drug efflux [49].

4 Notes

1. The ability of a sample to align is very dependent on the molar ratio of DMPC:DHPC ("*q*-value"). If the q-value is significantly below 3.2 the bicelle will be isotropic.
2. The final volume of the sample should be 125 μL to ensure that the total lipid concentration is at least 25% (w/v). If the lipid concentration is too low the bicelles will not align.

3. The sample will readily align at 37 °C with the bilayer normal perpendicular with respect to the direction of the magnetic field (i.e., unflipped bicelles). Some proteins that undergo rapid uniaxial rotational diffusion can be characterized under these conditions [50]. However, our experiments show that EmrE's rotational diffusion is not sufficiently rapid to average out the residual anisotropy [51]. For this reason, the bicelle normal is "flipped" by 90° using $YbCl_3$, which orients the bilayer normal parallel to the magnetic field. These paramagnetic ions bind to the phosphate within the lipid headgroup and change the magnetic susceptibility of the bicelle [19, 21].
4. It is important to note that these spectra do not guarantee that the protein will be properly aligned in the magnetic field. However, when the bicelle containing protein is unaligned or partially aligned, our experience is that the protein's ^{15}N spectrum will give a powder pattern or be of low quality.
5. The solid-state NMR probe for PISEMA spectroscopy needs to include a sensitive detection coil optimized for ^{15}N detection. The probe designed by Gor'kov et al. [29] is the most sensitive available and uses a loop-gap resonator for the ^1H channel that minimizes sample heating and uses a ^{15}N solenoid for detection. The inner solenoid coil is positioned closest to the sample in order to maximize the filling factor and sensitivity for ^{15}N detection.
6. It is important to acquire a full nutation curve when first optimizing power levels to ensure accurate measurement of the 360° pulse.
7. The optimal match condition for the single crystal is typically not the best for the protein sample; however, it should be optimized to ensure sufficient signal-to-noise for calibration of the ^{15}N pulse and PISEMA experiment. The contact time can also be adjusted iteratively, but typically ~2 ms will give the highest signal-to-noise.
8. Note that the applied nutation frequency can be calculated by $\omega_1/2\pi = 250/t_{90}$, where t_{90} is equal to the length of the 90° pulse in μsec and $\omega_1/2\pi$ is given in kHz. For example, a 6 μs pulse corresponds to a frequency of 41.7 kHz.
9. We commonly employ an effective ^1H frequency of $\omega_{\mathrm{eff}}/2\pi = 41.7$ kHz, an offset $\Delta\omega/2\pi = 24.1$ kHz, and an applied frequency of $\omega_1/2\pi = 34.0$ kHz. Unlike most multidimensional NMR experiments, the dwell time during the t_1 evolution period cannot be set independently from the FSLG or PMLG parameters. Specifically, the minimum t_1 dwell time is the application of two 360° pulses. This works out to give a minimum dwell time equal to $8 \times (250/(\omega_{eff}/2\pi))$. For example, with an effective frequency of $\omega_{\mathrm{eff}}/2\pi = 41.7$ kHz, the dwell time is 48 μs.

10. The indirect ^{1}H-^{15}N dipolar coupling is scaled due to the Lee-Goldburg sequence and needs to be adjusted by the scaling factor (0.82) to give the correct couplings [26]. For example, a dwell time of 48 μs during FSLG or PMLG gives a corrected spectral width of 25.4 kHz.
11. For selectively labeled samples, the signal-to-noise often prevents detailed optimization of pulse parameters. However, since the buffer composition is the same as that of the uniformly ^{15}N-labeled sample, the optimal values typically do not vary significantly.

Acknowledgments

This work was supported by NIH (R01AI108889) and NSF (MCB1506420). M.L. acknowledges support from a Margaret-Strauss Kramer Fellowship.

References

1. Nikaido H (2009) Multidrug resistance in bacteria. Annu Rev Biochem 78:119–146. https://doi.org/10.1146/annurev.biochem.78.082907.145923
2. Nikaido H, Pages JM (2012) Broad-specificity efflux pumps and their role in multidrug resistance of Gram-negative bacteria. FEMS Microbiol Rev 36(2):340–363. https://doi.org/10.1111/j.1574-6976.2011.00290.x
3. Allen HK, Donato J, Wang HH, Cloud-Hansen KA, Davies J, Handelsman J (2010) Call of the wild: antibiotic resistance genes in natural environments. Nat Rev Microbiol 4:251–259
4. Du D, van Veen HW, Murakami S et al (2015) Structure, mechanism and cooperation of bacterial multidrug transporters. Curr Opin Struct Biol 33:76–91. https://doi.org/10.1016/j.sbi.2015.07.015
5. Traaseth NJ, Shi L, Verardi R et al (2009) Structure and topology of monomeric phospholamban in lipid membranes determined by a hybrid solution and solid-state NMR approach. Proc Natl Acad Sci U S A 106 (25):10165–10170. https://doi.org/10.1073/pnas.0904290106
6. Vostrikov Vitaly V, Grant Christopher V, Opella Stanley J et al (2011) On the combined analysis of (2)H and (15)N/(1)H solid-state NMR data for determination of transmembrane peptide orientation and dynamics. Biophys J 101(12):2939–2947. https://doi.org/10.1016/j.bpj.2011.11.008
7. Marassi FM, Opella SJ (2000) A solid-state NMR index of helical membrane protein structure and topology. J Magn Reson 144 (1):150–155. https://doi.org/10.1006/jmre.2000.2035
8. Cross TA (1986) A solid state nuclear magnetic resonance approach for determining the structure of gramicidin a without model fitting. Biophys J 49(1):124–126
9. Wang J, Denny J, Tian C et al (2000) Imaging membrane protein helical wheels. J Magn Reson 144(1):162–167. https://doi.org/10.1006/jmre.2000.2037
10. Buffy JJ, Traaseth NJ, Mascioni A et al (2006) Two-dimensional solid-state NMR reveals two topologies of sarcolipin in oriented lipid bilayers. Biochemistry 45(36):10939–10946. https://doi.org/10.1021/bi060728d
11. Opella SJ, Marassi FM (2004) Structure determination of membrane proteins by NMR spectroscopy. Chem Rev 104(8):3587–3606. https://doi.org/10.1021/cr0304121
12. Clark NA, Rothschild KJ, Luippold DA et al (1980) Surface-induced lamellar orientation of multilayer membrane arrays. Theoretical analysis and a new method with application to purple membrane fragments. Biophys J 31 (1):65–96. https://doi.org/10.1016/S0006-3495(80)85041-7
13. Moll F 3rd, Cross TA (1990) Optimizing and characterizing alignment of oriented lipid bilayers containing gramicidin D. Biophys J

57(2):351–362. https://doi.org/10.1016/S0006-3495(90)82536-4

14. Sanders Ii CR, Hare BJ, Howard KP et al (1994) Magnetically-oriented phospholipid micelles as a tool for the study of membrane-associated molecules. Prog Nucl Magn Reson Spectrosc 26(Part 5):421–444. https://doi.org/10.1016/0079-6565(94)80012-X
15. Song Z, Kovacs FA, Wang J et al (2000) Transmembrane domain of M2 protein from influenza a virus studied by solid-state 15N polarization inversion spin exchange at magic angle NMR. Biophys J 79(2):767–775. https://doi.org/10.1016/S0006-3495(00)76334-X
16. Traaseth NJ, Verardi R, Torgersen KD et al (2007) Spectroscopic validation of the pentameric structure of phospholamban. Proc Natl Acad Sci U S A 104(37):14676–14681. https://doi.org/10.1073/pnas.0701016104
17. De Angelis AA, Opella SJ (2007) Bicelle samples for solid-state NMR of membrane proteins. Nat Protoc 2(10):2332–2338
18. Sanders CR, Schwonek JP (1992) Characterization of magnetically orientable bilayers in mixtures of dihexanoylphosphatidylcholine and dimyristoylphosphatidylcholine by solid-state NMR. Biochemistry 31(37):8898–8905. https://doi.org/10.1021/bi00152a029
19. Marcotte I, Auger M (2005) Bicelles as model membranes for solid- and solution-state NMR studies of membrane peptides and proteins. Concepts Magn Reson A 24A(1):17–37. https://doi.org/10.1002/cmr.a.20025
20. Warschawski DE, Arnold AA, Beaugrand M et al (2011) Choosing membrane mimetics for NMR structural studies of transmembrane proteins. Biochim Biophys Acta Biomembr 1808(8):1957–1974. https://doi.org/10.1016/j.bbamem.2011.03.016
21. Prosser RS, Hwang JS, Vold RR (1998) Magnetically aligned phospholipid bilayers with positive ordering: a new model membrane system. Biophys J 74(5):2405–2418. https://doi.org/10.1016/S0006-3495(98)77949-4
22. Yamamoto K, Pearcy P, Ramamoorthy A (2014) Bicelles exhibiting magnetic alignment for a broader range of temperatures: a solid-state NMR study. Langmuir 30(6):1622–1629. https://doi.org/10.1021/la404331t
23. Yamamoto K, Pearcy P, Lee D-K et al (2015) Temperature-resistant bicelles for structural studies by solid-state NMR spectroscopy. Langmuir 31(4):1496–1504. https://doi.org/10.1021/la5043876
24. Triba MN, Devaux PF, Warschawski DE (2006) Effects of lipid chain length and unsaturation on bicelles stability. A phosphorus NMR study. Biophys J 91(4):1357–1367. https://doi.org/10.1529/biophysj.106.085118
25. Wu C, Ramamoorthy A, Opella S (1994) High-resolution heteronuclear dipolar solid-state NMR spectroscopy. J Magn Reson 109A:270–272
26. Ramamoorthy A, Wei Y, Lee D-K (2004) PISEMA solid-state NMR spectroscopy. In: Annual reports on NMR spectroscopy, vol 52. Academic Press, London, pp 1–52. https://doi.org/10.1016/S0066-4103(04)52001-X
27. Gayen A, Banigan JR, Traaseth NJ (2013) Ligand-induced conformational changes of the multidrug resistance transporter EmrE probed by oriented solid-state NMR spectroscopy. Angew Chem Int Ed 52(39):10321–10324. https://doi.org/10.1002/anie.201303091
28. Cho M-K, Gayen A, Banigan JR et al (2014) Intrinsic conformational plasticity of native EmrE provides a pathway for multidrug resistance. J Am Chem Soc 136(22):8072–8080. https://doi.org/10.1021/ja503145x
29. Gor'kov PL, Chekmenev EY, Li C et al (2007) Using low-E resonators to reduce RF heating in biological samples for static solid-state NMR up to 900 MHz. J Magn Reson 185(1):77–93. https://doi.org/10.1016/j.jmr.2006.11.008
30. Delaglio F, Grzesiek S, Vuister GW et al (1995) NMRPipe: a multidimensional spectral processing system based on UNIX pipes. J Biomol NMR 6(3):277–293. https://doi.org/10.1007/bf00197809
31. Fung BM, Khitrin AK, Ermolaev K (2000) An improved broadband decoupling sequence for liquid crystals and solids. J Magn Reson 142(1):97–101. https://doi.org/10.1006/jmre.1999.1896
32. Bielecki ACK, De Groot HJM, Griffin RG, Levitt MH (1990) Frequency-switched Lee--Goldburg sequence in solids. Adv Magn Reson 14:111–150
33. Goldburg WG, Lee M (1965) Nuclear magnetic resonance line narrowing by a rotation RF field. Phys Rev 140:1261–1271
34. Vinogradov E, Madhu PK, Vega S (1999) High-resolution proton solid-state NMR spectroscopy by phase-modulated Lee–Goldburg experiment. Chem Phys Lett 314(5–6):443–450. https://doi.org/10.1016/S0009-2614(99)01174-4
35. Fu R, Tian C, Cross TA (2002) NMR spin locking of proton magnetization under a

frequency-switched Lee–Goldburg pulse sequence. J Magn Reson 154(1):130–135. https://doi.org/10.1006/jmre.2001.2468

36. Yamamoto K, Lee DK, Ramamoorthy A (2005) Broadband-PISEMA solid-state NMR spectroscopy. Chem Phys Lett 407 (4–6):289–293. https://doi.org/10.1016/j.cplett.2005.03.082
37. Gopinath T, Traaseth NJ, Mote K et al (2010) Sensitivity enhanced heteronuclear correlation spectroscopy in multidimensional solid-state NMR of oriented systems via chemical shift coherences. J Am Chem Soc 132 (15):5357–5363. https://doi.org/10.1021/ja905991s
38. Veglia TGaG (2009) Sensitivity enhancement in static solid-state NMR experiments via single- and multiple-quantum dipolar coherences. J Am Chem Soc 131(16):5754–5756
39. Gopinath T, Veglia G (2010) Improved resolution in dipolar NMR spectra using constant time evolution PISEMA experiment. Chem Phys Lett 494(1–3):104–110. https://doi.org/10.1016/j.cplett.2010.05.078
40. Koroloff SN, Nevzorov AA (2015) Optimization of cross-polarization at low radiofrequency fields for sensitivity enhancement in solid-state NMR of membrane proteins reconstituted in magnetically aligned bicelles. J Magn Reson 256:14–22. https://doi.org/10.1016/j.jmr.2015.03.016
41. Tang W, Nevzorov AA (2011) Repetitive cross-polarization contacts via equilibration-re-equilibration of the proton bath: sensitivity enhancement for NMR of membrane proteins reconstituted in magnetically aligned bicelles. J Magn Reson 212(1):245–248. https://doi.org/10.1016/j.jmr.2011.06.028
42. Bertram R, Quine JR, Chapman MS et al (2000) Atomic refinement using orientational restraints from solid-state NMR. J Magn Reson 147(1):9–16. https://doi.org/10.1006/jmre.2000.2193
43. De Angelis AA, Howell SC, Nevzorov AA et al (2006) Structure determination of a membrane protein with two trans-membrane helices in aligned phospholipid bicelles by solid-state NMR spectroscopy. J Am Chem Soc 128 (37):12256–12267. https://doi.org/10.1021/ja063640w
44. Traaseth NJ, Gopinath T, Veglia G (2010) On the performance of spin diffusion NMR techniques in oriented solids: prospects for resonance assignments and distance measurements from separated local field experiments. J Phys Chem B 114(43):13872–13880. https://doi.org/10.1021/jp105718r
45. Nevzorov AA (2008) Mismatched Hartmann–Hahn conditions cause proton-mediated intermolecular magnetization transfer between dilute low-spin nuclei in NMR of static solids. J Am Chem Soc 130 (34):11282–11283. https://doi.org/10.1021/ja804326b
46. Fleishman SJ, Harrington SE, Enosh A et al (2006) Quasi-symmetry in the cryo-EM structure of EmrE provides the key to modeling its transmembrane domain. J Mol Biol 364 (1):54–67. https://doi.org/10.1016/j.jmb.2006.08.072
47. Chen Y-J, Pornillos O, Lieu S et al (2007) X-ray structure of EmrE supports dual topology model. Proc Natl Acad Sci U S A 104 (48):18999–19004. https://doi.org/10.1073/pnas.0709387104
48. deAzevedo ER, Bonagamba TJ, Schmidt-Rohr K (2000) Pure-exchange solid-state NMR. J Magn Reson 142(1):86–96. https://doi.org/10.1006/jmre.1999.1918
49. Gayen A, Leninger M, Traaseth NJ (2016) Protonation of a glutamate residue modulates the dynamics of the drug transporter EmrE. Nat Chem Biol 12(3):141–145. https://doi.org/10.1038/nchembio.1999. http://www.nature.com/nchembio/journal/v12/n3/abs/nchembio.1999.html—supplementary-information
50. Lu GJ, Opella SJ (2014) Resonance assignments of a membrane protein in phospholipid bilayers by combining multiple strategies of oriented sample solid-state NMR. J Biomol NMR 58(1):69–81. https://doi.org/10.1007/s10858-013-9806-y
51. Banigan JR, Gayen A, Traaseth NJ (2015) Correlating lipid bilayer fluidity with sensitivity and resolution of polytopic membrane protein spectra by solid-state NMR spectroscopy. Biochim Biophys Acta 1848(1, Part B):334–341. https://doi.org/10.1016/j.bbamem.2014.05.003

Chapter 7

Generation of Conformation-Specific Antibody Fragments for Crystallization of the Multidrug Resistance Transporter MdfA

Frank Jaenecke, Yoshiko Nakada-Nakura, Kumar Nagarathinam, Satoshi Ogasawara, Kehong Liu, Yunhon Hotta, So Iwata, Norimichi Nomura, and Mikio Tanabe

Abstract

A major hurdle in membrane protein crystallography is generating crystals diffracting sufficiently for structure determination. This is often attributed not only to the difficulty of obtaining functionally active protein in mg amounts but also to the intrinsic flexibility of its multiple conformations. The cocrystallization of membrane proteins with antibody fragments has been reported as an effective approach to improve the diffraction quality of membrane protein crystals by limiting the intrinsic flexibility. Isolating suitable antibody fragments recognizing a single conformation of a native membrane protein is not a straightforward task. However, by a systematic screening approach, the time to obtain suitable antibody fragments and consequently the chance of obtaining diffracting crystals can be reduced. In this chapter, we describe a protocol for the generation of Fab fragments recognizing the native conformation of a major facilitator superfamily (MFS)-type MDR transporter MdfA from *Escherichia coli*. We confirmed that the use of Fab fragments was efficient for stabilization of MdfA and improvement of its crystallization properties.

Key words Multidrug resistance transporter, Crystallization, Antibody fragment, MFS transporter, Stabilization

1 Introduction

Multidrug resistance (MDR) transport is a major obstacle to the successful treatment of infectious diseases, and it is often mediated by the expression of multiple MDR transporters that recognize and export a variety of chemically different toxic compounds [1, 2]. The major facilitator superfamily (MFS) is ubiquitously expressed throughout the bacterial kingdom, with many MDR transporters. Although MFS-type MDR transporters have long been

Frank Jaenecke and Yoshiko Nakada-Nakura contributed equally to this work.

Akihito Yamaguchi and Kunihiko Nishino (eds.), *Bacterial Multidrug Exporters: Methods and Protocols*, Methods in Molecular Biology, vol. 1700, https://doi.org/10.1007/978-1-4939-7454-2_7, © Springer Science+Business Media, LLC 2018

investigated, the mechanisms through which they recognize and transport drugs are still enigmatic [3, 4].

To understand the efflux mechanism in MFS-type MDR transporters, high-resolution structural data potentially can provide invaluable information for the future defense strategy against pathogenic bacteria. MDR transporters are membrane proteins; the generation of well diffracting crystals of these proteins is very challenging, but there is no doubt that obtaining more structural information is desirable to understand their transport mechanism. First, they have to be solubilized using detergent for purification purpose. In this procedure, the detergent covers the hydrophobic surface of the transporter, and the purified protein is generally crystallized as a complex with the detergent micelle. There are two problems, (1) the protein is often unstable in detergent micelles, and (2) the exposed area of proteins from detergent micelles, which are important for crystal–crystal contacts are masked.

Stabilizing the transporter with an antibody fragment potentially solves these two problems. The use of antibody fragments to fix the transporter in a defined conformation increases its stability in solution. This approach also enhances the surface area exposed from detergent micelles, which is often thought to be critical for producing crystal contacts, thereby increasing the chances for generation of protein crystals [5, 6]. For successful cocrystallization, a stable complex between an antibody fragment and a conformational epitope present in the native 3D structure of the target membrane protein is required. Sequence-specific antibodies are less favorable to use as crystallization chaperones as they may, in fact, destabilize the protein. The protocol described here is an efficient and reliable screening system exemplified in our search for conformation-specific antibodies against *E. coli* MFS MDR transporter MdfA.

MdfA is an MFS-type MDR, which has close homologues in many pathogenic bacteria. It is composed of 410 amino acid residues corresponding to 12-transmembrane helices, and is capable of transporting a number of lipophilic, cationic, zwitterionic and neutral antibiotics and toxic compounds such as chloramphenicol, erythromycin, ethidium, tetraphenylphosphonium (TPP), in exchange for protons [7]. It has been recently found that MdfA is a proton/(Na^+)(K^+) antiporter involved in pH regulation and knocking out this transporter restricts cell growth in strongly alkaline conditions [8]. The first structure of MdfA was recently determined, in an inward facing form [9]. To understand the efflux mechanism in MFS-type MDR, alternative states of high-resolution structural data would be invaluable information for the future defense strategy from pathogenic bacteria.

We expressed and purified MdfA, and Fab fragments against MdfA reconstituted in liposomes to ensure it adopted a native conformation. In the course of multiple screening processes, we identified Fab fragments that cause an increase in the stability of the transporter as measured using N-[4-(7-diethylamino-4-methyl-3-coumarinyl)phenyl] maleimide (CPM) thermostability assays [10]. Here we describe a detailed protocol for the generation of Fab fragments suitable for use in structural biology.

2 Materials

2.1 Expression and Purification

1. Shaker incubator.
2. FPLC (e.g. ÄKTA system—GE Healthcare).
3. Superdex 200 10/300 GL (GE Healthcare) for size exclusion chromatography (SEC).
4. Ultrafiltration devices MWCO 100,000 Da (Millipore).
5. HiTrap ProteinG HP (GE Healthcare) for IgG purification.
6. NHS-activated Sepharose 4 Fast Flow (GE Healthcare) for papain digestion.
7. Protein A Sepharose Fast Flow (GE Healthcare) for Fab purification.
8. Poly-prep column.

2.2 Generation of Hybridoma

1. CO_2 incubator.
2. T225 culture flask.
3. 96-Well cell culture plate.

2.3 ELISAs, CPM Assays, Crystallization Trials

1. Bio-Beads SM-2 (Bio-Rad).
2. Fluorescence spectrometer.
3. 96-Well formatted Microplate Reader and Microplate washer, for ELISA.
4. Streptavidin-coated 96-well plates (e.g. Nunc Immobilizer Streptavidin—Thermo Scientific).
5. 96-Well ELISA plates (e.g. Nunc Immunoplate Maxisorp—Thermo Scientific).
6. Rotor Gene Q cycler (QIAGEN) qPCR system for CPM assays, with Rotor Gene Q Series Software Version 2.1.0; emission is recorded at 460 nm (excitation: 365 nm).
7. Lipidic cubic phase (LCP) mixer kit (e.g. Jena Bioscience).
8. Crystallization robot for LCP (e.g. TPP labtech, Zinsser, NT8, Glyphon).
9. Crystallization screens: MemPlus, MemGold2 (Molecular dimensions) and other screening kits.

2.4 Vector and Oligonucleotides, Strains

1. Expression vector pWaldo-GFPe [11] is a modified pET28a-vector including the GFP gene sequence and a TEV protease cleavage site corresponding to (ENLYFQ\S) before the GFP sequence.
2. Oligonuceotides for cloning of the *mdfA* gene for expression (5′-TCG CAC GAA GGG GGT ACC TAT GGA TCCGAA AAC CTG TAC-3′ and 5′-GTA CAG GTT TTC GGA TCC ATA GGT ACC CCC TTC GTG CGA-3′).
3. *Escherichia coli* C43(*DE3*) the host strain for MdfA expression.

2.5 Reagents and Buffers, Lipids, Enzymes, Media

1. Lipid solubilization buffer: 1.4% sodium cholate in phosphate-buffered saline (PBS), pH 7.4.
2. *Escherichia coli* polar lipid extract.
3. Lipid A.
4. Biotinyl phosphatidylethanolamine (Biotin-PE).
5. Hybridoma culture medium: 25% S-Clone (Eidia Co., Ltd.), 15% Fetal bovine serum (FBS), 5% BM condimed (Roche), 50 units/ml penicillin, 50 μg/ml streptomycin, 40 μg/ml gentamicin, 1 mM sodium pyruvate, RPMI1640 (Nacalai) containing 100 μM hypoxanthine, 0.4 μM aminopterin and 16 μM thymedine (HAT).
6. Red blood cell lysis buffer (Roche).
7. 50% (w/v) PEG1500, 17 mM HEPES, pH 8.0.
8. Reaction buffer: 1% bovine serum albumin (BSA) in PBS.
9. Denaturation buffer: 1% sodium dodecyl sulfate (SDS) in PBS.
10. 3,3′,5,5′-Tetra-methylbenzidine (TMB) in an acidic buffer.
11. TMB stop solution: 2 N sulfuric acid.
12. Papain.
13. Resin wash solution: 1 mM HCl.
14. Coupling buffer: 200 mM sodium hydrogen carbonate, 500 mM NaCl, pH 8.3.
15. Blocking buffer: 1 M Tris–HCl pH 9.0.
16. Wash buffer: 100 mM sodium acetate, 500 mM NaCl, pH 4.0.
17. Binding buffer: 20 mM sodium phosphate, 100 mM NaCl, pH 7.0.
18. Elution buffer: 100 mM Gly-HCl, pH 2.7.
19. Neutralization buffer: 1 M Tris–HCl, pH 9.0.
20. Digestion buffer: 20 mM cystein, 20 mM sodium phosphate, 5 mM EDTA, pH 7.0.
21. Buffer A: 10 mM MES, 20 mM NaCl, 0.02% DDM, pH 7.0.
22. *N*-[4-(7-Diethylamino-4-methyl-3-coumarinyl)phenyl] maleimide (CPM).

2.6 Mouse Strain and Myeloma Cell Line

1. BALB/cAJcl (CLEA) mice.
2. P3/NS1/1-Ag4-1 myeloma cell line (ATCC TIB-18).

3 Methods

Cloning, Expression and Purification of MdfA: His-tagged MdfA-GFP fusion protein was expressed in E. coli C43(*DE3*) by adding 0.5 mM IPTG at 28 °C for 6 h. The purification protocol essentially follows Drew et al. [11]. Briefly, the cells were disrupted by sonication, and the crude extract was fractionated by metal affinity chromatography using Ni-NTA resin. Nontagged MdfA was recovered after TEV protease treatment by a second metal affinity chromatography step. Approximately 0.3–0.4 mg of purified MdfA was obtained routinely from 1 L of 2×YT medium.

3.1 Reconstitution of MdfA Protein into Liposome

1. Solubilize 0.5 mg of *E. coli* polar lipid extract with 0.1 mg of the adjuvant lipid A in 0.1 ml of lipid solubilization buffer and add 0.1 mg of purified DDM-solubilized MdfA.
2. Add Bio-Beads SM-2 to the MdfA-lipid mixture to remove the detergents and reconstitute MdfA in phospholipid vesicles.
3. Incubate at 4 °C for 24 h.
4. Centrifuge for 10 min at 700 × *g* at 4 °C.
5. Transfer the turbid supernatant containing the MdfA-reconstituted liposomes to a new 1.5 ml-microcentrifuge tube.
6. Sonicate the proteoliposome solution to produce small vesicles for four cycles of 20–30 s. Incubate the solution on ice in between the cycles.
7. Incubate for 1 h on ice.
8. Centrifuge for 10 min with 3000 × *g* at 4 °C to remove aggregates. Transfer the supernatant to a new 1.5 ml-ultracentrifuge tube.
9. Centrifuge for 1.5 h at 50,000 × *g* to pellet the MdfA-reconstituted liposomes. Remove the supernatant.
10. Resuspend the proteoliposomes to perform analysis of the TPP binding activity by fluorescence quenching experiments (Fig. 1) [12]. Proteoliposomes containing MdfA capable of TPP binding and Lipid A were used for immunization and generation of antibodies (*see* below Subheading 3.2).
11. Typical example for TPP binding measurement, (fluorescence, excitation: 280 nm, emission: 300–400 nm), Take 50 μl (=50 μg MdfA) for measurement.
12. Add 200 μl of reaction buffer and record spectra (="without TPP").
13. Add 5 μl of TPP (5 mM) to the cuvette and mix well, record new spectra (="sample with 100 μM TPP").

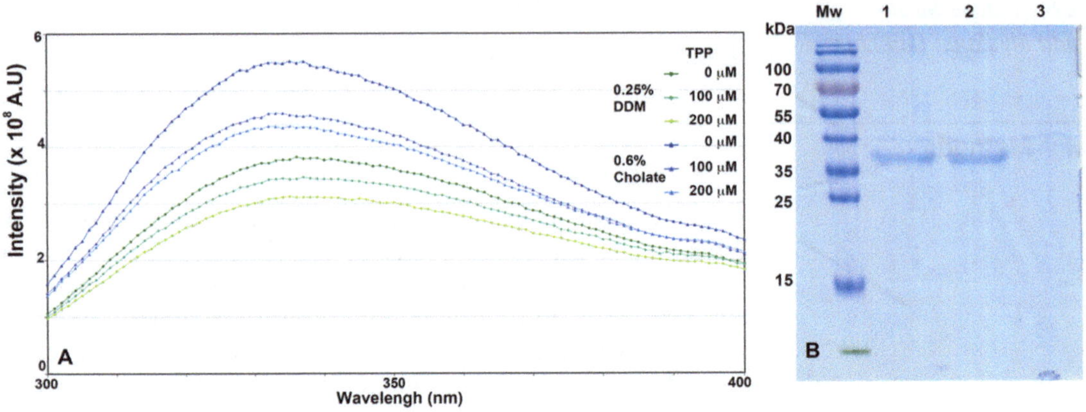

Fig. 1 Preparation of MdfA-liposome. (**a**) The substrate TPP-induced fluorescence quenching of MdfA-liposomes. The proteoliposomes were produced using either DDM-solubilized MdfA or cholate-solubilized MdfA. As previously shown by Fluman et al. [12], fluorescence quenching induced by substrate binding was observed in a concentration-dependent manner. (**b**) Sedimentation test of MdfA-liposome. *Lane 1*: purified MdfA protein, *lane 2*: supernatant fraction after low-speed centrifugation, *lane 3*: supernatant fraction after ultracentrifugation

3.2 Generation of Hybridoma-Producing Antibodies Against MdfA

Antibody-producing hybridoma cell lines are generated based on the established PEG methods [13, 14]. One should follow relevant guidelines for the ethics and use of laboratory animals. Cells were cultured at 37 °C in a humidified atmosphere of 5% CO_2 and 95% air.

1. Immunize two BALB/c mice initially with 100 μg of the proteoliposome antigen and 0.1 μg of pertussis toxin, followed by multiple booster injections of 50 μg of the proteoliposome antigen at 2-week intervals until the antiserum titer is increased.
2. Bleed the animal 7 days postinjection. A typical amount of blood to obtain from a single bleed is 30–50 μl. Centrifuge the blood sample for 10 min at 5000 × *g* to separate the serum from debris and clotted cells.
3. Apply the serum to liposome ELISA for titering immune serum. *See* below (Subheading 3.3 Liposome ELISA). To prepare microplates for liposome ELISA, incubate an appropriate number of wells of Nunc Immobilizer Streptavidin with proteoliposomes containing biotin-PE at a concentration between 0.2 and 2 μg/well in PBS.
4. Anesthetize the mice with a suitable inhalation anesthetic after 72 h of the final booster and sacrifice by cervical dislocation. Rinse the mouse in a bucket filled with 70% ethanol and lay it on a dissection tray with the face down and the left side up. Open the abdominal cavity with scissors, pull out the spleen with forceps and remove the attached connective tissue with

scissors. Squeeze the spleen and transfer into RPM1640 with forceps.

5. Pass the cell suspensions through a cell strainer. Pellet the cells at 400 × *g* for 5 min at RT and discard the supernatant.
6. Add 10 ml of red blood cell lysis buffer and suspend the cells gently. Pellet the cells at 400 × *g* for 5 min at RT and discard the supernatant. Add 20 ml of RPMI1640 and resuspend gently to wash the cells. Repeat this washing step twice.
7. Collect myeloma cells that grow in two T225 flasks. Pellet the cells at 400 × *g* for 5 min at RT and discard the supernatant.
8. Add 20 ml of RPMI1640 and suspend gently. Repeat this washing step twice.
9. Measure the volume of each cell mass and resuspend with RPMI1640. Add myeloma cells to the same volume of spleen cells. Mix the cells by inverting the tube gently and pellet cells at 400 × *g* for 5 min at RT. Discard supernatant and loosen the pellet by tapping.
10. Perform the cell fusion at 37 °C water with continuous stirring. Add 2 ml of 50% PEG 1500 (w/v) to the cells slowly at constant speed for 2 min. Keep stirring for 1 min. Add 4 ml of RPMI1640 slowly at a constant speed for 2 min. Dilute the cell suspension to the volume of 45 ml with RPMI1640 and pellet the cell at 400 × *g* for 5 min at RT. Discard supernatant and loosen pellet by tapping.
11. Dilute the cell suspension to the volume of 300 ml by adding hybridoma culture medium and pipette 300 μl/well to ten 96-well-culture plates. Incubate plates in a 5% CO_2 incubator at 37 °C for 1 week.
12. Dilute hybridoma cells in 3.0 ml of the culture medium at concentrations of 80, 40, 20, 10 cells/ml. Dispense 50 μl/well of each dilution into three columns of a 96-well culture plate containing 200 μl/well hybridoma culture medium. Incubate the plate 6–8 days and observe isolated clones.
13. Transfer the positive hybridoma cell lines to wells of 24-well-culture plates after the culture supernatants are tested by the liposome ELISA and denatured MdfA-targeted ELISA (*see* Subheading 3.3).
14. Incubate for 10–14 days until a sufficient amount of antibodies are secreted.

3.3 Liposome ELISA and Denatured MdfA-Targeted ELISA

Antibodies that bind to the native conformation of the hydrophilic portion of MdfA are positively selected by liposome ELISA as described [15] (**steps 1–4**). Furthermore, antibodies that bind to linear epitopes in MdfA are negatively selected by denatured MdfA-targeted ELISA (**steps 5–8**).

1. Dilute the liposomes containing MdfA and biotin-PE at the optimized concentration and dispense 50 μl each to the wells of ten streptavidin-coated 96-well plates. Incubate the plates overnight at 4 °C.
2. Discard the liposome solution and dispense 50 μl/well of the mixture of the hybridoma culture supernatant and reaction buffer (1:3 in vol/vol) and shake the plates for 1 h at 4 °C.
3. Wash the plates with PBS, add 50 μl/well of the secondary antibody solution diluted 1:60,000 in reaction buffer and shake the plates for 1 h at 4 °C.
4. Wash the plates, add 50 μl/well of TMB solution. Stop the reaction by adding 50 μl/well of TMB stop solution and read the absorbance at 450 nm.
5. Denature 5 μg of MdfA with 50 μl of denaturation buffer for 1 h at RT. Dilute the denatured MdfA solution with 5 ml of PBS, dispense 50 μl/well in a 96-well ELISA plate and incubate overnight at 4 °C. Discard the solution, block the remaining adsorption sites by adding 100 μl/well of reaction buffer and incubate for 1 h at RT.
6. Dispense 50 μl/well of the mixture of the hybridoma culture supernatant and reaction buffer (1:3 in vol/vol) and shake the plates for 1 h at 4 °C.
7. Wash the plates with PBS, add 50 μl/well of the secondary antibody solution diluted 1:60,000 in PBS and shake the plates for 1 h at 4 °C.
8. Wash the plates, add 50 μl/well of TMB solution. Stop the reaction by adding 50 μl/well of TMB stop solution and read the absorbance at 450 nm.

3.4 Preparation of Fab Fragments

1. Apply the mixture of the monoclonal hybridoma culture supernatant and binding buffer (1:1 in vol/vol) to a Protein G affinity column and wash the column with binding buffer. Elute antibodies (IgG) with elution buffer and neutralize the IgG fraction with a 10% volume of neutralization buffer.
2. Digest the purified antibodies with Papain to generate Fab fragments. Conjugate papain to NHS-activated Sepharose 4 Fast Flow resin according to the manufacture's protocol. Change the buffer containing IgG by dialysis with digestion buffer and concentrate IgG to the protein concentration of ca. 10 mg/ml. Add an equal volume of 50% suspension of papain conjugated resin and shake it for 4 h at 37 °C. Pass over an empty Poly-prep column to remove the resin.
3. Purify the Fab fragments on Superdex 200 gel filtration (SEC) followed by Protein A affinity chromatography. Perform the SEC to change the buffer to binding buffer and to remove

Table 1
Representative results of liposome ELISA and denatured MdfA-targeted ELISA

	ELISA signal ($A_{450\ nm}$)				
Clone number	MdfA liposome (A)	Empty liposome (B)	(A/B)	Denatured MdfA	GFP
YN1010	2.19	0.071	30.85	0.24	0.069
YN1014	2.218	0.068	32.62	0.18	0.069
YN1074	2.361	0.076	31.07	0.19	0.055
YN1082	1.683	0.076	22.14	0.2	0.068
YN1001	1.751	0.068	25.75	2.32	0.078
YN1053	2.532	0.073	34.68	1.59	0.808

GFP and empty liposome were also used as a background control. YN1010, 1014, 1074, and 1082 were candidates for further screening. YN1001 is a typical example of antibody-recognizing amino acid sequence but not 3D conformation of MdfA. YN1053 is also an example possibly recognizing amino acid sequence and noncleaved MdfA-GFP contaminant

undigested IgG and F(ab′)$_2$. Collect fractions of the Fab fragment contaminated with the Fc portion and apply to Protein A Affinity chromatography to remove the Fc portion. Collect the flow through fractions.

After two times of immunization with the MdfA-liposomes, a sufficient titer of antiserum was present. As a result of the screening of MdfA-specific monoclonal antibodies, we identified that 117 out of 960 clones (12.2%) are positive in the liposome ELISA, 59 clones (6.1%) also positive with denatured MdfA-targeted ELISA (Table 1, *see* **Note 1**).

3.5 Purification of the MdfA–Fab Complexes

In this study, we obtained four Fab fragments (clones YN1006, YN1010, YN1074, and YN1082). Each Fab was incubated with purified MdfA, and the MdfA–Fab complexes were purified by SEC.

1. Concentrate the purified MdfA and Fab fragments to the concentrations of approximately 5 mg/ml (For Fab, higher concentrations may also be used).
2. Dilute 1 mg of MdfA (~200 μl) into 9.5 ml buffer A. Add 1.5 mg of isolated Fab fragments (~300 μl) and invert the tube or pipette slowly up and down to mix.
3. Incubate for 1 h on ice.
4. Concentrate to a volume of ~200 μl and perform SEC in the presence of buffer B (Fig. 2, *see* **Note 2**).
5. Collect fractions containing MdfA–Fab complexes and concentrate to approx. 2 mg/ml for analysis of thermostability or higher concentration for protein crystallization (5 mg/ml recommended).

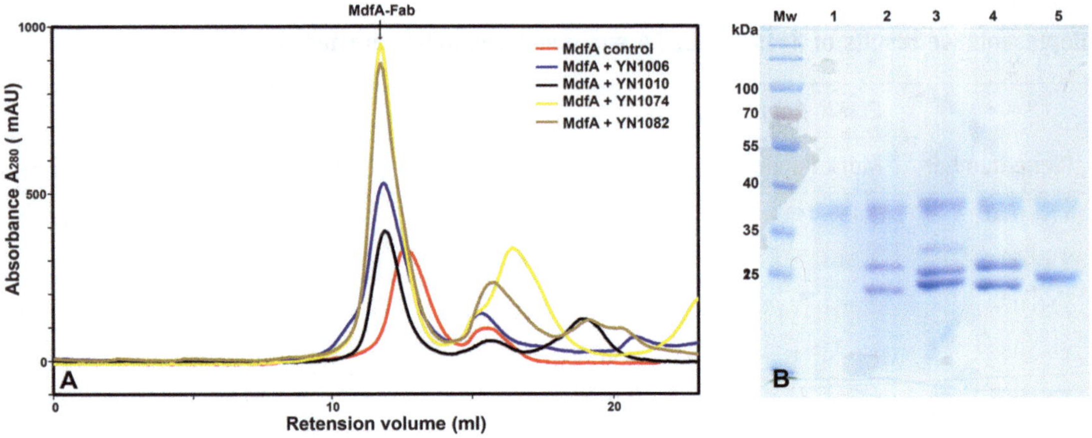

Fig. 2 Isolation of MdfA–Fab complexes. (**a**) Size exclusion chromatography of MdfA with Fab fragment. (**b**) Peak fractions (show *arrow* in **a**) were applied to 12% SDS-PAGE. *Lane 1*: MdfA apo, *lane 2*: MdfA-YN1006Fab, *lane 3*: MdfA-YN1010Fab, *lane 4*: MdfA-YN1074Fab, *lane 5*: MdfA-YN1082Fab. SEC profiles at pH 7.0 revealed a significant shift of the mono-disperse elution peak from ~12.7 ml for MdfA apo to small retention volumes (~12 ml) for all tested MdfA–Fab complexes. SDS-PAGE analysis also demonstrated that the main elution peak fractions contain MdfA and two more distinct polypeptides that correspond to the heavy chain V_H-C_H1 and light chain V_L-C_L of the Fab fragment. These four Fab fragments are able to tightly bind to the native structure of MdfA

3.6 Thermostability Analysis of MdfA–Fab Complexes

CPM assay was performed as described by Alexandrov et al. [10] with slight modifications. A qPCR machine is used to allow this to be done in a high-throughput manner.

1. Mix 12 μl MdfA–Fab complexes (2 mg/ml) with 45.6 μl of buffer A.
2. Add 2.4 μl of CPM dye (5 mg/ml).
3. Transfer 25 μl of this mixture to a clean PCR tube.
4. Analyze thermostability in a Rotor Gene Q cycler (QIAGEN). Heat the sample with 1 °C per min starting at room temperature (25 °C) and ending at 90 °C (Fig. 3). Set excitation wavelength at 365 nm and record emission at 460 nm (*see* **Note 3**).

3.7 Crystallization Trials of MdfA–Fab Complexes

1. MdfA–Fab complexes at 5 mg/ml is used for initial LCP crystallization (*see* **Note 4**). Centrifuge sample at 15,000 × *g* for 30 min to remove insoluble aggregates before reconstituting the protein into the lipid.
2. Two gas-tight syringes are coupled upon loading one syringe with the lipid, such as 9.9 MAG (monoolein) and the other containing MdfA–Fab respectively at a ratio between 6:4 and 7:3 (v/v). Push the syringe plungers alternately to move the lipid and protein through the inner needle of the coupler, back and forth, until the lipid mesophase becomes homogeneous.

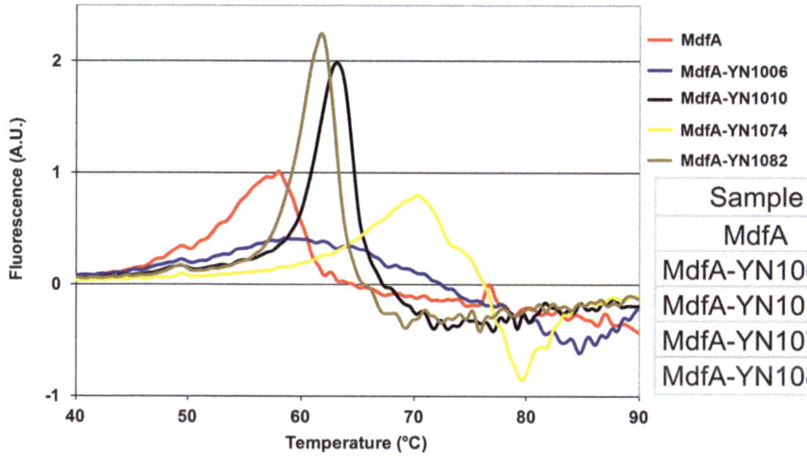

Sample	Tm (°C)
MdfA	59.4±1.7
MdfA-YN1006	63.6±6.2
MdfA-YN1010	63.6±0.5
MdfA-YN1074	71.8±1.6
MdfA-YN1082	62.5±0.7

Fig. 3 CPM assay. (**a**) Thermal transition temperatures (Tm) of MdfA wild-type, MdfA with respective Fab fragment were evaluated by the first derivation of the melting curve by thermal denaturation process. It was tested whether the Fab fragments are capable to stabilize MdfA. Analysis of the melting curves for MdfA with YN1010Fab, YN1074Fab and YN1082Fab showed a remarkable increase of Tm temperature. Compared to MdfA apo, MdfA–Fab were able to raise the Tm temperature by 3 °C (YN1010Fab, YN1082Fab)—12 °C (YN1074Fab) at pH 7.0

3. MdfA–Fab reconstituted into LCP is now ready for the crystallization setup. Protein laden phase is transferred completely to one of the syringes and a fine purging nozzle is attached to dispense the LCP as a bolus. The syringe is mounted on a movable arm in the crystallization robot (Zinsser analytics). One hundred nanoliters of the bolus is extruded onto each well of the 96 well Laminex glass base (200 μm) and 1 μl precipitant solution is pipetted over for crysallization. The Laminex glass base is sealed air-tight with a glass film cover to avoid dehydration.
4. Store the crystallization plate at 20 °C (Formulatrix Rock Imager 54) and evaluate drops by means of visible imaging and cross polarizers at several time points (e.g. at 0 h, 12 h, 1, 2, 4, 7, 10, 2 weeks ...) (Fig. 4) (*see* **Note 5**).
5. In the rest of steps for observing and harvesting crystals, conventional methods were used. This is described elsewhere [16].

4 Notes

1. For further screening or to analyze the affinity of the antibody to the target protein, surface plasmon resonance (SPR) can be used. However, the screening of antibodies using SPR is not absolutely required.
2. MdfA–Fab complex formation and SEC analysis was performed between pH 5.5 and pH 7.0 in this study. Binding of

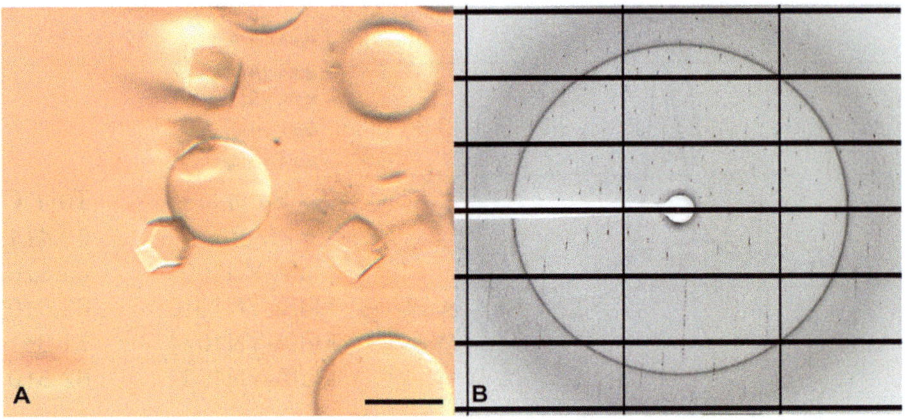

Fig. 4 Crystals of MdfA-YN1074Fab complex. (**a**) Crystals of MdfA–Fab complex appeared within 2 weeks in LCP crystallization method. The scale bar represents 100 μm. (**b**) The MdfA-Fab crystals diffracted to 3.5 Å. In contrast, crystals of MdfA apo have not diffracted more than 6 Å so far

MdfA by Fab fragments seems not to be affected by pH as we observed a shift to shorter retention times in all MdfA–Fab samples compared to MdfA without Fab fragments.

3. Theoretically, the increase of the measured fluorescence should give a maximum slope at the transition temperature/melting temperature of the protein. According to that the first derivation of the melting curve (which is conveniently done with Rotor Gene Q Series Software; Version 2.1.0) should show a maximum at this position, which allows an easy and accurate determination of the proteins Tm temperature.
4. A conventional setup of crystallization recommends the protein concentration is >10 mg/ml. The concentration limit here is due to the propensity of the complex forming insoluble aggregates.
5. We tested crystallization screens available from manufacturers such as Molecular Dimensions, Rigaku, and Hampton Research. We were able to obtain crystals from many conditions using the vapor diffusion method, yet could not able to obtain reasonable diffraction. We could successfully obtained 3.4 Å diffracting crystals of MdfA–Fab in LCP [17].

Acknowledgements

We thank Dr. Alexander Cameron for critically reading the manuscript. The work was supported by the Bundesministerium für Bildung und Forschung (BMBF) ZIK program (FKZ 03Z2HN21), by the ERDF (1241090001) (M.T.), by the ERATO Human Receptor Crystallography Project of the Japan

Science and Technology Agency (JST) (S.I.), by the Research Acceleration Program of the JST (S.I.), by the Targeted Proteins Research Program of the Ministry of Education, Culture, Sports, Science and Technology (MEXT) of Japan (S.I.), and by Grants-in-Aids for Scientific Research from the MEXT (No. 22570114 to N. N.). The crystallographic data were tested at Swiss light source (SLS, Villingen) with supported by the funding from the European Community's Seventh Framework Programme (FP7/2007–2013) under BioStruct-X (grant agreement N°283570, project ID: BioStructx_5450). The authors declare no conflict of interest. Frank Jaenecke and Yoshiko Nakada-Nakura have contributed equally to this work.

References

1. Paulsen IT, Brown MH, Skurray RA (1996) Proton-dependent multidrug efflux systems. Microbiol Rev 60:575–608
2. Nikaido H (2009) Multidrug resistance in bacteria. Annu Rev Biochem 78:119–146
3. Pao SS, Paulsen IT, Saier MH (1998) Major facilitator superfamily. Microbiol Mol Biol Rev 62:1–34
4. Shi Y (2013) Common folds and transport mechanisms of secondary active transporters. Annu Rev Biophys 42:51–72
5. Bukowska MA, Grütter MG (2013) New concepts and aids to facilitate crystallization. Curr Opin Struct Biol 23:409–416
6. Lieberman RL, Culver JA, Entzminger KC, Pai JC, Maynard JA (2011) Crystallization chaperone strategies for membrane proteins. Methods 55:293–302
7. Edgar R, Bibi E (1997) MdfA, an Escherichia coli multidrug resistance protein with an extraordinarily broad spectrum of drug recognition. J Bacteriol 179:2274–2280
8. Lewinson O, Padan E, Bibi E (2004) Alkalitolerance: a biological function for a multidrug transporter in pH homeostasis. Proc Natl Acad Sci U S A 101:14073–14078
9. Heng J, Zhao Y, Liu M, Liu Y, Fan J, Wang X, Zhao Y, Zhang XC (2015) Substrate-bound structure of the E. coli multidrug resistance transporter MdfA. Cell Res 25:1060–1073
10. Alexandrov AI, Mileni M, Chien EY, Hanson MA, Stevens RC (2008) Microscale fluorescent thermal stability assay for membrane proteins. Structure 16:351–359
11. Drew DE, von Heijne G, Nordlund P, de Gier JW (2001) Green fluorescent protein as an indicator to monitor membrane protein overexpression in Escherichia coli. FEBS Lett 507:220–224
12. Fluman N, Cohen-Karni D, Weiss T, Bibi E (2009) A promiscuous conformational switch in the secondary multidrug transporter MdfA. J Biol Chem 284:32296–32304
13. Köhler G, Milstein C (1975) Continuous cultures of fused cells secreting antibody of predefined specificity. Nature 256:495–497
14. Pontecorvo G (1976) Polyethylene glycol (PEG) in the production of mammalian somatic cell hybrids. Birth Defects Orig Artic Ser 12:399–400
15. Hino T, Iwata S, Murata T (2013) Generation of functional antibodies for mammalian membrane protein crystallography. Curr Opin Struct Biol 23:563–568
16. Li D, Boland C, Aragao D, Walsh K, Caffrey M (2012) Harvesting and cryo-cooling crystals of membrane proteins grown in lipidic mesophases for structure determination by macromolecular crystallography. J Vis Exp:e4001
17 Nagarathinam K, Jaenecke F, Nakada-Nakura Y, Hotta Y, Liu K, Iwata S, Stubbs MT, Nomura N, Tanabe M (2017) The multidrug-resistance transporter MdfA from: crystallization and X-ray diffraction analysis. Acta Crystallogr F Struct Biol Commun 73:423–430

Part II

Biochemical and Bioengineering Analysis of Bacterial Multidrug Exporters

Chapter 8

Biochemical Reconstitution and Characterization of Multicomponent Drug Efflux Transporters

Martin Picard, Elena B. Tikhonova, Isabelle Broutin, Shuo Lu, Alice Verchère, and Helen I. Zgurskaya

Abstract

Efflux pumps are the major determinants in bacterial multidrug resistance. In Gram-negative bacteria, efflux transporters are organized as macromolecular tripartite machineries that span the two-membrane cell envelope of the bacterium. Biochemical data on purified proteins are essential to draw a mechanistic picture of this highly dynamical, multicomponent, efflux system. We describe protocols for the reconstitution and the in vitro study of transporters belonging to RND and ABC superfamilies: the AcrAB–TolC and MacAB–TolC efflux systems from *Escherichia coli* and the MexAB–OprM efflux pump from *Pseudomonas aeruginosa*.

Key words Membrane protein purification, Proteoliposomes, Transport kinetics

1 Introduction

1.1 Multicomponent Drug Transporters

Gram-negative bacteria are notoriously more resistant to antibiotics than Gram-positive bacteria. For a long time, the outer membrane was believed to be the major reason why Gram-negative pathogens are recalcitrant to antibiotic treatment [1]. The seminal finding that *Escherichia coli* cells contain multidrug efflux transporters changed dramatically our view on how these bacteria resist antibiotics [2, 3]. Current models postulate that intrinsic multidrug resistance of Gram-negative bacteria is the result of synergy between reduced uptake of drugs across the outer membrane and active drug efflux from the inner membrane [4–6]. This synergy is possible because multidrug efflux transporters of Gram-negative bacteria function together with proteins belonging to the Membrane Fusion Protein (MFP) and the Outer Membrane Factor (OMF) families [7]. MFPs are located in the periplasm and act on both membranes to enable drug efflux across the whole cell envelope directly into the medium. On the inner membrane (IM), MFPs associate with drug efflux

Akihito Yamaguchi and Kunihiko Nishino (eds.), *Bacterial Multidrug Exporters: Methods and Protocols*, Methods in Molecular Biology, vol. 1700, https://doi.org/10.1007/978-1-4939-7454-2_8, © Springer Science+Business Media, LLC 2018

transporters and stimulate their activities [8–10]. On the outer membrane (OM), they recruit OMF channels that enable expulsion of drugs into the medium [11, 12]. The three components form large multiprotein assemblies that traverse both the inner and outer membranes of Gram-negative bacteria. Working together as a well-coordinated team, they achieve the direct extrusion of substrates from the cytoplasm and/or the periplasm into the medium.

Drug transporters that associate with MFPs and OMF channels can belong to any of the three major superfamilies of proteins: RND (Resistance-Nodulation-Cell Division), ABC (ATP-Binding Cassette) and MF (Major Facilitator) superfamilies [13–16]. These transporters are structurally and mechanistically very diverse. ABC transporters are driven by ATP hydrolysis, whereas drug efflux by RND and MF pumps is coupled to transport of protons. MF transporters are thought to function as monomers, whereas ABC and RND transporters are dimers and trimers, respectively. Surprisingly, MFPs that associate with these transporters are structurally very similar. Furthermore, in *E. coli* and other enterobacteria, multidrug transporters share the same OM channel. How transporters with different architectures and molecular mechanisms assemble into tri-partite complexes with MFPs and OM channels and how transporter activities are coupled to efflux across the OM are the main questions of our research programs.

This chapter is focused on reconstitution and in vitro activities of transporters belonging to RND and ABC superfamilies that include the best-studied RND transporters *Escherichia coli* AcrB and *Pseudomonas aeruginosa* MexB functioning with MFPs AcrA/MexA and the OM channels TolC/OprM, respectively [10, 17] and the ABC transporter MacB with MFP MacA and TolC [18–20].

1.2 Reconstitution of Proteoliposomes and Transport Across One Membrane

Reconstitution of proteoliposomes is a tried-and-tested methodology that requires expertise and skills. It allows incorporation of the purified proteins into lipid membranes of known composition that create a closed compartment necessary for the vectorial transport of substrates. Such reconstituted proteoliposomes enable specific characterization of membrane protein transporters outside of the complexity of their native environment.

Zgurskaya and Nikaido achieved the first successful functional reconstitution of AcrB in 1999 [10], where they showed that proteoliposomes containing AcrB catalyze the extrusion of fluorescent phospholipids to protein-free acceptor vesicles in the presence of a proton gradient. This study provided the first in vitro experimental evidence that AcrB is a proton antiporter and that its activity is catalyzed in the presence of AcrA and Mg^{2+}. Proteoliposomes can also be used to extract dynamic and kinetic information: for example, the functional reconstitution of CzcA allowed characterization of the kinetics of heavy metal extrusion by this transporter, which is responsible for heavy metal resistance in *Ralstonia* sp. [21], whereas reconstitution of MacAB led to the proposal of a kinetic mechanism of this ATP-dependent transporter, which depends on MFP MacA [18, 20, 22]. Lastly, proteoliposome reconstitution can also be used to address the molecular mechanism of transport. Indeed, upon reconstitution of AcrD from *E. coli*, Aires and Nikaido reported that gentamicin can be transported from either side of the membrane, thereby providing experimental evidence in support of a controversial cytoplasmic access of the drug toward its transport side [8]. At the atomic level, the group of Edward Yu obtained a high-resolution structure of the RND CusA from the efflux pump CusCBA *E. coli* [23, 24]. They designed a functional assay for characterization of the proton transport pathway suggested by their X-ray crystal structure and confirmed that CusA mutants with substitutions at amino acid residues involved in the putative proton pathway were transport defective. More recently, Verchère and colleagues have described the first successful reconstitution of the whole tripartite efflux pump MexA-MexB-OprM from *P. aeruginosa*. This study highlights the mutual interplay between the various partners of the pump [17].

1.3 Energizing the System

Bacterial efflux pumps catalyze the active transport of substrates at the expense of proton counter transport or ATP hydrolysis. In the latter case, the system is energized upon addition of ATP and the activity of ABC transporters can be indirectly investigated by titrating the release of Pi by colorimetry [25] or by ^{32}P γ-phosphate counting [20]. Using such methodologies, it was shown that the MFP MacA does stimulate the basal activity of the macrolide transporter MacB [20]. Note that the use of transition state analogs such as aluminum fluoride, magnesium fluoride, or sodium orthovanadate or nonhydrolyzable nucleotide analogs can provide very useful mechanistic insights [18].

In the case of pmf-dependent transporters, several strategies can be followed to trigger the function of a transporter. Generation of a proton gradient can be achieved simply upon incubation of the proteoliposome suspension in a buffer which pH is different from that used to prepare the liposomes [10] or even more simply by adding HCl to the proteoliposome suspension [17]. However, such procedures are believed to generate gradients that are not very stable. More stable gradients can be obtained with proteoliposomes loaded with NH_4Cl that are subsequently diluted into an iso-osmotic buffer containing KCl, which leads to diffusion of uncharged NH3 and acidification of the proteoliposome lumen [21]. Proton gradient can also be generated in the other direction (alkaline-inside) by diluting proteoliposomes containing potassium acetate in a potassium acetate-free buffer [26]. The most stable gradients are obtained when the protein under study is co-reconstituted together with bacteriorhodopsin, the famous archaebacterial proton transporter from *Halobacterium salinarium* [27, 28].

Another classical strategy involves the use of valinomycin, a potassium-selective ionophore. Addition of valinomycin to KCl-containing liposomes that have been diluted into a buffer containing NaCl allows potassium to leak out of the vesicle down its concentration gradient. This creates a membrane potential across the membrane that is spontaneously compensated by an influx of protons, hence a proton gradient [8, 29, 30].

1.4 Fluorescent, Nucleic Acid, and Radioactive Probes

The choice of a quantitative readout for the monitoring of transport is the most critical parameter. In case of amphiphilic and hydrophobic substrates, such as most antibiotics and drugs, assays relying on the transmembrane transport are often difficult to design and interpret because substrates can bind nonspecifically to the membrane or leak out of the vesicles. In such cases, it is possible to deduce the transport activity from the associated variations of the pH inside the liposome or ATP hydrolysis rates. For example, dissipation of a proton gradient is often monitored by using pH-sensitive probes such as pyranine [8, 29, 30]. However in multidrug efflux transporters, proton translocation or ATP hydrolysis processes not always are stoichiometrically coupled to the substrate transfer. Therefore, energy depletion assays may not serve as a quantitative measure of the substrate transport.

The direct transport of substrates can sometimes be deduced from the distribution of the substrate itself. In that case, studies often resort to the use of radioactively labeled substrates such as heavy metals [21], or antibiotics [8] but assays with fluorescent substrates can also be performed [10]. Fluorescently labeled lipids have been demonstrated to be useful reporters of activities of multidrug efflux pumps [10]. In addition, Hoechst 33,342 or ethidium bromide are commonly used because they become highly fluorescent once intercalated into lipid bilayers or into DNA molecules that can be loaded into the proteoliposome lumen [26, 27] or into the lumen of an acceptor vesicle [17]. Note that we demonstrated that incorporation of stable RNA scaffolds is much more efficient than incorporation of DNA [17].

2 Materials

2.1 Purification of Inner Membrane Transporters

2.1.1 AcrB^{His} Purification

1. Luria-Bertani (LB) broth or 2×TY (16 g Bacto-tryptone, 10 g Yeast extract, 5 g NaCl for 1 L of medium) were used as growth media.
2. Unless otherwise noticed, 100 μg/ml ampicillin were added to the growth medium for plasmid selection.
3. Syringes and syringe acetate 0.22 μm filters.
4. Rotors JLA8.1, 70 Ti, TLA 100 rotors (Beckman Coulter).
5. Beckman Coulter centrifuges Avanti J-20 XP and Optima LE-80K.

6. AcrB Cell Buffer: 20 mM Tris–HCl (pH 8.0), 0.1 M NaCl, 1 mM phenylmethylsulfonyl fluoride (PMSF) in MilliQ water.
7. AcrB Cell Lysis Buffer: AcrB cell buffer supplemented with 1 mM $MgCl_2$ and 0.1 mg/ml DNase I.
8. French Pressure Cell Press (SLM Instruments, Inc.).
9. 0.5 M disodium ethylenediaminetetraacetate (Na_2EDTA) (pH 8.0) stock solution in MilliQ water.
10. 6 M Urea solution in MilliQ water.
11. 10% Triton X-100 (TX) prepared in buffers as indicated for each protein purification protocol.
12. AcrB Binding Buffer: 20 mM Tris–HCl (pH 8.0), 0.5 M NaCl, 1 mM PMSF, 5 mM imidazole in MilliQ water.
13. Metal-chelating Sepharose resin charged with Cu^{2+}.
14. Column buffers: AcrB Binding Buffer supplemented with 0.2% TX and a linear imidazole gradient of 5–200 mM.

2.1.2 MexB^{His} Purification

1. Syringes and syringe acetate 0.22 μm filters.
2. Rotors JLA8.1, 45Ti, SW32 rotors (Beckman Coulter).
3. Beckman Coulter high performance centrifuge (Avanti JXN-26), Beckman ultracentrifuge (Optima XPN).
4. MexB Cell Buffer: 20 mM Tris–HCl (pH 7.0), 0.15 M NaCl, supplemented with a protease inhibitor tablet (Roche) in MilliQ water.
5. MexB Membrane Buffer: 20 mM Tris–HCl (pH 7.4), 20% glycerol, 10 mM imidazole, 500 mM NaCl.
6. Cell-disrupter (Cell-D from Constant LTD).
7. 10% dodecyl maltoside (DDM) in MilliQ water.
8. MexB Binding Buffer: 10 mM Bis-Tris 10 (pH 7.4), 0.2% DDM, 20% glycerol, 10 mM imidazole and 500 mM NaCl in MilliQ water.
9. MexB storage buffer: 10 mM Bis-Tris (pH 7.4), 20% glycerol (w/v), 500 mM NaCl, 0.2% DDM (w/v).
10. BCA Reagent A, containing sodium carbonate, sodium bicarbonate, bicinchoninic acid and sodium tartrate in 0.1 M sodium hydroxide.
11. BCA Reagent B containing 4% cupric sulfate.
12. A plastic Bio-Rad column for gravity, step purification.
13. Column buffers: MexB Binding Buffer supplemented with imidazole for a concentration step-gradient as follows: 10, 50, 300 mM.
14. A PD-10 column (GE Healthcare).
15. Ni-NTA resin (Macherey Nagel).

2.1.3 MacB^{His} Purification

1. 20% L-arabinose in MilliQ water.
2. MacB Cell Washing Buffer: 10 mM Tris–HCl (pH 8.0), 0.1 M NaCl in Milli Q water.
3. MacB Cell Lysis Buffer: 10 mM Tris–HCl (pH 8.0), 5 mM Na_2EDTA, 1 mM PMSF and 100 μg/ml lysozyme in MilliQ water.
4. Sonifier 450 (Branson).
5. MacB Membrane Buffer: 20 mM Tris–HCl (pH 7.0), 200 mM NaCl, 1 mM PMSF, 0.05 mM β-mercaptoethanol in MilliQ water.
6. Cu^{2+}-charged NTA column (Novagen).
7. Column washing and elution buffers: MacB Membrane Buffer supplemented with 0.2% TX and either one of 5, 20, 40, 100 and 250 mM of imidazole.
8. MacB Storage Buffer: 20 mM HEPES-KOH (pH 7.7), 200 mM NaCl, 1 mM PMSF, 0.2% TX, 2 mM DTT, 50% glycerol in MilliQ water.

2.2 Purification of Accessory Proteins

2.2.1 AcrA^{His} Purification

1. 1 M isopropyl β-D-1-thiogalactopyranoside (IPTG) stock prepared in MilliQ water.
2. AcrA Binding Buffer: 50 mM Tris–HCl (pH 8.0), 150 mM NaCl, 1 mM PMSF, 10% glycerol with 5 mM imidazole and lysozyme (100 μg/ml) in MilliQ water.
3. Sonifier 450 (Branson).
4. Cu^{2+}-charged His-binding resin (Novagen).
5. Washing and elution buffers: AcrA Binding Buffer supplemented with imidazole for a concentration step-gradient as follows: 5, 20, 50 and 500 mM.
6. AcrA Storage Buffer: 50% glycerol, 20 mM HEPES-KOH, (pH 7.7), 500 mM NaCl, 1 mM PMSF.

2.2.2 MexA^{His} Purification

1. 20% L-arabinose in MilliQ water.
2. Selection marker: 25 μg/ml chloramphenicol in sterile LB medium, filter sterilized.
3. Syringes and syringe acetate 0.22 μm filters.
4. Rotors used in all protocols: JLA8.1, 45Ti, SW32 rotors (Beckman).
5. Beckman high performance Centrifuges (Avanti JXN-26), Beckman ultracentrifuge (Optima XPN).
6. MexA Cell Buffer: 20 mM Tris–HCl (pH 7.0), 0.15 M NaCl, supplemented with a protease inhibitor tablet (Roche) in MilliQ water.

7. Cell-disrupter (Cell-D from Constant LTD).
8. 10% β-octyl glucopyranoside (β-OG) in MilliQ water.
9. 10% N-laurylsarcosine sodium salt solution from Sigma Aldrich.
10. MexA Membrane Buffer: 10 mM Tris-HCl (pH 8.0), 10% glycerol, 15 mM imidazole in MilliQ water.
11. MexA Binding Buffer (B): 10 mM Tris–HCl (pH 8.0), ß-OG 0.9%, 10% glycerol, 15 mM imidazole in MilliQ water.
12. MexA storage Buffer: 10 mM Tris–HCl (pH 8.0), 10% glycerol, in MilliQ water.
13. Bicinchoninic acid (BCA) from Sigma Aldrich.
14. Plastic Bio-Rad columns.
15. Column buffers: MexA Binding Buffer supplemented with 0.9% β-OG and increasing imidazole concentrations (10, 50, 300 mM).

2.2.3 MacA^{His} Purification

1. MacA Membrane Buffer: 20 mM HEPES-KOH (pH 7.7), 500 mM NaCl, 5 mM imidazole and 1 mM PMSF in MilliQ water.
2. Column washing and elution buffers: MacA Membrane Buffer containing 0.2% TX and 5, 50, 500 mM imidazole.
3. MacA Dialysis Buffer: polyethylene glycol 20,000 in 20 mM HEPES-KOH (pH 7.7), 200 mM NaCl, 1 mM PMSF, 0.2% TX in MilliQ water.
4. MacA Storage Buffer: 20 mM HEPES-KOH (pH 7.7), 200 mM NaCl, 1 mM PMSF, 0.2% TX, 50% glycerol in MilliQ water.

2.3 Purification of Outer Membrane Channels

2.3.1 TolC^{His} Purification

1. 1 M IPTG prepared in MilliQ water.
2. TolC Cell Washing Buffer: 20 mM Tris–HCl (pH 7.5) prepared in MilliQ water.
3. TolC Cell Lysis Buffer: 20 mM Tris–HCl (pH 7.5), 5 mM $MgCl_2$, 1 mM PMSF and 0.05 mg/ml DNAase I (Sigma) prepared in MilliQ water.
4. French Pressure Cell Press (SLM Instruments, Inc.).
5. TolC Membrane Buffer: 20 mM Tris–HCl (pH 7.5), 100 mM NaCl and 1 mM PMSF prepared in MilliQ water.
6. Solubilization buffer: 5% TX, 20 mM Tris–HCl (pH 7.5), 1 mM PMSF and 100 mM NaCl prepared in MilliQ water.
7. Protein Assay (Bio-Rad).
8. His-Binding resin (Novagen).

9. TolC Binding Buffer: 20 mM Tris–HCl (pH 7.5), 100 mM NaCl, 1 mM PMSF, 0.2% TX and 5 mM imidazole prepared in MilliQ water.
10. Column washing buffers: TolC Binding Buffer supplemented with imidazole for a step-gradient of 20, 50, and 100 mM concentrations.
11. $TolC^{His}$ Elution Buffer: TolC Binding Buffer supplemented with 500 mM imidazole.
12. $TolC^{His}$ Storage Buffer: 20 mM Tris–HCl (pH 7.5), 1 mM PMSF, 500 mM NaCl, and 0.2% TX prepared in MilliQ water.

2.3.2 $OprM^{His}$ Purification

1. 20% L-arabinose in MilliQ water.
2. Selection marker: 25 μg/ml chloramphenicol in sterile LB medium, filter sterilized.
3. Syringes and syringe acetate 0.22 μm filters.
4. Rotors used in all protocols: JLA8.1, 45Ti, SW32 rotors (Beckman Coulter).
5. Beckman Coulter high performance centrifuge (Avanti JXN-26), Beckman Coulter ultracentrifuge (Optima XPN).
6. OprM Cell Buffer (A): 20 mM Tris–HCl (pH 7.0), 0.15 M NaCl, supplemented with a protease inhibitor tablet (Roche) in MilliQ water.
7. Cell-disrupter (Cell-D from Constant LTD).
8. 10 Octyl-polyoxyethylene (C8POE) in MilliQ water.
9. 10% β-OG in MilliQ water.
10. OprM Binding Buffer (B): 10 mM Tris–HCl 20 (pH 8), 0.9% β-OG, 10% glycerol, 15 mM imidazole in MilliQ water.
11. BCA Reagent A, containing sodium carbonate, sodium bicarbonate, bicinchoninic acid and sodium tartrate in 0.1 M sodium hydroxide.
12. BCA Reagent B containing 4% cupric sulfate.
13. Plastic Bio-Rad columns.
14. Column buffers: OprM Binding Buffer supplemented with imidazole for a step-gradient of imidazole (washing step at 25 mM imidazole, elution at 300 mM imidazole).

2.4 Purification of the tRNA Scaffold

1. 1 M IPTG prepared in MilliQ water.
2. Selection marker: 100 μg/ml of ampicillin in sterile LB medium, filter sterilized.
3. Syringes and syringe acetate 0.22 μm filters.
4. Rotors used in all protocols: JLA8.1, 45Ti, SW32 rotors (Beckman).

5. RNA Cell Buffer: 20 mM Tris, 200 mM NaCl pH 7.5.
6. RNA Lysis Buffer: 40 mM of $MgSO_4$, 50 mM Na_3Citrate pH 5.6.
7. Phenol.
8. Ethanol.
9. Resource Q column (GE Healthcare, 50 ml).

2.5 Preparation of Proteoliposomes

2.5.1 AcrB Reconstitution into Unlabeled and Fluorescent Vesicles

1. Sepharose G-25 prepacked column (GE Healthcare).
2. Prepacked metal-chelating Sepharose column, charged with Cu^{2+} (GE Healthcare).
3. 10% β-OG stock prepared in AcrB Binding Buffer (*see* Subheading 2.1.1).
4. Column buffers: 20 mM HEPES-KOH (pH 7.0), 500 mM KCl, 1 mM PMSF, 1.2% β-OG, 10% glycerol, 5 or 100 mM imidazole in MilliQ water.
5. *Escherichia coli* polar lipids (Avanti).
6. 1,2-Dipalmitoyl N-(7-nitrobenz-2-oxa-1,3-diazol-4-yl) phosphatidylethanolamine (N-NBD-PE) (Avanti).
7. 1,2-Dipalmitoyl N-(lissamine rhodamine B sulfonyl) phosphatidylethanolamine (N-Rh-PE) (Avanti).
8. Reconstitution Buffer: 25 mM HEPES-KOH (pH 7.0), 1 mM DTT, 100 mM KCl in MilliQ water.
9. Water bath sonicator (Branson, Model 1510).
10. SM-2 Adsorbent Bio-Beads (Bio-Rad).

2.5.2 MexB Reconstitution into Vesicles

1. PD-10 prepacked column (GE Healthcare).
2. 10% β-OG stock prepared in MexB Binding Buffer (Subheading 2.1.2).
3. MexB Reconstitution Buffer: 25 mM HEPES-KOH (pH 7.0), 100 mM K_2SO_4, 2 mM $MgSO_4$ in MilliQ water.
4. 1,2-Dioleoyl-sn-glycero-3-phosphocholine (DOPC) (Avanti).
5. Mini-extruder (Avanti).
6. Cholesterol, powder (Sigma Aldrich).
7. Pyranine 100 mM (Sigma Aldrich) prepared in MilliQ water.
8. Sonifier (GM 3100 generator, MS73 sonotrode).
9. SM-2 Adsorbent Bio-Beads (Bio-Rad).

2.5.3 Reconstitution of MacAB and TolC into Vesicles

1. *Escherichia coli* polar lipids (Avanti).
2. MacAB reconstitution buffer: 20 mM HEPES-KOH (pH 7.0), 5 mM DTT in MilliQ water.
3. 10% TX stock solution in MacAB reconstitution buffer.

4. MacAB reconstitution buffer supplemented with 100 mM KCl.
5. SM-2 Adsorbent Bio-Beads (Bio-Rad).
6. TLA 100 rotor (Beckman Coulter).
7. Proteoliposome Storage Buffer: 20 mM HEPES-KOH (pH 7.0), 5 mM DTT and 50 mM KCl in MilliQ water.
8. Bovine Serum Albumin (BSA) (Sigma Aldrich).

2.5.4 ATP Hydrolysis

1. 100 mM Adenosine-5′-Triphosphate (ATP) (Sigma Aldrich).
2. ^{32}P γ-phosphate-labeled ATP (3000 Ci $mmol^{-1}$, Amersham).
3. Reaction buffer: 20 mM HEPES-KOH (pH 7.0), 5 mM DTT, 50 mM KCl, 2 mM $MgCl_2$ in MilliQ water. Chill to 4 °C before use.
4. 10 mM Mg-ATP mix: 10 mM ATP, 10 mM $MgCl_2$, 1 μl ^{32}P γ-phosphate-labeled ATP in 20 μl reaction buffer.
5. Stop buffer: 50 mM Tris–HCl (pH 8.0), 20 mM EDTA (pH 8.0), 0.5% SDS, 200 mM NaCl, 0.5 mg/ml proteinase K in MilliQ water.
6. PEI-F cellulose (Millipore).
7. TLC running buffer: 10% formic acid, 0.5 mM LiCl in MilliQ water.
8. Storm PhosphoImager (GE Healthcare).
9. ImageQuant Software (Molecular Dynamics).

2.5.5 Reconstitution of OprM

1. 10% β-OG stock prepared in OprM Binding Buffer (*see* Subheading 2.3.2).
2. OprM Reconstitution Buffer: 25 mM HEPES-KOH (pH 7.0), 100 mM K_2SO_4, 2 mM $MgSO_4$ in MilliQ water.
3. L-α-Phosphatidylcholine (Egg-PC) and 1,2-dioleoyl-snglycero-3-phosphoethanolamine (DOPE) (Avanti Polar lipids).
4. Mini-extruder (Avanti).
5. Sonifier (GM 3100 generator, MS73 sonotrode).
6. SM-2 Adsorbent Bio-Beads (Bio-Rad).

2.6 Energy Consumption Assays

1. AcrB, MexB, and MacB proteoliposomes prepared as described in Subheading 2.5.
2. Purified AcrA (sAcrA^{His}), MexA and MacA (*see* Subheading 2.2).
3. Pyranine (Molecular Probes, 100 mM in methanol).
4. Nigericin (Molecular Probes, 1 mM in methanol).
5. Buffer 25 mM HEPES-KOH (pH 7.0), 5 mM DTT and 0.1 M KCl.

6. Buffer 25 mM HEPES-KOH (pH 6.0), 5 mM DTT and 0.1 M NaCl.
7. 25 mM 2-(*N*-morpholino)ethane-sulfonic acid (MES) buffers with pH ranges of 4.5 to 6, 5 mM DTT and 0.1 NaCl.
8. Erythromycin (Sigma Aldrich) 200 mM stock in methanol.
9. Chloramphenicol (Sigma Aldrich) 200 mM stock in methanol.
10. Cloxacillin (Sigma Aldrich) 200 mM stock in methanol.
11. Sodium glycolate (Sigma Aldrich) 200 mM stock in water.
12. Sodium taurocholate (Sigma Aldrich) 200 mM stock in water.
13. Valinomycin (Sigma Aldrich) 10 mM stock in methanol.
14. 1 M $MgCl_2$ prepared in MilliQ water.
15. 0.5 M Na_2EDTA (pH 8.0) prepared in MilliQ water.
16. 100 mM ATP (Sigma Aldrich).
17. Trypsin (Sigma, 10 mg/ml in water).

2.7 In Vitro AcrB and MexB-Dependent Transport Assays

1. AcrB and MexB proteoliposomes prepared as described in Subheading 2.5.
2. Purified AcrA and MexA (*see* Subheading 2.2).
3. Purified tRNA scaffold (*see* Subheading 2.4).
4. Reconstituted TolC and OprM (*see* Subheading 2.5).
5. Buffer 25 mM HEPES-KOH (pH 7.0), 5 mM DTT and 0.1 M KCl.
6. Buffer 25 mM HEPES-KOH (pH 6.0), 5 mM DTT and 0.1 M NaCl.
7. 25 mM 2-(*N*-morpholino)ethane-sulfonic acid (MES) buffers with pH ranges of 4.5–6.0, 5 mM DTT and 0.1 NaCl.
8. Erythromycin (Sigma) 200 mM stock in methanol.
9. Chloramphenicol (Sigma) 200 mM stock in methanol.
10. Cloxacillin (Sigma) 200 mM stock in methanol.
11. Sodium glycolate (Sigma) 200 mM stock in water.
12. Sodium taurocholate (Sigma) 200 mM stock in water.
13. Valinomycin (Sigma) 10 mM stock in DMSO.
14. Ethidium Bromide (Sigma) 2 mM stock in water.
15. 1 M $MgCl_2$ prepared in MilliQ water.
16. 0.5 M Na_2EDTA (pH 8.0) prepared in MilliQ water.
17. 10% DDM stock prepared in MilliQ water.

3 Methods

3.1 Purification of Inner Membrane Transporters

3.1.1 AcrB^{His} Purification

1. Grow *E. coli* DH5α cells with pUC151A plasmid in LB medium supplemented with ampicillin (100 mkg/ml) to induce until OD_{600} reaches 1.8 (*see* **Note 1**).
2. Harvest the cells by centrifugation at 3220 × *g* for 20 min at 4 °C and wash once in AcrB Cell Washing Buffer.
3. Resuspend the cell pellet in the same buffer supplemented with 1 mM $MgCl_2$ and 0.1 mg/ml DNase I. Break the cells once by French press at 12,000 psi. After 10 min incubation on ice add 2 mM Na_2EDTA (pH 8.0).
4. Remove unbroken cells by centrifugation at 3220 × *g* for 20 min at 4 °C.
5. Isolate membrane fractions by high speed centrifugation at 200,000 × *g* for 1 h at 4 °C.
6. Resuspend the membrane pellet in AcrB Cell Washing Buffer. Slowly add an equal volume of 6 M urea at room temperature. Incubate for 10 min and collect the membrane fraction as described above.
7. Resuspend the resulting membrane pellet in AcrB Binding Buffer.
8. Extract AcrB^{His} overnight by slow addition of 10% TX prepared in AcrB Binding Buffer.
9. Remove insoluble fraction by centrifugation and load the supernatant onto a column packed with metal-chelating Sepharose charged with Cu^{2+} and pre-equilibrated with AcrB Binding Buffer supplemented with 0.2% TX (*see* **Note 2**).
10. Wash the column with AcrB Binding Buffer containing 0.2% TX and then with a linear imidazole gradient (5–200 mM). AcrB was found to elute with 100 mM imidazole. The protein was stable at 4 °C for at least 1 week.

3.1.2 MexB^{His} Purification

1. Grow *E. coli* C43 (DE3) ΔAcrB cells with pET22b-MexB^{His} [31] plasmid in 2xTY medium supplemented with 100 μg/ml ampicillin until OD_{600} reaches 0.6. Induce proteins expression with 1 mM IPTG overnight at 20 °C.
2. Harvest the cells by centrifugation at 9000 × *g* for 20 min at room temperature and wash once in ice-cold MexB Cell Buffer.
3. Resuspend the cell pellet in MexB Cell Buffer. Break the cells by a cell-disrupter (Cell-D from Constant LTD) at 4 °C, two passages at 2.4 kbar.
4. Remove unbroken cells by centrifugation at 9000 × *g* for 20 min at 4 °C. Repeat the centrifugation step 2–3 times, until no pellet is visible after the centrifugation.

5. Isolate membrane fractions by high speed centrifugation at 100,000 × *g* for 1 h at 4 °C.
6. Resuspend the membrane pellet in 10 ml MexB Membrane Buffer.
7. Determine the total membrane protein concentration by the BCA test (usually around 15 mg/ml).
8. Gently add a DDM solution to the membrane suspension under agitation. The final DDM concentration should be 2% (w/v) in a final volume, so that the total amount of the detergent is twice of that of the total membrane proteins. Solubilize the membrane fraction overnight at 4 °C under gentle stirring on an end-over-end rotator.
9. Remove insoluble fraction by centrifugation (100,000 × *g* for 1 h, 4 °C) and incubate the supernatant for 2 h with 3 ml of Ni-NTA resin (previously equilibrated with MexB Binding Buffer.
10. Transfer the resin into a plastic Bio-Rad column and collect the flow-through (FT).
11. Wash the resin with 50 ml of MexB Binding Buffer supplemented with 10 mM imidazole and then repeat the washing with 50 ml of MexB Binding Buffer supplemented with 50 mM imidazole.
12. Elute MexB with 10 ml of MexB Binding Buffer supplemented with 300 mM imidazole.
13. Remove imidazole by gel filtration on PD-10 column equilibrated with 10 mM Bis-Tris (pH 7.4), 20% glycerol (w/v), 500 mM NaCl, 0.2% DDM (w/v).

3.1.3 MacB^{His} Protein Purification

1. Grow *E. coli* cells carrying pBB^{His} [20] until $OD_{(100\ \mu g/ml)}$ ~ 0.6 at 37 °C in LB medium supplemented with ampicillin. Induce proteins expression at 37 °C with 0.1% arabinose for 3 h.
2. Collect cells by centrifugation at 3220 × *g* for 20 min at 4 °C. Wash cells once with the ice-cold MacB Cell Washing Buffer.
3. Resuspend the cell pellet in the ice-cold MacB Cell Lysis Buffer. Incubate cells on ice for 30 min and then sonicate six times 30 s each on ice using Sonifier 450 (Branson).
4. Remove unbroken cells by centrifugation at 3220 × *g* for 20 min at 4 °C.
5. Pellet the membrane fractions by centrifugation at 200,000 × *g* for 1.5 h at 4 °C.
6. Resuspend the membrane fraction pellet in MacB Membrane Buffer. Slowly add an equal volume of the same buffer containing 10% TX (*see* **Notes 2** and **3**). Leave the membrane fraction for solubilization overnight at 4 °C with stirring.

7. Remove an insoluble material by centrifugation at 200,000 × *g* for 40 min at 4 °C and load solubilized membrane fractions onto the Cu^{2+}-charged NTA column pre-equilibrated with MacB Membrane Buffer supplemented with 0.2% TX.
8. Wash the column using a step-gradient of imidazole in the same buffer at 5, 20, 40, 100 and 250 mM of imidazole (*see* **Note 3**).
9. Concentrate the purified protein if needed against polyethylene glycol 20,000 in MacB Dialysis Buffer following by dialysis in the same buffer containing 50% glycerol instead of polyethylene glycol. Protein can be stored at −20 °C.

3.2 Purification of Membrane Fusion Proteins sAcrA^{His}, MexA^{His} and MacA^{His}

3.2.1 sAcrA^{His} Purification

1. Grow *E. coli* BL21 cells harboring pEM13 [30] plasmid in LB medium with ampicillin (100 μg/ml) until OD_{600} reaches 0.3–0.4.
2. Induce the protein expression with 1 mM IPTG for 3 h at 37 °C.
3. Collect the cells at 3220 × *g* for 15 min at 4 °C.
4. Lyse the cells in AcrA Binding Buffer supplemented with lysozyme (100 μg/ml) for 30 min on ice.
5. Break the cells by 4–6 rounds of sonication on ice using Sonifier 450 (Branson).
6. Remove unbroken cells by centrifugation at 3,220 × *g* for 15 min at 4 °C.
7. Remove membrane fraction by centrifugation at 200,000 × *g* for 40 min at 4 °C.
8. Load the supernatant onto a Cu^{2+}-charged His-binding column (Novagen) equilibrated with AcrA Binding Buffer.
9. Wash the column with AcrA Binding Buffer and an imidazole concentration step-gradient of 5, 20 and 50 mM imidazole in AcrA Binding Buffer.
10. Elute the protein by 500 mM imidazole-containing AcrA Binding Buffer.
11. To remove imidazole, dialyze the eluted AcrA first in AcrA Binding Buffer without imidazole.
12. For storage at −20 °C perform dialysis in AcrA Storage Buffer.

3.2.2 MexA Purification

1. Grow *E. coli* BL21 cells carrying pBAD33-GFPUv-MexA^{His} [32] plasmid in LB medium supplemented with 25 μg/ml chloramphenicol until OD_{600} reaches 0.6. Induce the protein expression with 0.02% L-arabinose for 3 h at 20 °C.
2. Harvest the cells by centrifugation at 9000 × *g* for 20 min at room temperature and wash once with the ice-cold MexA Cell Buffer.

3. Resuspend the cell pellet in the same buffer. Break the cells by a cell-disrupter (Cell-D from Constant LTD) at 4 °C, two passages at 2.4 kbar.
4. Remove unbroken cells by centrifugation at 9000 × *g* for 20 min at 4 °C. Repeat centrifugation 2–3 times until no pellet is visible upon centrifugation.
5. Isolate membrane fractions by a high speed centrifugation at 100,000 × *g* for 1 h 4 °C.
6. Resuspend the membrane pellet in 10 ml of MexA Membrane Buffer.
7. Determine the total membrane protein concentration with Bicinchoninic acid (BCA) test (usually around 10 mg/ml).
8. Gently add β-OG to the membrane suspension under agitation. The final β-OG concentration should be 2% (w/v) in a final volume, so that the total amount of the detergent is 40 times that of the membrane proteins. Solubilize the membranes overnight at room temperature under gentle stirring on an end-over-end rotator.
9. Remove insoluble fraction by centrifugation (50,000 × *g* for 45 min, 4 °C) and add 0.2% of N-lauryl sarkosyl to minimize binding of contaminants.
10. Incubate the supernatant for 2 h with 3 ml of Ni-NTA resin (previously equilibrated with MexA Binding Buffer supplemented with 0.2% N-lauryl sarkosyl (w/v)).
11. Transfer the resin into a plastic Bio-Rad column and collect the flow-through.
12. Wash the resin with 50 ml of MexA Binding Buffer supplemented with 0.2% N-lauryl sarkosyl (w/v)).
13. Elute protein with MexA binding buffer containing 0.2% N-lauryl sarkosyl (w/v) and 300 mM imidazole.
14. Remove imidazole by gel filtration on a PD-10 column equilibrated with MexA Storage Buffer.

3.2.3 MacA^{His} Purification

1. Grow *E. coli* cells lacking OmpT protease (*see* **Note 4**) and harboring pBMA^{His} [20]. To produce MacA^{His} and purify membrane fractions use the same protocol described in Subheading 3.1.3.
2. Resuspend isolated membranes in MacA Membrane Buffer and extract the protein with 5% TX as described in Subheading 3.1.3.
3. Centrifuge the sample at 200,000 × *g* for 30 min at 4 °C to remove insoluble fraction.

4. Load solubilized membrane proteins onto a Cu^{2+}-charged NTA column prewashed with MacA Membrane Buffer containing 0.2% TX.
5. Wash the column with 10 volumes of MacA Membrane Buffer containing 0.2% TX and then with 10 volumes of the same buffer supplemented with 50 mM imidazole.
6. Elute His-tagged MacA with MacA Membrane Buffer containing 0.2% TX and 500 mM imidazole.
7. Dialyze and concentrate eluted protein against MacA Concentrating Buffer. For storage dialyze protein samples in MacA Storage Buffer containing 50% glycerol. Protein can be stored at −20 °C.

3.3 Purification of Outer Membrane Channels

3.3.1 Purification of $TolC^{His}$

1. Grow freshly transformed with $pTolC^{His}$ *E. coli* ZK796 cells [20, 33] at 30 °C until OD_{600} reaches 0.4, then induce $TolC^{His}$ expression by 0.2 mM IPTG and continue cell growth overnight.
2. Pellet the cells by centrifugation at 3220 × *g* for 20 min at 4 °C. Wash cell once with TolC Cell Washing Buffer.
3. Resuspend cells in TolC Cell Lysis buffer.
4. Break the cells by passing three times through French Pressure cell at 12,000 psi.
5. Remove unbroken cells by repeating centrifugation at 3220 × *g* for 20 min at 4 °C.
6. Isolate total membrane fractions by ultracentrifugation at 200,000 × *g* for 1.5 h at 4 °C.
7. Resuspend membranes in TolC membrane buffer. Measure the total protein concentration with Bio-Rad Protein Assay.
8. Dilute the membrane fractions with TolC Membrane buffer supplemented with 0.5% TX and 20 mM $MgCl_2$ to have the final protein concentration of 5 mg/ml. Centrifuge for 1 h as described in **step 6**.
9. Repeat washing of the membrane pellet with the same buffer one more time.
10. Resuspend the pellet, containing mostly the outer membrane proteins, in TolC solubilization buffer. Keep the protein concentration of 5 mg/ml. Stir the membrane sample at 4 °C overnight, then remove insoluble fraction by centrifugation at 200,000 × *g* for 1 h at 4 °C. The supernatant is ready for metal affinity purification.
11. Equilibrate Cu^{2+}-charged His-binding resin with TolC Binding Buffer.
12. Load the solubilized $TolC^{His}$ sample (**step 10**), wash the column with 10 volumes of TolC Binding Buffer and then

with the same buffer containing imidazole step gradient (20, 50 and 100 mM).

13. Elute $TolC^{His}$ with 5 volumes of TolC Binding Buffer containing 500 mM imidazole. At this step, collect one column volume fractions.
14. Remove imidazole and increase NaCl concentration to 500 mM by dialysis in TolC Storage Buffer. Protein will be stable on ice for a few weeks.

3.3.2 Purification of OprM

1. Grow *E. coli* BL21 cells with pBAD33-GFPUv-$OprM^{His}$ [34] plasmid in LB medium supplemented with 25 μg/ml chloramphenicol until OD_{600} reaches 0.6. Induce proteins expression with 0.02% Arabinose 3 h at 20 °C.
2. Harvest the cells by centrifugation at 9000 × *g* for 20 min at 4 °C and wash once in ice-cold OprM Cell Buffer.
3. Resuspend the cell pellet in the same buffer. Break the cells by a cell-disrupter (Cell-D from Constant LTD) at 4 °C, two passages at 2.4 kbar.
4. Remove unbroken cell by centrifugation at 9000 × *g* for 20 min at 4 °C. The suspension is centrifuged 2–3 times, until no pellet is visible after centrifugation.
5. Isolate membrane fractions by high speed centrifugation at 100,000 × *g* for 1 h 4 °C.
6. Resuspend the membrane pellet in 10 ml of 20 mM Tris–HCl (pH 8.0), 10% glycerol, 15 mM imidazole.
7. Add C_8POE (2% final) to the membrane suspension and incubate the mixture for 30 min at 37 °C. C_8POE is a detergent that specifically solubilizes bacterial inner membranes (*see* **Note 5**).
8. Centrifuge the resulting suspension at 100,000 × *g* for 1 h at 4 °C and resuspend the membranes in 10 ml of OprM Binding Buffer without β-OG.
9. Determine the total membrane protein concentration with Bicinchoninic acid (BCA) test (usually around 12 mg/ml).
10. Gently add β-OG on the membrane suspension under agitation. The final β-OG concentration should be 2% (w/v) in a final volume, so that the total amount of the detergent is 40 times that of the membrane proteins. Solubilize the membranes overnight at room temperature under gentle stirring in a conical flask (200 ml) with a stir bar.
11. Remove insoluble fraction by centrifugation (100,000 × *g* for 1 h, 4 °C).
12. Incubate the supernatant for 2 h with 3 ml of Ni-NTA resin (previously equilibrated with OprM Binding Buffer).

13. Transfer the resin into a plastic Bio-Rad column and collect the flow-through (FT).
14. Wash the resin with 50 ml of 10 mM Tris–HCl (pH 8.0), 10% glycerol (w/v), 25 mM imidazole, 0.9% β-OG (w/v)).
15. Elute protein with 10 mM Tris HCl (pH 8.0), 10% glycerol (w/v), 300 mM imidazole, 0.9% β-OG (w/v)).
16. Remove imidazole by gel filtration on a PD-10 column equilibrated with 10 mM Bis-Tris (pH 7.4), 20% glycerol (w/v), 500 mM NaCl, 0.2% DDM (w/v).

3.4 Purification of the tRNA Scaffold (Adapted from [35])

1. Grow *E. coli* XL1 cells carrying the pBSTNAV-AtRNA [35] plasmid in LB medium supplemented with 100 μg/ml ampicillin until OD_{600} reaches 0.6. Induce the protein expression with 1 mM IPTG 3 h at 37 °C.
2. Harvest the cells by centrifugation at 7000 × *g* for 20 min at room temperature and wash once in tRNA Cell Buffer (*see* Subheading 2.4).
3. Resuspend the cell pellet in the Cell Lysis Buffer. Extract RNA upon incubation with 40 ml of phenol during 2 h under gentle stirring at room temperature. The solution should turn from green to white.
4. Centrifuge the mixture 15 min at 3200 × *g*.
5. Incubate the supernatant with two volumes of ethanol and 1/20 volume of NaCl 5 M. A precipitate forms.
6. Centrifuge 5 min at 3200 × *g*.
7. Dry the pellet under the fume hood for 5 min and add 10 ml of distilled water.
8. Purify the RNA by anion chromatography on a Resource Q column equilibrated in 20 mM potassium phosphate (pH 6.5), wash with potassium phosphate (pH 6.5), 400 mM NaCl, and elute with a gradient of NaCl.
9. Dialyze the eluate against 10 mM potassium phosphate (pH 6.5) and 50 mM KCl.

3.5 Preparation of Proteoliposomes

3.5.1 Reconstitution of AcrB into Proteoliposomes

1. First, remove imidazole by gel filtration on a Sepharose G-25 column equilibrated with AcrB Binding Buffer with 0.2% TX.
2. Bind the purified AcrB on Cu^{2+}-column again. To exchange TX with β-OG, wash the column extensively with a buffer containing 20 mM HEPES-KOH (pH 7.0), 500 mM KCl, 5 mM imidazole, 1 mM PMSF, 1.2% β-OG, and 10% glycerol.
3. Elute AcrB in the same buffer containing 100 mM imidazole. Use AcrB for reconstitution into proteoliposomes immediately (*see* **Note 6**).

4. Resuspend *E. coli* polar lipids in AcrB Reconstitution Buffer to the final concentration 20 mg/ml. Briefly sonicate the mixture using Water Bath Sonicator (Branson 1510).
5. Mix 50 μg of AcrB and 4.5 mg of lipids. Adjust the volume of mixture to 500 μl with reconstitution buffer and 1.1% β-OG.
6. Dilute the protein–lipid mixture with 3 ml of prechilled AcrB Reconstitution Buffer. Remove the detergent by dialysis against 1 L of AcrB Reconstitution Buffer containing 2 g of SM-2 Adsorbent Bio-Beads overnight at 4 °C.
7. To prepare control vesicles, substitute the protein sample with the same volume of AcrB Reconstitution Buffer.

3.5.2 Preparation of Fluorescent AcrB-Containing Donor Vesicles

1. Dissolve *E. coli* phospholipids, N-NBD-PE and N-Rh-PE lipids in chloroform.
2. Mix *E. coli* phospholipids, N-NBD-PE and N-Rh-PE lipids in 99:0.5:0.5 molar ratio to the final concentration of 20 mg/ml. Aliquot 0.25 ml of the lipid mixture into glass tubes, dry it down under nitrogen, and then under reduced pressure for 2 more hours.
3. Rehydrate the mixed lipid pellets in AcrB Reconstitution Buffer.
4. To reconstitute AcrB into fluorescent vesicles follow the procedure described in Subheading 3.5, **step 1**.
5. To prepare unlabeled vesicles use the same technique without addition of fluorescent phosphatidylethanolamine derivatives.

3.5.3 Reconstitution of MexA and MexB

1. Dissolve 3 mg cholesterol in 800 μl DOPC suspension in chloroform (at 25 mg/ml). Let the chloroform evaporate overnight under a vacuum bell (*see* **Note 7**).
2. Resuspend the latter in MexB Reconstitution Buffer supplemented with 3 mM pyranine, to the final concentration 20 mg/ml.
3. Incubate for 10 min at 37 °C and subject to sonication for 10 min with 30 s pulse/30 s pause cycles (power: 40 W).
4. Extrude the lipid suspension through 200 and 100 nm membranes (21 pass for each type of membrane), using an Avanti mini-extruder. Before extruding the liposome, make sure the membrane is moisturized by rinsing the whole set-up with the MexB Reconstitution Buffer (*see* **Note 8**).
5. Aliquot 100 μl of liposomes at 3 mM of lipids into a 96-well plate (cuvettes could be used for a smaller number of samples). Add increasing concentrations of β-OG and measure the optical density at 550 nm in each well. Determine the Rsat, which corresponds to the maximum optical density at a saturating level (*see* **Note 9**).

6. Solubilize the liposomes with β-OG at Rsat for 30 min at 4 °C.
7. Add the purified MexB at the 2.5:1 lipid to protein ratio w/w and the purified MexA at the 20:1 protein-to-protein ratio.
8. Remove the detergent by addition of Bio-Beads at a Bio-Bead-to-detergent ratio equal to 30 (w/w) and incubation overnight at 4 °C with gentle stirring.
9. Remove the nonencapsulated pyranine using a PD-10 desalting column previously equilibrated with MexB Reconstitution Buffer.
10. To prepare control vesicles, substitute the protein sample with the same volume of a protein buffer (this buffer should have the exact same detergent concentration that the protein, so one should use the buffer used to equilibrate the PD-10 column at the end of the protein purification).
11. Filter the liposomes through a 0.22 μm filter.

3.5.4 Reconstitution of MacA and MacB Proteins into Proteoliposomes

1. Resuspend *E. coli* polar lipids in MacAB Reconstitution Buffer to the final concentration 20 mg/ml. Briefly sonicate the mixture using Water Bath Sonicator. Dilute 250 μl of the lipid solution with the same volume of the MacAB Reconstitution Buffer containing 0.45% TX.
2. Dilute 25 μg of MacB only or mixture of MacA and MacB proteins (with molar ratio MacA to MacB being 3:1 and *see* **Note 10**) in 500 μl of MacAB Reconstitution Buffer supplemented with 100 mM KCl. Adjust concentration of TX to 0.45% (*see* **Note 11**).
3. Slowly add 500 μl of the protein sample to 500 μl of the lipid solution and incubate for 30 min at the room temperature.
4. Wash 40 mg of SM-2 Adsorbent Bio-Beads three times in methanol, then in water and MacAB Reconstitution Buffer (*see* **Note 12**).
5. To remove TX, incubate protein–lipid mix at room temperature for 1 h with prewashed beads. Briefly spin down the beads and transfer the supernatant into another tube with fresh beads. Repeat the procedure two more times with fresh beads for 1 h at room temperature and 1 h at 4 °C.
6. To decrease the residual concentration of TX, dilute the sample with 2 ml of cold MacAB Reconstitution Buffer. Collect the proteoliposomes by ultracentrifugation at 250,000 × *g* for 1 h at 4 °C using TLA 100 rotor (Beckman).
7. Resuspend the pellet in MacAB Proteoliposome Storage Buffer (*see* **Note 13**) and briefly sonicate.
8. Determine the protein concentration in the resulting proteoliposomes by a quantitative SDS-PAGE. On average, up to 40% of the protein will be recovered in proteoliposomes.

9. Measure total phosphate concentration of phospholipids according to Ames [36]. The molar protein:lipid ratio varied between 2:1 and 4:1.
10. To prepare control vesicles substitute the protein sample by the same volume of MacAB Reconstitution Buffer.
11. Store the sample on ice. The ATPase activity of reconstituted MacAB complex is stable and reproducible up to 1 week of storage on ice.

3.5.5 Reconstitution of OprM in Proteoliposomes

1. Mix 24.4 mg of Egg PC and 12 mg of DOPE. Let the chloroform evaporate overnight under a vacuum bell.
2. Resuspend the latter in 2 ml of OprM Reconstitution Buffer supplemented with 3.2 mg/ml RNA (*see* Subheading 3.4 and **Note 14**).
3. Incubate 10 min at 37 °C and subject to five freeze–thaw cycles (freeze in liquid nitrogen and thaw in water at 37 °C).
4. Extrude the lipid suspension through 200 and 100 nm membranes (21 pass for each type of membrane), using an Avanti mini-extruder. Before extruding the liposome, make sure the membrane is moisturized by rinsing the whole set-up with the buffer.
5. Aliquot 100 μl of liposomes at 3 mM of lipids into a 96-well plate (cuvettes could be used for a smaller number of samples). Add increasing concentrations of ß-OG and measure the optical density at 550 nm in each well. Determine the Rsat, i.e. the optical density reaches a maximum.
6. Solubilize the liposomes with β-OG at Rsat during 30 min at 4 °C.
7. Add OprM at a 20: 1 lipid-to-protein ratio (w/w).
8. Remove the detergent by addition of Bio-Beads at a Bio-Bead-to-detergent ratio equal to 30 (w/w) overnight at room temperature under gentle stirring.
9. Liposomes are concentrated using a Millipore concentrator with the cut off 100,000 kDa to the final volume of 1 ml.
10. Remove the nonencapsulated RNA by anion exchange chromatography. Equilibrate a Sephadex Q column with OprM Reconstitution Buffer and load liposomes.
11. Liposomes loaded with RNA are found in the flow-through, while the nonencapsuated RNA remains bound to the column. It can be subsequently eluted with 1 M NaCl.
12. To prepare control vesicles, substitute the protein sample with the equal volume of the protein buffer.
13. The efficiency of the RNA encapsulation and the removal of the nonencapsulated RNA are analyzed by the agarose-sybersafe

gel. Nonencapsulated RNA migrates into the gel during electrophoresis (bands around 40 bp) but encapsulated RNA remains in wells because liposomes are too large to enter the gel matrix. If liposomes are now solubilized with detergent, the encapsulated material migrates into the gel.

3.6 Loading of Fluorescent, Nucleic Acid, and Radioactive Probes into Proteoliposomes

3.6.1 Entrapment of Probes

1. Add a probe to the final concentration of 1 mM in case of pyranine to the reconstituted proteoliposomes. Freeze the sample in a dry ice/ethanol bath and thaw at room temperature twice. Briefly bath-sonicate liposomes between the cycles.
2. To remove untrapped probe, load the sample onto a gel filtration P-40 column (Pharmacia). Use the protein reconstitution buffer to wash the column. Collect the fractions containing proteoliposomes.

3.6.2 Loading of ATP Inside of Vesicles (See Note 15)

1. Mix 1 μl of proteoliposomes and ATP probe in a 10 μl reaction volume on ice. Immediately freeze the tube in liquid N_2.
2. Thaw proteoliposomes on ice and sonicate briefly in water bath sonicator. Repeat the procedure two more times.
3. Minimize the exposure time of proteoliposomes mixed with ATP to room temperature to delay ATP hydrolysis.

3.7 ATP Hydrolysis by Reconstituted MacAB (See Note 17)

1. Prepare 10 mM unlabeled Mg-ATP solution by mixing equal volumes of 20 mM ATP and 20 mM $MgCl_2$ diluted in the reaction buffer.
2. Add 5% ^{32}P γ-phosphate labeled Mg-ATP (3000 Ci/mmol, Amersham).
3. Prepare 10 μl reaction mix containing 1–2 μl of MacAB proteoliposomes (0.3–0.4 μM of MacB and 0.9–1.9 μM MacA depending on proteoliposome batch), reaction buffer and ATP in desired concentrations (*see* **Note 17**). Keep all solutions on ice while mixing.
4. Immediately after addition of ATP, freeze the sample in liquid N_2. Load the radioactive ATP inside of vesicle as described in Subheading 3.6.2.
5. After the last round of sonication, start the reaction by placing the sample into a designated 37 °C water bath.
6. To study the kinetics of the ATP hydrolysis perform a time course experiment (Fig. 1). Transfer 1 μl of reaction at different time points into a tube containing 10 μl of Stop Buffer. Immediately place the tube with a time point sample into a designated 50 °C water bath. To ensure complete digestion of the protein incubate the tube for 20 min at 50 °C.
7. Spot 1 μl of stop reaction on prewashed PEI-F cellulose plate and separate the products of the reaction by thin-layer chromatography (TLC) in 10% formic acid and 0.5 M LiCl [37].

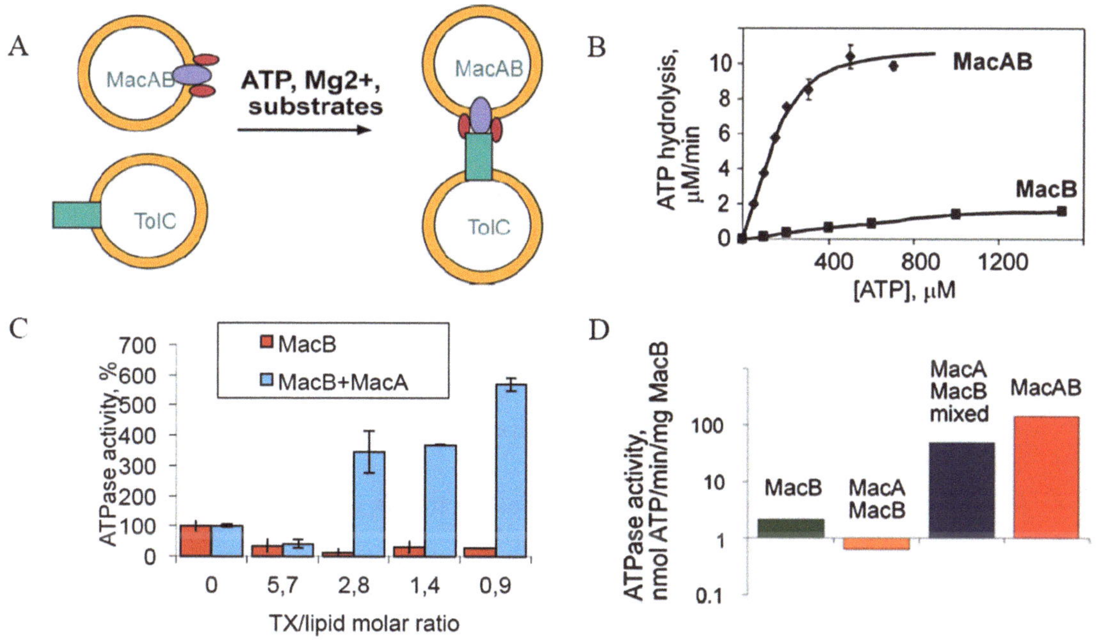

Fig. 1 Assay schematics and reconstitution of the ABC-type efflux pump MacAB. (**a**) Proteins are reconstituted to mimic the arrangement of the pump within the two-membrane envelope of *E. coli*. For this purpose, MacB alone or in combination with MacA is reconstituted into one population of proteoliposomes, whereas TolC into a different population of proteoliposomes and they are mixed in the presence of cofactors such as ATP and Mg^{2+} ions. (**b**) ATP hydrolysis rates of MacB-proteoliposomes and MacAB-proteoliposomes. (**c**) ATPase activities of MacB and MacAB measured at different detergent:lipid molar ratios. (**d**) ATPase activity of MacB proteoliposomes alone or mixed with MaxA-proteoliposomes with and without sonication. The activity of the co-reconstituted MacAB-proteoliposomes is shown for comparison

8. After overnight exposure scan dried TLC plates using Storm PhosphoImager (GE Health Sciences). Quantify the amounts of generated Pi using ImageQuant Software. Based on the concentration of MacB and ATP in the reaction, calculate the specific ATPase activity of MacB in mol ATP/min/mol MacB.
9. To evaluate the effect of substrates, inhibitors or accessory proteins on the ATPase activity of MacB, perform the reaction in the presence of different concentration of additives ([20] and *see* **Note 18**).

3.8 Proton Translocation by Reconstituted AcrB and MexB

1. Unload AcrB/MexB proteoliposomes in reconstitution buffer pH 7.0 and 100 mM KCl with 1 mM pyranine and remove the access of the probe as described in Subheading 3.6.1. Use empty liposomes as a negative control.
2. Calibrate the fluorescent response of pyranine trapped inside the vesicles by evaluating the pyranine fluorescence at different pH in the presence of 0.1 µM nigericin (*see* **Note 19**). Add 20 µl of AcrB/MexB-pyranine vesicles in 2 ml buffers with

0.1 M NaCl and different pH (*see* Subheading 2.6). Measure pyranine fluorescence with excitation and emission wavelengths as 455 and 509 nm correspondingly and plot it against the ΔpH value.

3. To study the effect of drugs on proton efflux through AcrB/MexB, mix 20 μl AcrB/MexB vesicles containing pyranine in 25 mM HEPES-KOH (pH 7.0), 5 mM DTT and 0.1 M KCl in 2 ml of iso-osmotic 0.1 M NaCl -buffer, then add substrates at 0.2 mM and allow equilibration for 1 min.
4. To generate ΔpH across the membrane, add 10 μM valinomycin (*see* **Note 20**) which causes fast decrease in pyranine fluorescence. ΔpH will be stable for at least 4 min with the half-life time of 5 min. Dissipation of ΔpH will be faster in the presence of AcrB substrates.
5. Perform control measurements with AcrB-free liposomes and in the absence of ΔpH.

3.9 Transport of Fluorescent Lipids by Reconstituted AcrB (See Note 21)

1. Prepare AcrB-containing donor vesicles with fluorescent lipids with the internal pH 7.0 as described in Subheading 3.5, **step 2**.
2. Mix 5 μl of donor vesicles (0.36 μg of AcrB in 40 nmol of total phospholipids) with 10 μl of unlabeled protein-free acceptor vesicles (650 nmol of phospholipids) (*see* Subheading 3.5.2). Dilute the vesicles into buffer with pH 6.0 to generate ΔpH.
3. Initiate the transport reaction by adding 5 mM $MgCl_2$. Monitor NBD fluorescence at 30 °C (*see* **Note 21**). To stop the reaction, add 10 mM EDTA (Fig. 2). Then solubilize the vesicles with 0.5% DDM to completely de-quench N-NBD-PE fluorescence.
4. To exclude nonspecific effect of Mg^{2+} on NBD fluorescence, perform the control reaction in the absence of ΔpH.
5. To evaluate the effect of accessory protein AcrA on the transport of lipids by reconstituted AcrB, add $sAcrA^{His}$ after addition of $MgCl_2$. To confirm that the effect of AcrA is specific, perform the transport reaction with different AcrA concentrations (from 15 to 90 μg/ml).
6. To confirm that the observed transport reaction depends on the presence of AcrB in vesicles, digest the AcrB-containing vesicles with 0.1 μg/ml trypsin for 15 min and then repeat the transport reaction. The treatment should result in 50% decrease of the AcrB-dependent lipid transport.

3.10 Transport of Ethidium Bromide by Reconstituted MexAB-OprM (See Note 22)

1. Mix 125 μl of OprM-proteoliposomes (Subheading 3.5, **step 5**) with 125 μl of MexAB-proteoliposomes (Subheading 3.5, **step 3**) and incubate for 20 min at room temperature, no stirring.

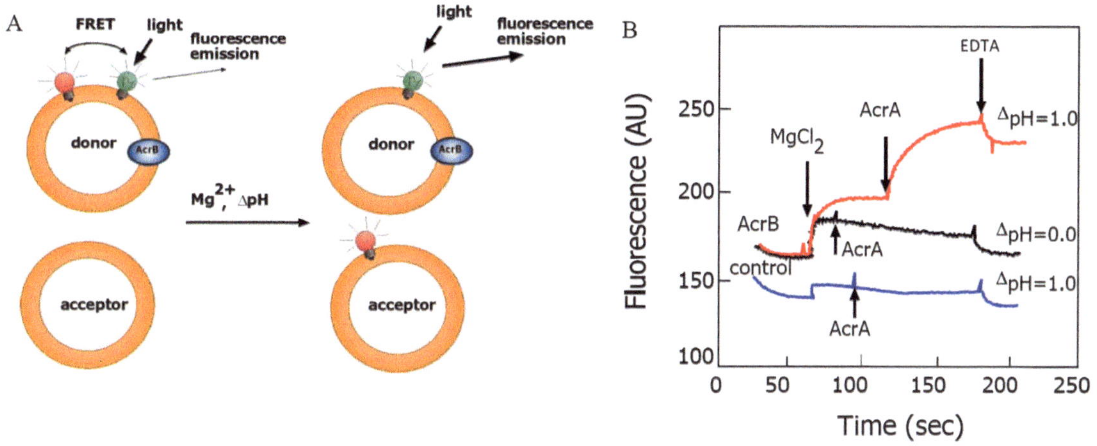

Fig. 2 Assay schematics and reconstitution of the RND-type efflux pump AcrAB. (**a**) Proteins are reconstituted to mimic their arrangement within the two-membrane cell envelope of *E. coli*. For this purpose, AcrB is reconstituted into proteoliposomes containing two types of fluorescent lipids that constitute a FRET pair (donor proteoliposomes). After mixing of donor proteoliposomes with the empty acceptor liposomes, the transport reaction is initiated by addition of cofactors such as AcrA, Mg^{2+} ions and by imposing a proton gradient. (**b**) Time-dependent changes in fluorescence intensity of empty donor liposomes mixed with acceptor vesicles (blue trace), and AcrB-containing donor proteoliposomes mixed with acceptor vesicles in the presence (*red trace*) and absence (*black trace*) of a proton gradient. The addition of Mg^{2+} ions and AcrA are shown by *arrows*. The reaction is stopped by EDTA, which chelates Mg^{2+} and separates the donor and the acceptor vesicles

Add substrate at 5 μM at the same time as liposomes (« time of incubation is the key », *see* **Note 23**).

2. Initiate the transport reaction by adding between 5 to 20 μl HCl 0.15 M (Fig. 3).
3. Monitor fluorescence of probes at 25 °C with continuous stirring using a SAFAS Xenius spectrofluorometer. Measurements are made using the "dual wavelengths" mode of the spectrofluorometer. Pyranine fluorescence ($\lambda_{ex} = 455$ nm and $\lambda_{em} = 509$ nm) and ethidium bromide fluorescence ($\lambda_{ex} = 300$ nm and $\lambda_{em} = 600$ nm) are monitored as a function of time (Fig. 3b and c, respectively).
4. To evaluate the effect of accessory protein, MexA, on the transport of substrates by reconstituted MexB, add palmitoylated MexA in the same time as MexB during the reconstitution procedure (*see* Subheading 3.5, **step 3**).
5. To confirm that the observed transport reaction depends on the presence of MexB in vesicles, perform the same procedure with an inactive mutant of MexB.

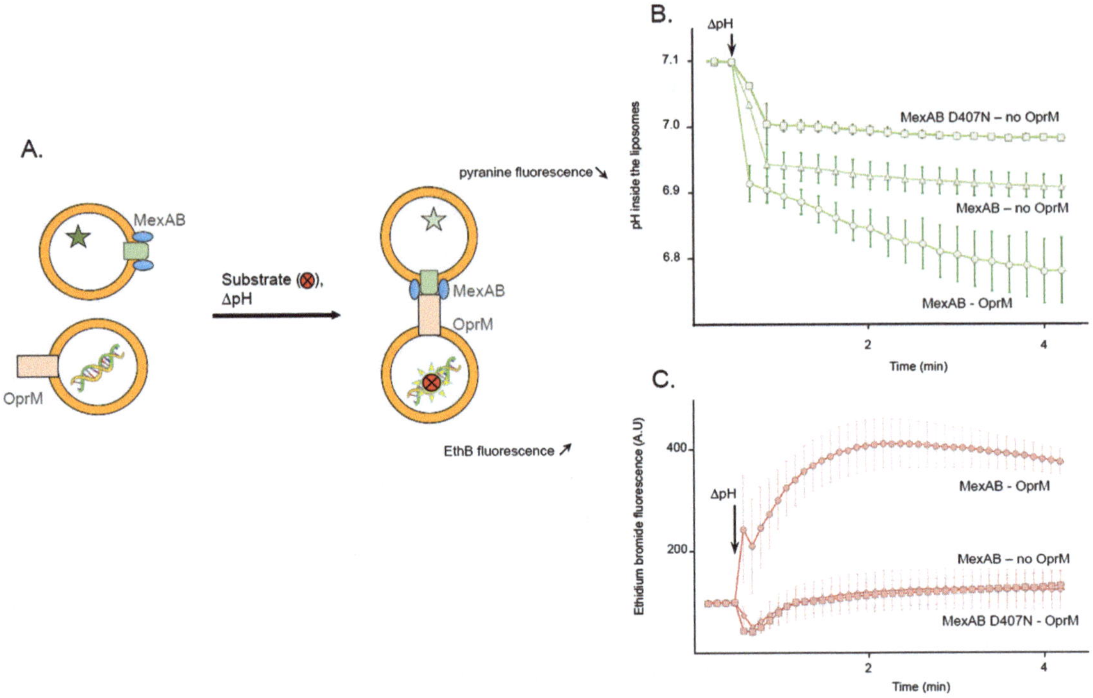

Fig. 3 Assay schematics and reconstitution of the tri-partite MexAB-OprM efflux pump. (**a**) Proteins are reconstituted to mimic the arrangement of the pump within the two-membrane envelope of *P. aeruginosa*. For this purpose, MexB alone or in combination with MexA is reconstituted into one population of proteoliposomes that contain pyranine, whereas OprM is reconstituted into a different population of proteoliposomes containing RNA. Both types of liposomes are mixed in the presence of Mg^{2+} ions and ethidium bromide. The transport reaction is initiated by imposing a proton gradient. (**b**) Time-dependent changes in pyranine fluorescence intensity of proteoliposomes containing a wild-type (*triangle*) or an inactive version (*squares*) of MexB. Effect of OprM on the MexAB-induced acceleration of acidification (circle). (**c**) Time-dependent changes in ethidium bromide fluorescence intensity of proteoliposomes containing OprM (*circle* and *square*), or of empty liposomes (*diamond*), mixed with liposomes containing MexAB. As a negative control, an inactive version of MexAB is also used (*square*)

4 Notes

1. Proteins encoded on pUC151A plasmid are constitutively expressed and do not require any induction for the overexpression of AcrB. Typical purification profiles can be found, e.g., in Fig. 1, Ref. [20], for MacAB and in Fig. 1, Ref. [40], as for MexAB.
2. 5% TX was used only to extract AcrB, MacA, MacB and TolC proteins from the *E. coli* membranes. During purification steps, storage and biochemical assays, concentration of detergent can be decreased according critical micelles concentration (CMC) value of the detergent. 0.2% TX was used to maintain purified

membrane proteins in soluble form. MexB was not stable enough in TX so we had to use DDM.

3. The choice of TX as a detergent for MacA and MacB proteins solubilization was determined by the value of the ATP-ase activity of purified and/or reconstituted MacB. We have found that the ATP-ase activity of MacB was irreversibly inhibited by DDM or OG, but not by TX. After exposure MacB to DDM the specific activity was as low as 0.035 mol ATP/min/mol MacB. In comparison, MacB purified using TX maintained two orders higher activity. Furthermore stimulating effect of MacA on the ATP-ase activity of MacB purified using DDM was no longer detected even after detergent removal. High concentration of imidazole inhibited the ATP-se activity of MacB therefore only 100 mM imidazole elution fractions containing MacB were taken for further assays.
4. We found that MacA is not structurally stable in *E. coli* containing chromosomal OmpT protease. The protein is cleaved by OmpT protease during the detergent solubilization step. To purify a full-length stable MacA, we deleted *ompT* gene from the chromosome of *E. coli* BW25113 cells by method described in [38].
5. C8POE is a detergent that is known to specifically solubilize membrane proteins from the inner membrane. Upon solubilization and ultracentrifugation, proteins from the outer membrane are found in the pellet. The latter must be washed extensively before being subjected to solubilization with the proper detergent. An alternative solution consists in separating the outer membrane from the inner membrane by subjecting them to a sucrose gradient: membranes are loaded on top of a two-layer sucrose cushion (1.5 M/0.5 M sucrose), ultracentrifuged for 1 h at 100,000 × *g* and the outer membranes are found in the 1.5 M sucrose phase.
6. After detergent exchange AcrB is stable only for a few hours and should be used for reconstitution into proteoliposomes immediately. Do not exceed the AcrB concentration at this step higher than 0.3 mg/ml.
7. Although not physiologically relevant, the use of cholesterol turned out to dramatically improve our reconstitution protocols, most probably thanks to the tightening effect of cholesterol on membranes.
8. Preparation of liposome samples as monodisperse and pure as possible is mandatory. A good quality control of the preparation helps very much the reproducibility of the procedures. We routinely use dynamic light scattering (DLS) to check the quality of the liposomes at every step of the reconstitution protocol. Sucrose gradients and cryo-electron microscopy are

also used to assess the dispersity of the final proteoliposome suspension. Following similar lines, the buffer used to prepare the liposomes and that of the buffer used to solubilize them should be isotonic. We systematically measure the osmolarity of our buffers.

9. Liposome solubilization upon addition of increasing concentrations of detergent can be described as a three-phase process. Phase I corresponds to detergent partitioning into the liposome membranes up to a threshold corresponding to the point where membranes are saturated with detergent, the Rsat value. During phase II, detergent is no longer incorporated into the liposomes but is instead found as free monomers in solution. The concentration of free detergent increases until formation of detergent micelles and lipid/detergent mixed micelles. Phase III corresponds to the situation where liposomes are completely solubilized, the Rsol value. Rsat and Rsol, which are specific for the type of lipid and detergent used, should be measured before any membrane protein reconstitution assay.

10. Mixing MacA and MacB proteins prior reconstitution into proteoliposomes resulted in reconstitution of highly active ATP-ase complex (Fig. 1b). To study the effect of MacA on the ATP-ase activity of MacB using single-protein vesicles requires intermixing of bilayers by additional sonication round [20]. The resulting activity will be significantly higher than for MacB alone or when MacA and MacB proteoliposmes are combined in the same reaction without intermixing of bilayers (Fig. 1d). We have concluded that ATP-ase activity of MacAB in membrane vesicles requires a physical interaction between MacA and MacB within a bilayer and that MacA activates MacB in proteoliposomes.

11. Molar ratio between lipids and detergent used during reconstitution reaction is very important for achieving of functionally active MacB ATP-ase (Fig. 1c). We prepared MacAB vesicles with constant concentration of TX (0.45%) and different concentration of *E. coli* phospholipids. Detergent was not removed in this experiment. The molar ratio of detergent:lipid varied from 5.7 to 2.8, 1.4 and 0.9. Based on previous observations [39] and our results, we concluded that at the detergent:lipid ratio 5.7 lipids were completely solubilized in TX and interfered with the activity of MacAB complex. At the detergent: lipid ratio of 2.8 and below, there is mixed population of lipids micelles and vesicles, which resulted in significant ATP-ase activity of MacAB vesicles. At the molar ratio 0.9, reconstituted lipid vesicles dominate, which coincides with the highest level of MacAB ATP-ase activity.

12. Detergent removal can be also achieved by extensive dialysis against the reconstitution buffer without detergent or simply diluting the concentration of detergent in sample below CMC level. We found that using SM-2 Adsorbent Bio-Beads simplifies this step.
13. To determine the optimal concentration of KCl we titrated the effect of KCl on the ATP-ase activity of reconstituted proteoliposomes. 50 mM KCl was the minimal concentration required for full activity of MacAB proteoliposomes.
14. We have tried to incorporate various types of nucleic acids (sonicated calf thymus DNA, PCR products, genomic DNA) but the resulting suspension turned out to be unsatisfactory (*see* **Note 8**). By opposite, RNA scaffolds are very stable and soluble. Their use improved very much the reconstitution protocol.
15. The described protocol of MacB reconstitution into proteoliposomes does not allow a control of the MacB orientation in prepared vesicles. Analysis of the ATP-ase activity of MacAB proteoliposomes loaded with ATP showed that MacB is predominantly inserted with NBD domains inside of vesicles. The protocol described for MexB has been optimized in order to favor a 100% incorporation inside-out. To do so, we take extreme care to solubilize the liposomes with a detergent concentration precisely equal to that needed to saturate the membranes. In such conditions, the protein is supposed to insert unidirectionally with its hydrophobic part first.
16. Purified $MacB^{His}$ has basal ATP-ase activity in detergent solution. To measure MacB activity in detergent micelles follow the procedure described in Subheading 3.7 with one modification. Include 0.2% TX in all buffers used in reaction. Substrates and accessory proteins should be also prepared in reaction buffer.
17. To characterize the kinetics of ATP hydrolysis by reconstituted MacAB complex, in reaction we used increasing concentrations of ATP up to 3 mM. We found that at these reaction conditions, ATP hydrolysis by MacAB complex reaches saturation at about 0.3–1 mM ATP depending on protein and liposomes preparations (Fig. 1b).
18. To verify the specificity of MacA stimulation of the MacB ATP-ase activity, we perform the preparation of MacAB proteoliposomes using different ratios between MacB and MacA during reconstitution. We evaluated the final concentration of MacA and MacB in proteoliposomes preparations by SDS-PAGE (*see* Subheading 3.5, **step 3**) and measured the ATP-ase activity. We found that MacA stimulation was concentration dependent with the half-maximum being achieved when MacB:MacA ratio was about 2:3. Furthermore, AcrA

protein, the accessory protein of another drug efflux pump AcrB, did not show any stimulation activity when used in the same experimental set up. Thus, we proved that the effect of MacA is specific to MacB protein. Addition of TolC-proteoliposomes into the reaction mixture did not further stimulate the activity of the pump.

19. The pH-sensitive fluorescent probe pyranine is water soluble and does not readily cross a lipid bilayer. Since the pyranine fluorescence changes with the pH value of environment in a linear fashion, this probe can be used to calibrate changes in ΔpH caused by active drug transport in AcrB-proteoliposomes. Dilution of vesicles loaded with pyranine into a buffer with pH 6.0 will create ΔpH across the membrane with alkaline inside and acidic outside. Only slow decrease in pyranine fluorescence will be observed due to spontaneous leakage of H^+. Addition of nigericin will make the membrane permeable to proton and cause immediate collapse of ΔpH.
20. Valinomycin is a lipid-soluble ionophore with high selectivity against potassium ions (K^+). It binds K^+ and facilitates their transport across lipid bilayers. Addition of valinomycin to K^+-lipid vesicles into an isosmotic NaCl-containing buffer leads to generation of an electrical potential across the membrane. Then transmembrane potential is converted to an interior-acid ΔpH in the presence of 0.1 M Cl^-. In the presence of AcrB and its substrates ΔpH dissipates rapidly due to the proton efflux coupled with inward transport of drugs.
21. AcrB expels multiple substrates from cells providing multidrug resistance to bacteria. Some of the substrates include lipophilic molecules capable of diffusion through the lipid barrier. Using the transport assays with fluorescent lipids as substrates for AcrAB pump (*see* Subheading 3.9), Zgurskaya and Nikaido [10] experimentally showed that AcrA stimulated the lipid transport by AcrB. Fluorescent derivatives of phospholipids, N-NBD-PE and N-Rh-PE, were included for AcrB vesicles preparation (Fig. 2a). When these two lipids are present at high concentrations in the same bilayer, the fluorescence of NBD is quenched due to the fluorescence energy transfer to the rhodamine-containing lipid (N-Rh-PE). With the AcrB-dependent efflux of fluorescent lipids, their concentration in AcrAB-proteoliposomes decreases by transfer into acceptor vesicles and de-quenching of NBD fluorescence can be observed and quantified. AcrA triggers faster de-quenching thus showing the stimulating role of the accessory protein during the active transport (Fig. 2b).
22. MexB is a *P. aeruginosa* homolog to AcrB. Using the transport assays described in Subheading 3.10, Verchère et al. reported a

system where the functional assembly of the tripartite efflux pump could be mimicked (Fig. 3). By using specific reporters incorporated into proteoliposomes containing either transporter or the outer membrane channel, the observed transport occurs through the pump and not by a simple passive diffusion [17]. On the one side, pyranine loaded into MexAB proteoliposomes allows the monitoring of pH variations upon transport through MexB (transport is coupled to the counter transport of protons, hence by the acidification of the liposome). On the other side, RNA is loaded into the OprM-proteoliposomes. The substrate, ethidium bromide, a RNA intercalating agent, is added in the outer medium, corresponding to the periplasmic space where the drug is supposed to reach its binding sites. Upon generation of a proton gradient, ethidium bromide is transported by MexB, at the expense of proton counter transport, through OprM, making it possible for its intercalation into RNA. The concomitance of the two signals (decrease of the fluorescence of pyranine/ increase of the fluorescence of ethidium bromide) is the evidence for transport through the pump.

23. Incubation is a key parameter. Indeed, we had noticed that the system had to equilibrate for a significant period of time for the liposome-to-liposome transport to occur. This was later confirmed in an assay specifically dedicated to the in vitro monitoring of the pump assembly [40].

References

1. Nikaido H, Vaara M (1985) Molecular basis of bacterial outer membrane permeability. Microbiol Rev 49:1–32
2. Lomovskaya O, Lewis K (1992) Emr, an Escherichia coli locus for multidrug resistance. Proc Natl Acad Sci U S A 89:8938–8942
3. Ma D, Cook DN, Alberti M, Pon NG, Nikaido H, Hearst JE (1993) Molecular cloning and characterization of *acrA* and *acrE* genes of *Escherichia coli*. J Bacteriol 175:6299–6313
4. Lomovskaya O, Zgurskaya HI, Totrov M, Watkins WJ (2007) Waltzing transporters and 'the dance macabre' between humans and bacteria. Nat Rev Drug Discov 6:56–65
5. Nikaido H (2001) Preventing drug access to targets: cell surface permeability barriers and active efflux in bacteria. Semin Cell Dev Biol 12:215–223
6. Zgurskaya HI, Krishnamoorthy G, Tikhonova EB, Lau SY, Stratton KL (2003) Mechanism of antibiotic efflux in Gram-negative bacteria. Front Biosci 8:s862–s873
7. Dinh T, Paulsen IT, Saier MH Jr (1994) A family of extracytoplasmic proteins that allow transport of large molecules across the outer membranes of gram-negative bacteria. J Bacteriol 176:3825–3831
8. Aires JR, Nikaido H (2005) Aminoglycosides are captured from both periplasm and cytoplasm by the AcrD multidrug efflux transporter of *Escherichia coli*. J Bacteriol 187:1923–1929
9. Krishnamoorthy G, Tikhonova EB, Zgurskaya HI (2008) Fitting periplasmic membrane fusion proteins to inner membrane transporters: mutations that enable *Escherichia coli* AcrA to function with *Pseudomonas aeruginosa* MexB. J Bacteriol 190:691–698
10. Zgurskaya HI, Nikaido H (1999) Bypassing the periplasm: reconstitution of the AcrAB multidrug efflux pump of *Escherichia coli*. Proc Natl Acad Sci U S A 96:7190–7195
11. Andersen C, Hughes C, Koronakis V (2001) Protein export and drug efflux through bacterial channel-tunnels. Curr Opin Cell Biol 13:412–416

12. Thanabalu T, Koronakis E, Hughes C, Koronakis V (1998) Substrate-induced assembly of a contiguous channel for protein export from *E. coli*: reversible bridging of an inner-membrane translocase to an outer membrane exit pore. EMBO J 17:6487–6496
13. Paulsen IT, Chen J, Nelson KE, Saier MH Jr (2001) Comparative genomics of microbial drug efflux systems. J Mol Microbiol Biotechnol 3:145–150
14. Saier MH Jr, Beatty JT, Goffeau A, Harley KT, Heijne WH, Huang SC, Jack DL, Jahn PS, Lew K, Liu J, Pao SS, Paulsen IT, Tseng TT, Virk PS (1999) The major facilitator superfamily. J Mol Microbiol Biotechnol 1:257–279
15. Tseng TT, Gratwick KS, Kollman J, Park D, Nies DH, Goffeau A, Saier MH Jr (1999) The RND permease superfamily: an ancient, ubiquitous and diverse family that includes human disease and development proteins. J Mol Microbiol Biotechnol 1:107–125
16. Saier MH Jr, Paulsen IT, Sliwinski MK, Pao SS, Skurray RA, Nikaido H (1998) Evolutionary origins of multidrug and drug-specific efflux pumps in bacteria. FASEB J 12:265–274
17. Verchere A, Dezi M, Adrien V, Broutin I, Picard M (2015) In vitro transport activity of the fully assembled MexAB-OprM efflux pump from *Pseudomonas aeruginosa*. Nat Commun 6:6890
18. Modali SD, Zgurskaya HI (2011) The periplasmic membrane proximal domain of MacA acts as a switch in stimulation of ATP hydrolysis by MacB transporter. Mol Microbiol 81:937–951
19. Tikhonova EB, Dastidar V, Rybenkov VV, Zgurskaya HI (2009) Kinetic control of TolC recruitment by multidrug efflux complexes. Proc Natl Acad Sci U S A 106:16416–16421
20. Tikhonova EB, Devroy VK, Lau SY, Zgurskaya HI (2007) Reconstitution of the *Escherichia coli* macrolide transporter: the periplasmic membrane fusion protein MacA stimulates the ATPase activity of MacB. Mol Microbiol 63:895–910
21. Goldberg M, Pribyl T, Juhnke S, Nies DH (1999) Energetics and topology of CzcA, a cation/proton antiporter of the resistance-nodulation-cell division protein family. J Biol Chem 274:26065–26070
22. Lu S, Zgurskaya HI (2012) Role of ATP binding and hydrolysis in assembly of MacAB-TolC macrolide transporter. Mol Microbiol 86:1132–1143
23. Su CC, Long F, Zimmermann MT, Rajashankar KR, Jernigan RL, Yu EW (2011) Crystal structure of the CusBA heavy-metal efflux complex of *Escherichia coli*. Nature 470:558–562
24. Su CC, Yang F, Long F, Reyon D, Routh MD, Kuo DW, Mokhtari AK, Van Ornam JD, Rabe KL, Hoy JA, Lee YJ, Rajashankar KR, Yu EW (2009) Crystal structure of the membrane fusion protein CusB from *Escherichia coli*. J Mol Biol 393:342–355
25. Lin HT, Bavro VN, Barrera NP, Frankish HM, Velamakanni S, van Veen HW, Robinson CV, Borges-Walmsley MI, Walmsley AR (2009) MacB ABC transporter is a dimer whose ATPase activity and macrolide-binding capacity are regulated by the membrane fusion protein MacA. J Biol Chem 284:1145–1154
26. Welch A, Awah CU, Jing S, van Veen Hendrik W, Venter H (2010) Promiscuous partnering and independent activity of MexB, the multidrug transporter protein from *Pseudomonas aeruginosa*. Biochem J 430:355–364
27. Verchère A, Broutin I, Picard M (2012) Photo-induced proton gradients for the in vitro investigation of bacterial efflux pumps. Sci Rep 2:306
28. Kapoor V, Wendell D (2013) Engineering bacterial efflux pumps for solar-powered bioremediation of surface waters. Nano Lett 13:2189–2193
29. Picard M, Verchère A, Broutin I (2012) Monitoring the active transport of efflux pumps after their reconstitution into proteoliposomes: caveats and keys. Anal Biochem 420:194–196
30. Zgurskaya HI, Nikaido H (1999) AcrA is a highly asymmetric protein capable of spanning the periplasm. J Mol Biol 285:409–420
31. Mokhonov V, Mokhonova E, Yoshihara E, Masui R, Sakai M, Akama H, Nakae T (2005) Multidrug transporter MexB of *Pseudomonas aeruginosa*: overexpression, purification, and initial structural characterization. Protein Expr Purif 40:91–100
32. Trépout S, Taveau JC, Benabdelhak H, Granier T, Ducruix A, Frangakis AS, Lambert O (2010) Structure of reconstituted bacterial membrane efflux pump by cryo-electron tomography. Biochim Biophys Acta Biomembr 1798:1953–1960
33. Gilson L, Mahanty HK, Kolter R (1990) Genetic analysis of an MDR-like export system: the secretion of colicin V. EMBO J 9:3875–3894
34. Phan G, Benabdelhak H, Lascombe MB, Benas P, Rety S, Picard M, Ducruix A, Etchebest C, Broutin I (2010) Structural and dynamical insights into the opening mechanism of *P. aeruginosa* OprM channel. Structure 18:507–517

35. Ponchon L, Catala M, Seijo B, El Khouri M, Dardel F, Nonin-Lecomte S, Tisné C (2013) Co-expression of RNA–protein complexes in *Escherichia coli* and applications to RNA biology. Nucleic Acids Res 41:e150
36. Ames BN (1966) Assay of inorganic phosphate, total phosphate and phospholipids. Methods Enzymol VIII:115–118
37. Randerath E, Randerath K (1967) Ion-exchange thin-layer chromatography: XVI. Techniques for preparation and analysis of oligonucleotides. J Chromatogr 31:485–499
38. Datsenko KA, Wanner BL (2000) One-step inactivation of chromosomal genes in *Escherichia coli* K-12 using PCR products. Proc Natl Acad Sci U S A 97:6640–6645
39. Lopez O, de la Maza A, Coderch L, Lopez-Iglesias C, Wehrli E, Parra JL (1998) Direct formation of mixed micelles in the solubilization of phospholipid liposomes by Triton X-100. FEBS Lett 426:314–318
40. Ntsogo Enguene VY, Verchère A, Phan G, Broutin I, Picard M (2015) Catch me if you can: a biotinylated proteoliposome affinity assay for the investigation of assembly of the MexA-MexB-OprM efflux pump from *Pseudomonas aeruginosa*. Front Microbiol 6:541

Chapter 9

Covalently Linked Trimers of RND (Resistance-Nodulation-Division) Efflux Transporters to Study Their Mechanism of Action: *Escherichia coli* AcrB Multidrug Exporter as an Example

Hiroshi Nikaido

Abstract

Transporters undergo large conformational changes in their functional cycle. RND (Resistance-Nodulation-Division) family efflux transporters usually exist as homotrimers, and each protomer was proposed to undergo a cycle of conformational changes in succession so that at any given time the trimer would contain three protomers of different conformations, the functionally rotating mechanism of transport. This mechanism implies that the inactivation of one protomer among three will inactivate the entire trimeric ensemble by blocking the functional rotation. We describe a biochemical approach to test this prediction by first producing a giant protein in which the three protomers of *Escherichia coli* AcrB efflux pump are covalently linked together through linker sequences, and then testing for its function by inactivation of a single protomer unit. Inactivation can be done permanently by mutating a residue involved in proton relay, or in "real time" by using a protein in which one protomer contains two Cys residues on both sides of the large cleft in the periplasmic domain and then by rapidly inactivating this protomer with a methanethiosulfonate cross-linker.

Key words RND family transporters, Antibiotics, Drug efflux, Disulfide bonds, Methanethiosulfonate, AcrB, Proton relay, Functionally rotating mechanism

1 Introduction

RND (Resistance-Nodulation-Division) family efflux transporters undoubtedly play a major role in the development of drug resistance in Gram-negative bacteria [1], because of their ability to pump out drugs directly into the external medium. Such transporters exist as trimers, as was first shown by the crystal structure of *Escherichia coli* AcrB pump in 2002 [2]. Several years later, AcrB was crystallized as an asymmetric trimer [3–5], and these structures suggested strongly that the transporter works by a functionally rotating mechanism, in which each protomer undergoes a cyclic conformational change, from the Access to Binding to Extrusion

Akihito Yamaguchi and Kunihiko Nishino (eds.), *Bacterial Multidrug Exporters: Methods and Protocols*, Methods in Molecular Biology, vol. 1700, https://doi.org/10.1007/978-1-4939-7454-2_9, © Springer Science+Business Media, LLC 2018

(or L (loose) to T (tight) to O (open)). This mechanism was further supported by the elucidation of the drug-AcrB cocrystal structure [3], in which only the Binding protomer was found to contain the drug in its distal (or deep) binding pocket in the periplasmic domain.

Although this hypothesis explains the mechanism of drug uptake, binding, and extrusion by AcrB, it requires direct confirmation because it is based on the static crystal structure of the transporter. This need is especially acute because there are pieces of evidence suggesting the existence, in *E. coli*, of AcrB trimers not conforming to the classical Access-Binding-Extrusion structure [6]. In this chapter we describe a method to test directly the functionally rotating mechanism, by creating a giant gene coding for a covalently linked functional trimer of AcrB, so that the effect of inactivating only one of the three linked protomers can be tested. If the sequential conformational alteration of each protomer within the trimeric complex is essential, this should result in the inactivation of the pumping function of the entire trimer. This approach was accomplished in my laboratory through the innovative and yet meticulous work of Yumiko Takatsuka, and the results were published in 2009 [7]. This chapter describes the details of the construction of such a fused gene, the methods used to test the functionality of the gene product as well as the real and potential problems in this approach. It should be mentioned that the linked giant trimer approach has since been successfully utilized in showing that the modulating effects of some ligands in AcrB are not caused by the modification of assembly rates of the trimer from monomer [8], and also in the study of the functional role of each protomer in $MdtB_2C$, an unusual RND pump that contains two different protomer types [9].

2 Materials

2.1 Construction of the Plasmids Containing Giant Gene Coding for Three Trimers with Linker Peptides and Introducing Mutations in It

1. Strains and Plasmids: Those needed at the outset are shown in Table 1. *E. coli* strains DH5α and DH10B are prepared as competent cells by Ca^{2+} treatment method [10] and stored at −80 °C. *E. coli* DH10B is mostly used for construction and propagation of large plasmids. Plasmid miniprep kit (Fermentas Inc. or Qiagen Inc) is used for the preparation of plasmid DNA.
2. Recombinant DNA: For manipulation follow the standard procedures [10].
3. Pfu Ultra high-fidelity (HF) DNA polymerase (Stratagene): For PCR amplification. 10× Pfu Ultra HF reaction buffer (Stratagene) is also needed. dNTP mixture, 2.5 mM each, is stored at −20 °C.

Table 1
Strains and plasmids needed at the outset

Strain or plasmid	Genotype or description	Reference or source
Strains		
DH5α	F^- *endA1 recA1* Φ80*lacZ*Δ*M15* Δ(*lacZYA-argF*) *U169 hsdR17*(r_K^-, m_K^+) *supE44 thi-1 gyrA96 relA1 phoA deoR* λ^-	[10]
DH10B	F^- *endA1 recA1 galE15 galK16 nupG rpsL* Δ*lacX74* Φ80*lacZ*Δ*M15 araD139* Δ(*ara,leu*)*7697 mcrA* Δ(*mrr-hsdRMS-mcrBC*) λ^-	Invitrogen
BLR	F^- *ompT* $hsdS_B$(r_B^-, m_B^-) *gal dcm lon* Δ(*srl-recA*)*306*::Tn*10*(Tc^R)	Novagen
RI90	MC1000 *phoR* Δ*ara714 leu*$^+$ *dsbA1*::*kan*	[11]
AG100	K-12 *argE3 thi-1 rpsL xyl mtl supE44* Δ(*gal-uvrB*)	[12]
AG100YB	AG100 Δ*acrB*::Spc^r	[13]
AG100YBR	AG100YB Δ(*srl-recA*)*306*::Tn*10*(Tc^R)	[7]
BL21	F^- *ompT* $hsdS_B$(r_B^-, m_B^-) *gal dcm lon*	Novagen
BL21YB	BL21 Δ*acrB*::Spc^r	[7]
BL21YBR	BL21YB Δ(*srl-recA*)*306*::Tn*10*(Tc^R)	[7]
BL21YBDR	BL21YB *dsbA1*::*kan* Δ(*srl-recA*)*306*::Tn*10*(Tc^R)	[7]
Plasmids		
pUC19	High-copy-number cloning vector; *amp*	
pGB2	Low-copy-number cloning vector; Spc^r	[14]
pSPORT1	Medium-copy-number cloning and expression vector; *amp*, *lac*-inducible expression	Gibco BRL
pUCK151A	6.5-kb BglII fragment containing the entire *acrAB* operon cloned into pUC19 vector in the opposite direction of *lac* promoter	[15]
pSAcrB^{His}	*acrB* gene with His_4 tag sequence at the 3′ end cloned into pSPORT1, producing AcB protein containing a hexahistidine C-terminal sequence (two intrinsic and four additional histidine residues) under the control of the *lac* promoter	[16]
pSCLBH	Derived from pSAcrB^{His}; codons for two intrinsic cysteines (Cys493 and Cys887) of the *acrB* gene converted to codons for serines, producing cysteineless and His_6-tagged AcrB	[13]
pSCLB^{10His}	pSPORT1 containing cysteineless *acrB* gene with His_8 tag sequence at the 3′ end, producing cysteineless and His_{10}-tagged AcrB	[7]
pSCL-F666C/Q830C	Derived from pSCLBH; producing His_6-tagged AcrB with double-Cys mutations, F666C and Q830C	[13]

Table 2
Primers for amplification of *acrB* sequence

Name	Sequence	Comments
For amplification of acrB sequence		
NSFw	5′-ATGCATGTCGACTCAGCCTGAACAGTCCA-3′	Covers nearly 50 nucleotides upstream of *acrB*, and introduces SalI site (underlined)
MetSacFw	5′-ATGCATGAGCTC*ATG*CCTAATTTCTTTATCGATC-3′	Begins with the Met start codon of acrB, and introduces SacI site (underlined)
MetBFw	5′-CGCGGATCC*ATG*CCTAATTTCTTTATCGATC-3′	Begins with the Met start codon of acrB, and introduces BamHI site (underlined)
CdHisXRv	5′-CGCTCTAGAATCGACAGTATGGCTGTGCT-3′	Deletes the two C-terminal His residues in AcrB; introduces XbaI site (underlined) right after the coding region
CH104Rv	5′-ATGCATAAGCTTTTATCCGTGGTTAATACTG-3′	Covers about 120 nucleotides downstream from *acrB* coding sequence, and introduces HindIII site (underlined)
For amplification of linker sequence		
linkXFw	5′-GGCTCTAGAATGCTGAAACCGATTGCCA-3′	Introduces XbaI site (underlined)
linkSacRv	5′-ATGCATGAGCTCACGCCCCGTACTGCGCAG-3′	Introduces SacI site (underlined)
linkBRv	5′-CGCGGATCCACGCCCCGTACTGCGCAG-3′	Introduces BamHI site (underlined)
linkSmHRv	5′-ATGCATAAGCTTCCCGGGACGCCCCGTACTGCGCAg-3′	Introduces SmaI-HindIII sites (underlined)

4. Tris-EDTA (TE) buffer: Contains 10 mM Tris–HCl and 1 mM EDTA, adjusted to pH 8.0. Autoclaved and stored at room temperature.
5. Oligonucleotide primers (Table 2): Dissolved in TE buffer and stored at −20 °C.
6. LB (liquid) medium: 1% Tryptone, 0.5% Yeast Extract, 0.5% NaCl, supplemented with 100 μg/ml ampicillin.

7. LB agar plates: Add 1.5% Bacto Agar to LB medium, and after autoclaving and cooling pour into plates right after the addition of 100 μg/ml ampicillin.
8. QuickChange site-directed mutagenesis system from Stratagene: Use for site-directed mutagenesis. DpnI restriction enzyme is from New England Biolabs.
9. Autoclaved distilled water: Stored at room temperature.

2.2 Expression and Localization of the Product(s) from $pS(CLB)_3$

1. Strains and plasmids: Shown in Table 1. *E. coli* AG100YBR and BL21YBR contain an *acrB* deletion in which 90% of the *acrB* gene is deleted and is replaced by *aadA*, the spectinomycin resistance gene from pGB2 plasmid [13]. For the stable maintenance of the plasmids carrying tandem-repeat *acrB* sequences, Δ*recA*::Tn*10* allele of strain BLR was transduced into the *acrB*::*spc* strains, to produce AG100YBR and BL21YBR. Grown in LB medium supplemented with spectinomycin (50 μg/ml) and tetracycline (10 μg/ml), prepared as competent cells, and stored at −80 °C.
2. LB media: Described in Subheading 2, **items 1.6** and **1.7**.
3. 10 mM HEPES-KOH buffer: pH was adjusted to 7.5, autoclaved, and stored at room temperature.
4. Complete, EDTA-free protease inhibitor cocktail (Roche).
5. 10% Sodium *N*-lauroylsarcosinate: This stock solution is in distilled water, and stored at 4 °C.
6. Sonicator: Gallenkamp Soniprep 150.
7. Ultracentrifuge: Beckman-Coulter TLA 100.2.
8. Electrode buffer for Blue Native (BN) PAGE [17]: Use the electrode buffer of the Laemmli system [18] with SDS omitted (25 mM Tris–192 mM glycine buffer, pH adjusted to 8.3 with HCl).
9. BN-sample buffer for Blue Native PAGE, 5× solution: 0.5 M 6-aminocaproic acid-50 mM BisTris buffer, pH adjusted to 7.0 with HCl, containing 30% (w/v) sucrose and 5% (w/v) Coomassie brilliant blue R-250.
10. Tris–HCl precast gel, 4–15% linear gradient (Bio-Rad): Used for BN PAGE.
11. Nitrocellulose membrane for Western Blotting: GE Water & Process Technologies.
12. Antibodies: polyclonal rabbit anti-AcrB [19] and tetra-His antibody (Qiagen) as primary antibodies. Alkaline phosphatase-conjugated anti-rabbit (Sigma) and anti-mouse (Bio-Rad), and horseradish peroxidase-conjugated anti-rabbit immunoglobulin G (Pierce), as secondary antibodies.

13. Nitrotetrazolium Blue chloride, 5-bromo-4-chloro-3-indolyl phosphate (both from Sigma-Aldrich), and Western lightning chemiluminescence reagent *plus* (Perkin-Elmer Life Sciences, Inc.): These are used for visualization of protein-antibody conjugates.

2.3 Evaluation of the Drug Efflux Function of the Linked AcrB Trimer

1. Strains: *E. coli* AG100YBR and BL21YBR, transformed by plasmid pSPORT1, pSCLBH (both in Table 1), $pS(CLB)_3$, $pS(CLB)_2$, $pS(B_2\text{-}D407A)$, or $pS(D407\text{-}B_2)$. Plasmids $pS(CLB)_3$ and $pS(CLB)_2$ express linked AcrB trimer and linked AcrB dimer, respectively. Plasmids $pS(B_2\text{-}D407A)$ and $pS(D407A\text{-}B_2)$ express the linked AcrB trimer containing the proton-relay-inactive monomeric unit as the first and the last unit, respectively. For detailed description of the plasmids, *see* Subheading 3.
2. LB agar plates and liquid medium, both supplemented with 100 μg/ml ampicillin.
3. LB agar, autoclaved and kept at 50 °C.
4. Sodium cholate hydrate, from ox or sheep bile, ≥99%: Obtained from Sigma.
5. Square dishes with grid, 100 × 100 × 15 mm: Fisher Scientific.
6. Disposable loops, 1 μl: Fisher Scientific.7. For details of gradient plate assay, please see reference [20].
7. For details of gradient plate assay, please see reference [20].

2.4 "Real-Time" Inactivation of the Entire Trimer by the Inactivation of One Component Monomer

1. Strain: *E. coli* strain BL21YBDR (Table 1), a Δ*acrB dsbA* Δ*recA* mutant generated from BL21. Grown in LB medium supplemented with 50 μg/ml spectinomycin, 35 μg/ml kanamycin, and 10 μg/ml tetracycline, prepared as competent cells, and stored at −80 °C.
2. Plasmids: $pS(CLB)_3$, $pS(FQ\text{-}B_2)$, pS(B-FQ-B), and $pS(B_2\text{-}FQ)$, the latter three containing double-Cys mutations F666C and Q830C in $pS(CLB)_3$ at first, second, and third *acrB* unit, respectively. For description in detail, please *see* Subheading 3.
3. LB agar plates and liquid medium, both supplemented with 100 μg/ml ampicillin.
4. Assay buffer: 50 mM sodium phosphate, 0.1 M NaCl, 0.1% (v/v) glycerol, pH 7.0. Autoclaved and stored at room temperature.
5. Spectrofluorometer: Shimadzu RF-5301PC (Shimadzu Scientific Instruments, Inc.).
6. Quartz cuvette for fluorometry, 10 mm path length: Starna Cells, Inc.
7. Micro stirring bar, 7 mm length × 2 mm diameter: Fisher Scientific.

8. Ethidium bromide, 2 mM stock solution in distilled water: Stored at −20 °C.
9. MTS reagents: 2.5 mM 1,2-ethanediyl bismethanethiosulfonate (MTS-2-MTS, approximately 5.2 Å spacer), and 2.5 mM pentyl MTS (5-MTS), from Toronto Research Chemicals. Freshly dissolved in dimethyl sulfoxide (DMSO)-ethyl acetate (3:1, vol/vol) (*see* **Note 1**), and kept on ice until use (*see* **Note 2**).
10. Proton conductor: Carbonyl cyanide *m*-chlorophenylhydrazone (CCCP) from Sigma-Aldrich. Stored as a 10 mM stock solution in ethanol at −20 °C. The solution is stable at least for a month.

3 Methods

3.1 Construction of the Giant Gene Coding for Three Trimers and Linker Peptides

N- and C-termini of AcrB protomers are both in the cytosolic face of the transmembrane domain. However, inspection of the crystal structure shows that the C-terminus of one protomer (for example the Binding protomer) is far away (by almost 50 Å) from the N-terminus of the neighboring protomer (in this case the Extrusion protomer). We need a linker peptide to cover this distance and allow the AcrB protomers to fold more or less correctly. We use, for this purpose, an internal sequence within AcrB that normally is located in the same cytosolic area and is lying parallel to the membrane surface, i. e. an α-helix (Iα) that normally connects N-terminal and C-terminal halves of AcrB as well as its flanking regions, corresponding to Met496 to Arg540 (5 kDa). The DNA sequence (135 bp) coding for this stretch was added as the "linker" between the individual sequences coding for each protomers.

The overall scheme for the construction of the giant plasmid is shown in Fig. 1. Because the real-time inactivation of the linked trimer relies on the crosslinking of Cys residues, all Cys codons in each protomer sequence was converted to Ser codons, in the early stage by site-directed mutagenesis [13], producing the CL (for *c*ysteine*l*ess) sequence.

The first stage (Step I, Fig. 1) involved the creation of three plasmids, each coding for a monomer unit (and also the C-terminal linker sequence in the first two), as well as the restriction sites needed (*see* **Note 3**).

1. For pU (for *pU*C19)CL$^{\Delta H}$ (no His at C-terminus)-linkXSac, first amplify the linker sequence by PCR with pSCLBH as the template, with primers linkXFw and linkBRv (Table 2), and insert the amplicon into pUC19 vector cut with XbaI and BamHI, generating pUlinkXB. Then cut this plasmid with SalI and XbaI, and insert here the amplicon, coding for the

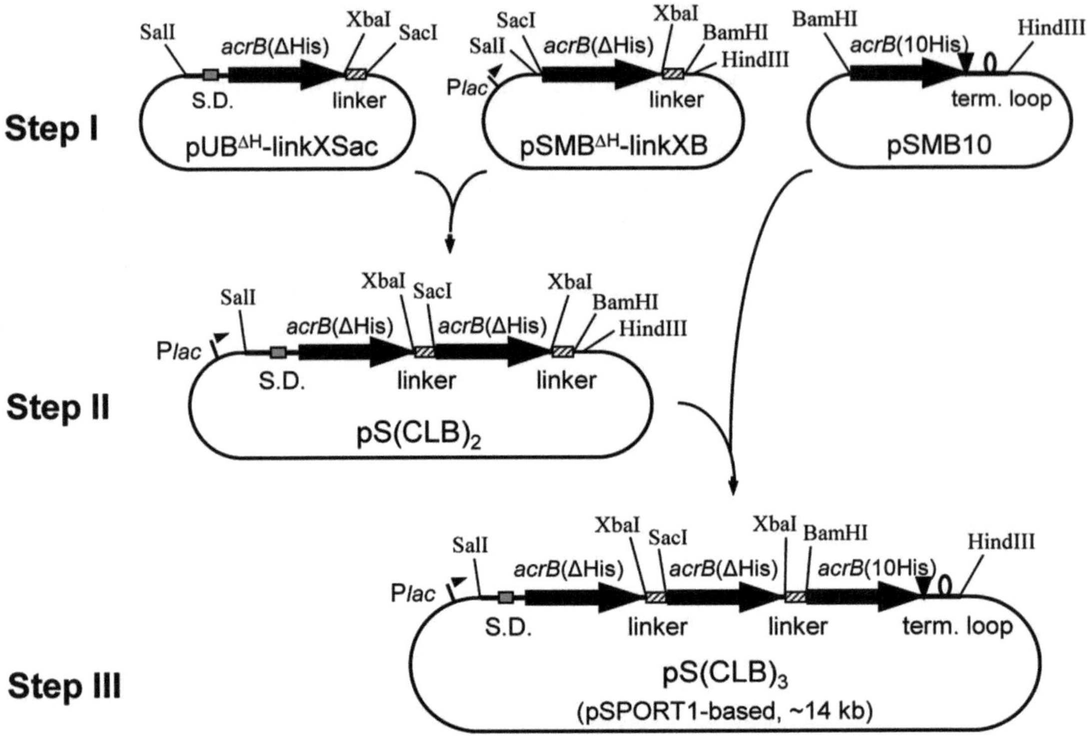

Fig. 1 Construction of the plasmid pS(CLB)$_3$ for the expression of the cysteineless linked trimer of AcrB. Construction of the giant gene containing three *acrB* sequences connected by linkers was achieved through three major steps. A 135-bp internal sequence of *acrB* was used as the linker. *See* text for details. *S.D.* Shine-Dalgarno sequence, *term. loop* termination loop. From [7]

cysteineless *acrB* (with upstream ca. 50 nucleotide sequence and with the deletion of C-terminal His-His) and obtained by PCR with the primers NSFw and CdHisXRv with pSCLBH as the template, generating pUCL$^{\Delta H}$-linkXB. In the final step, exchange the linker sequence with the one containing SacI site at the 3′-end, by inserting linkXSac sequence obtained by PCR with the primers linkXFw and linkSacRv (Table 2) by using pU19$^{\Delta H}$-linkXSmH (*see* **Note 4** for construction of this plasmid) as the template, into the above plasmid cut with XbaI and SacI.

2. For pS (for *pS*PORT1) M (for *acrB* sequence beginning with the initiation *M*et codon) B (for the CL derivative of *acr*B)$^{\Delta H}$-linkXB, first PCR-amplify the linker sequence by using the primers linkXFw and linkBRv (Table 2), with pSCLBH as the template. Insert the amplicon into pSPORT1 cut with XbaI and BamHI, generating an intermediate plasmid pSlinkXB. Then cut this plasmid with SacI and XbaI, and insert into it the PCR amplicon CLMet$^{\Delta H}$ generated by using primers

MetSacFw and CdHisXRv (Table 2), using pU19$^{\Delta H}$-linkXSmH (see above) as the template.

3. For pSMB10 (for the terminal His10 sequence), PCR-amplify the *acrB* sequence together with the terminal His10 sequence as well as the transcription termination loop sequence with primers MetSmFw (similar to MetSacFw (Table 2) but with the SmaI site) and CSac104Rv (similar to CH104Rv (Table 2) but with SacI site), using pSCLB^{10His} as the template. Then insert the amplicon into pUC19 vector cleaved with SmaI and SacI. Use the resultant plasmid as the template for amplification of *acrB* with His10 sequence as well as the 122-nucleotide downstream sequence with primers MetBFw and CH104Rv (Table 2), and insert the amplicon into the pSPORT1 vector cleaved with BamH1 and HindIII, producing pSMB10 (Fig. 1).

In the second stage (Step II, Fig. 1), insert the 3.3-kb SalI-SacI fragment from pUB$^{\Delta H}$-linkXSac into pSMB$^{\Delta H}$-linkXB, cut with these enzymes. This will result in the insertion of the first protomer unit, with the upstream sequence and downstream linker, in front of the *acrB*(ΔHis)-linker sequence in pSMB$^{\Delta H}$-linkXB in frame, generating pS(CLB)$_2$ that codes for a linked cysteineless AcrB dimer.Finally, in the last stage (Step III, Fig. 1), insert the 3.3-kb BamHI-HindIII fragment from pSMB10 into pS(CLB)$_2$ cut with the same enzymes. This will result in the addition of the third AcrB protomer, followed by His10 tag sequence, behind the second linker sequence in frame, creating the pS(CLB)$_3$ (pSPORT1 based; 14 kb). This plasmid codes for a covalently linked trimer of cysteineless AcrB. The gene expression is under the control of Plac promoter of the vector pSPORT1. Importantly, we have tried to construct a similar plasmid based on the high copy number vector pUC19, but this attempt was not successful presumably because elevated production of this "unnatural" membrane protein is deleterious for *E. coli*. The medium copy number vector pSPORT1 seems to allow production of the linked trimer, at least up to a moderate level.

3.2 Construction of Plasmids Containing Mutations Within the Giant Gene

To construct the plasmid containing the giant gene with one of the three protomers containing point mutation(s), one cannot use the site-directed mutagenesis in a conventional manner, as there are three identical coding sequences in the giant gene. Thus we have to introduce mutations into a plasmid coding for a single protomer, and then assemble the giant gene from this and other plasmids.

For a linked trimer containing the D407A mutation (which abolishes proton conduction and therefore the efflux activity [21]) in the first or the third protomer [pS(D407A-B$_2$) or pS(B$_2$-D407A)] or the double-Cys mutation (F666C and Q830C) in the third protomer [pS(B$_2$-FQ)], mutation(s) were introduced by site-directed mutagenesis into the plasmids at step I (Fig. 1). For double-Cys mutation in the first or the second protomer [pS(FQ-B_2) or pS(B-FQ-B)], *acrB*(ΔHis) sequences with mutations and without linkers were amplified by PCR with pSCL-F666C/Q830C [13] as the template, using the primers NSFw (or MetSacFw) and CdHisXRv (Table 2), and were put into the SalI (or SacI)-XbaI sites of the plasmids at step I (Fig. 1). After the sequences were confirmed, the 3.3-kb SalI-SacI, SacI-BamHI, or BamHI-HindIII fragments were replaced into corresponding sites of pS$(CLB)_3$ or inserted into pS$(CLB)_2$. The region(s) containing mutation(s) in the resulting linked-trimer plasmids were sequenced, and it was confirmed that they contained the expected sequences.

3.3 Expression and Localization of the Product from pS$(CLB)_3$

The giant plasmid pS$(CLB)_3$ was introduced into *E. coli* AG100YBR [13], a Δ*acrB*::Spc^r derivative of AG100 [12], which also contained the *recA* mutation, in order to prevent homologous recombination between tandemly repeated *acrB* sequences. Since we noticed that the products were more stable in the presumably protease-deficient host strains (see below), the *acrB recA* strain BL21YBR generated from BL21 was also used. Western blotting analysis of the whole-cell proteins from the exponential-phase cells harboring pS$(CLB)_3$ without induction showed the presence of a large protein of 300 kDa which reacted with anti-AcrB antibody (not shown). The 300-kDa protein was also detected with anti-His_4 antibody (data not shown), indicating that the giant gene was translated completely up to the C-terminal end containing the histidine tag sequence.

The total cell proteins were further fractionated, after sonication, into the soluble fraction, inner membrane (extractable with 1.5% sodium *N*-laurylsarcosinate), and insoluble residue (containing the outer membrane) [22]. SDS-PAGE (carried out by using the Laemmli system [18]) followed by Western blotting with anti-AcrB antibody (Fig. 2a) showed that the 300-kDa protein was mainly localized in the inner membrane fraction, similar to the monomeric cysteineless and hexahistidine-tagged AcrB (CL-AcrBHis) expressed from pSCLBH. In addition to the 300-kDa protein, smaller amounts of an approximately 220-kDa protein, presumably corresponding to a linked dimer of AcrB, were also detected in the cells harboring pS$(CLB)_3$. In the host strain AG100YBR, the monomeric form of AcrB (110 kDa) was also detected at faint levels. The intensities of the 220- and 110-kDa bands in the inner membrane fraction from pS$(CLB)_3$-harboring AG100YBR were 20% and 11%, respectively, of the intensity of the 300-kDa band, as estimated by the use of the NIH ImageJ program, while they were 10% and 2% in BL21YBR (Fig. 2a), suggesting that the giant 300-kDa protein is more stable in the BL21YBR host.

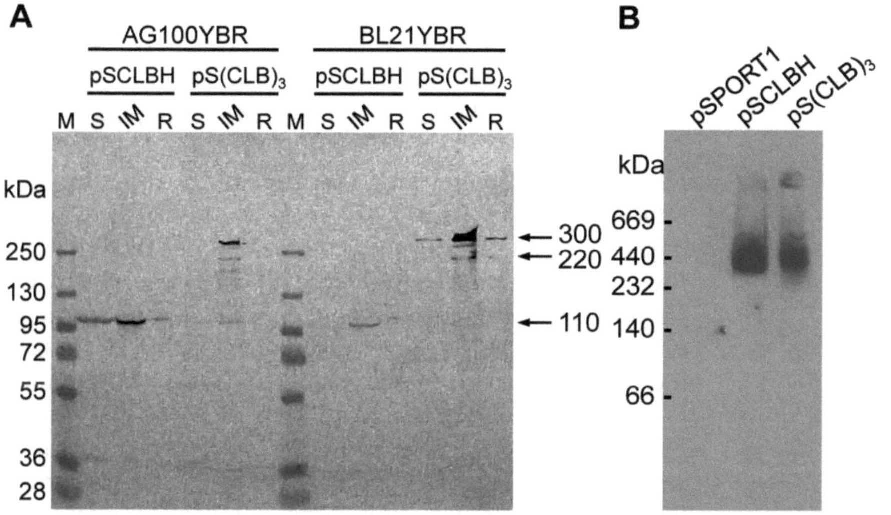

Fig. 2 Expression and localization of the linked-trimer AcrB. AcrB proteins were analyzed by Western blotting using a polyclonal anti-AcrB antibody. (**a**) Total cell proteins from noninduced *acrB recA* host strains AG100YBR and BL21YBR containing each plasmid were fractionated into three fractions and separated by SDS-PAGE, as described in Subheadings 2 and 3. The amount of protein in each lane corresponds to that from ~3.6×10^8 cells. Fractions: *S* soluble fraction, *IM* inner membrane proteins, *R* insoluble residue (containing the outer membrane). (**b**) Blue native PAGE. Inner membrane proteins were prepared from BL21YBR (*acrB recA*) cells containing pSPORT1-derived plasmids and separated by Blue native PAGE. The amount of protein in each lane corresponds to that from 7×10^9 cells for pSPORT1 and pSCLBH and 1×10^9 cells for pS(CLB)$_3$. AcrB was detected with anti-AcrB polyclonal antibody. The product from pS(CLB)$_3$ migrates at the same rate as wild-type AcrB trimer. From [7]

The proteins in the inner membrane fractions from BL21YBR cells harboring pSPORT1, pSCLBH, or pS(CLB)$_3$ were also separated on Blue native gel and analyzed by Western blotting using anti-AcrB antibody (Fig. 2b) as described [7]. Because of the different expression levels of AcrB between pSCLBH and pS (CLB)$_3$ in BL21YBR, about seven-times-larger amounts of proteins were applied for pSPORT1 and pSCLBH lanes than for pS (CLB)$_3$. The product from pS(CLB)$_3$ migrated at the same rate as the wild-type AcrB trimer (360 kDa), suggesting that the quaternary structure of the 300-kDa protein is similar to that of the wild-type AcrB trimer.

3.4 Evaluation of the Drug Efflux Function of the Linked AcrB Trimer

The transport activity of the linked trimer was evaluated by examining the drug susceptibilities of plasmid-containing *acrB recA* strains AG100YBR and BL21YBR. Two different methods, MIC determination by the broth dilution method and gradient plate assay, were employed [7]. AcrB proteins were expressed without IPTG induction. We previously observed variation in resistance levels following storage of the transformed cells of another host strain, HNCE1a [13, 23]. However, this was not much of a problem with the *recA* null host strains AG100YBR and BL21YBR, and reproducible results were obtained with the transformed colonies stored at 4 °C for up to 5 days. The measurement of MICs against cholic acid, erythromycin, ethidium bromide, and novobiocin showed that the AG100YBR cells harboring pS(CLB)$_3$ were resistant to these compounds at about the same level as the pSCLBH-harboring cells, suggesting that the linked trimer functions fully in intact cells. With BL21YBR as the host, the drug resistance produced by the linked trimer was often higher than that produced by the monomeric *acrB* gene in pSCLBH. For example, the MIC for cholic acid, that increased from the 2.5 mg/ml in the pSPORT1-containing cells to 5.0 mg/ml in pSCLBH-containing cells, increased further to 10 mg/ml in pS(CLB)$_3$-containing cells.

Higher cholate resistance levels of pS(CLB)$_3$-containing cells could be observed even more clearly in the gradient plate assay, which gives more precise and quantitative data on drug resistance levels [13, 16, 24]. A linear concentration gradient of cholate is prepared in square LB agar plates containing 10,000, and 3000–5000 μg/ml of sodium cholate in the lower layer for AG100YBR and BL21YBR, respectively, and the upper layer consisting of LB agar without cholate is added. Mid-exponential cultures of AG100YBR and BL21YBR containing various plasmids are diluted to OD660 of 0.1, and 1 μl aliquots are streaked across the plate using the disposable loop. The relative activity of each linked or mutated AcrB protein is calculated by dividing the length of growth of each strain tested (minus that of the strain carrying the vector alone) by the length of growth of the strain containing pSCLBH (minus that of the strain containing the vector alone), and then multiplying the result by 100. Thus, full efflux activity and no activity should produce values of 100% and 0%, respectively.

In this assay, AG100YBR containing pS(CLB)$_3$ is found to produce the same activity as the strain expressing monomeric AcrB from pCLBH (Fig. 3a). In confirmation of the MIC data, in the host strain BL21YBR, the relative activity of the linked trimer was 2.6 times higher than that of the monomeric CL-AcrBHis (produced by pSCLBH) (Fig. 3b). The reason for this influence of the host strains could be the greater stability of the linked trimer in the presumably protease-deficient strain BL21YBR (but *see* **Note 5**), as well as the enhanced expression level of the linked trimer in comparison with the expression of the monomer in BL21YBR. Western blot analysis of the whole cell proteins revealed that several proteins smaller than 300 kDa, especially the apparently monomeric form of 110 kDa, were seen in AG100YBR expressing the linked trimer. In contrast, in BL21YBR containing pS(CLB)$_3$, only one 220-kDa band was detected other than the main 300-kDa band (not shown). However, especially in AG100YBR (Fig. 3a), plasmids that are not expected to produce active linked trimers, such as pS(CLB)$_2$ and pS(B$_2$-D407A), produced considerable levels of cholate efflux activity. This problem is discussed in the next section.

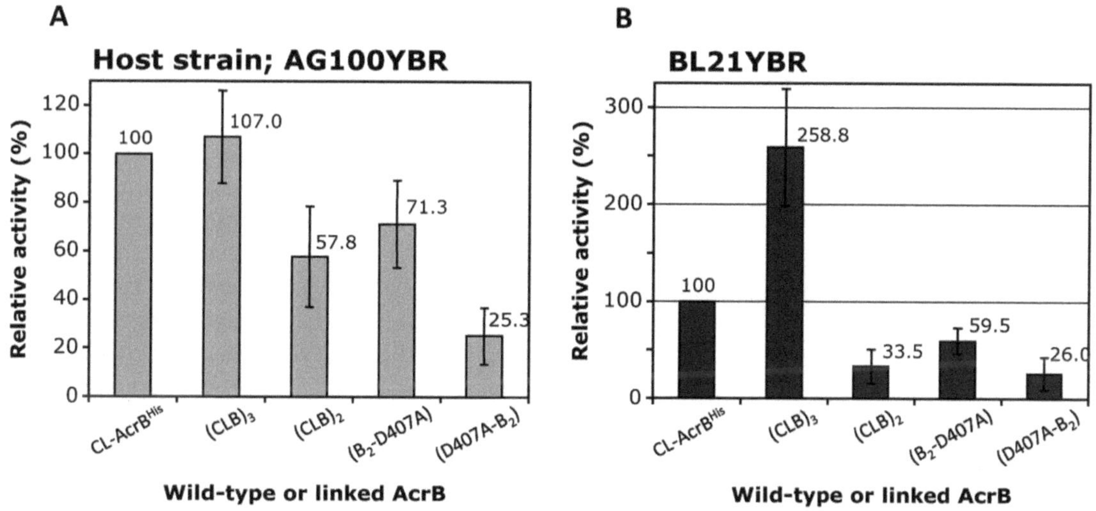

Fig. 3 Activities of linked AcrB proteins. Wild-type cysteineless AcrB (CL-AcrBHis) and linked AcrB proteins were expressed in the *acrB recA* host strains with different backgrounds, AG100YBR and BL21YBR, and their efflux activities were estimated from their levels of resistance to cholate by using the gradient plate method. Error bars show standard deviations. From [7]

3.5 Problems Caused by the Monomeric AcrB Produced by Proteolysis or Other Mechanisms

It is well known that "unnatural" proteins made in *E. coli* tend to become degraded by intracellular protease machineries, such as Lon [25]. Indeed the appearance of 220- and 110-kDa fragments, that correspond to an AcrB dimer and monomer, respectively, in cells containing p(CLB)$_3$ suggests that either the linker region or the connection between the linker and the proper AcrB sequence might be the weak spot for such proteolytic degradation. This production of the monomeric AcrB unit causes a major problem in the interpretation of experimental results using p(CLB)$_3$, because what is thought initially as the activity of linked trimer could be due to the trimeric assembly of the monomeric units generated by proteolysis.

We use two controls to evaluate the contribution of such monomeric units. One is the plasmid p(CLB)$_2$, which should be inactive in transport, except through the reassembly of the monomers generated by proteolysis or other mechanisms. The other is the plasmid coding for a linked trimer containing a monomeric unit that should inactivate the entire transport process, such as the proton translocation pathway mutation D407A.

1. Examine the efflux functions of AG100YBR cells containing the control plasmids, by assessing the extent of cholate resistance with the gradient plate method (Fig. 3a). In the K-12 background of AG100YBR, both of these controls will produce very significant activities (24–54% of the activity of p (CLB)$_3$) in cholate efflux activity (Fig. 3a), suggesting that

the generation of monomeric AcrB units is a real problem in this host.

2. Examine the efflux functions created by the control plasmids in an *E. coli* strain of a totally different lineage BL21YBR, in exactly the same manner. The results will be quite different in this host (Fig. 3b). First, pS(CLB)$_3$ will produce a much stronger cholate resistance than the plasmid coding for a monomeric unit, pSCLBH. Second, even more importantly, the control plasmids produce only very low levels of resistance in comparison with pS(CLB)$_3$, essentially indistinguishable from the level seen in cells containing only the pSPORT1 vector. This result is important not only because it indicates that the linked trimer is stable in this host, but also because it shows that the inactivation of only one protomer within the linked trimer essentially inactivates the entire trimer (see the low activities seen with (B$_2$-D407A) and (D407A-B$_2$) in comparison with (CLB)$_3$, in Fig. 3b).
3. The above results suggest that there is much less cleavage of the linked trimer in BL21YBR. Examine this hypothesis more directly by carrying out the SDS-PAGE analysis of whole cell proteins from cells containing pS(CLB)$_3$, followed by staining for AcrB proteins with anti-AcrB antibody. The results will show that much less dimers and monomers are found in BL21YBR than in AG100YBR (*see* Fig. 4 of [7]). Although these results suggest that proteolysis is less of a problem in BL21YBR (but *see* **Note 5**), careful inspection of the data in Fig. 3 suggests the operation of other mechanism(s) for the generation of AcrB monomer. Thus when we compare the activity of (B$_2$-D407A) and (D407A-B$_2$), the plasmid producing the former always shows more activity than that coding for the latter. That is, the functional monomeric AcrB tends to become generated more frequently from the N-terminal portion of the linked trimer, than from its C-terminal portion. This suggests that there may be additional (minor) mechanism (s) for generating a monomer from the linked trimer gene, such as premature termination of transcription or translation.

3.6 "Real-Time" Inactivation of the Entire Trimer by the Inactivation of One Component Monomer

The main purpose of this study is to see if the inactivation of one component monomer will inactivate the entire trimeric complex by blocking the functional rotatory mechanism. This has already been achieved by the experiment in the previous section, as pS(B$_2$-D407A) and pS(D407A-B$_2$) plasmids, inactivated in one protomer by the mutation in D407 residue involved in proton relay, were unable to produce cholate resistance (Fig. 3b), at least in the host BL21YBR. That this is not due to the poor expression of these constructs can be ascertained by the SDS-PAGE followed by the Western blot using anti-AcrB antibody (*see* Fig. 4 of [7]).

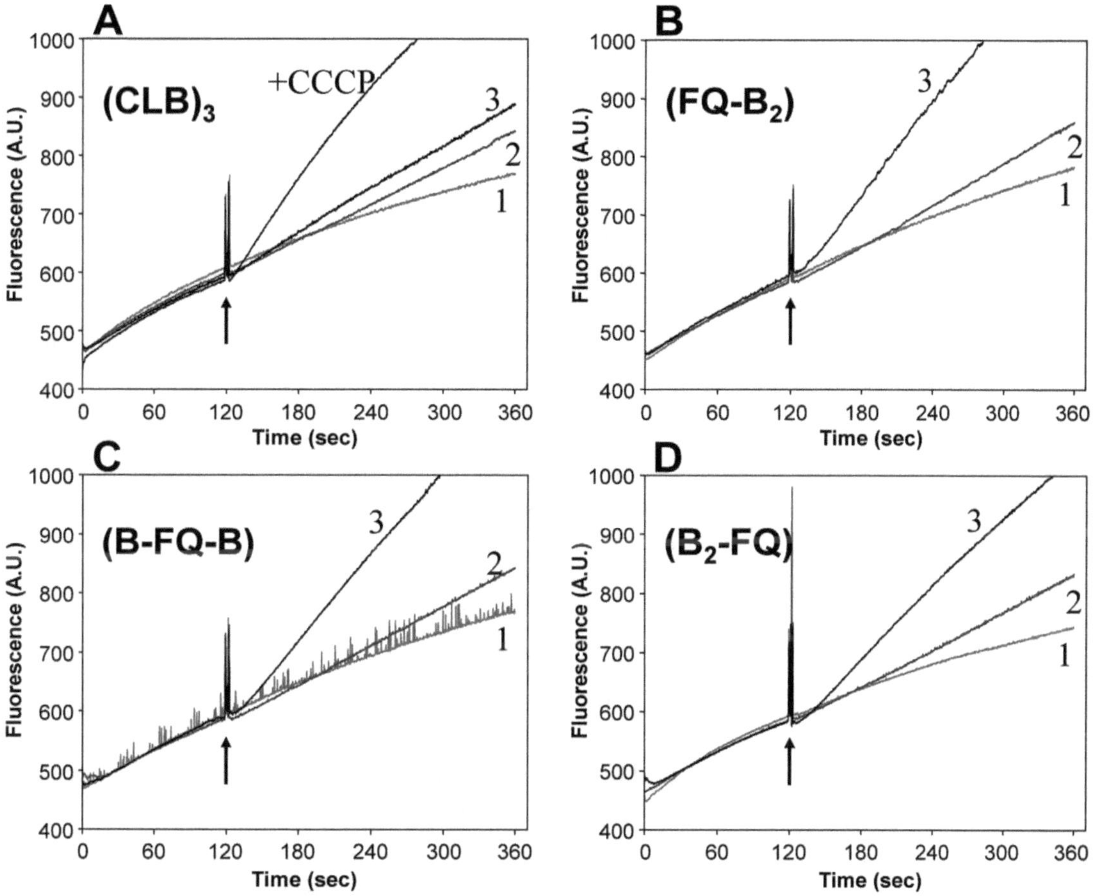

Fig. 4 Effect of cross-linker addition on ethidium accumulation in BL21YBDR cells expressing linked-heterotrimer AcrB with double-Cys mutation in only one protomer. Cellular accumulation of ethidium was monitored continuously by measuring the fluorescence of the ethidium-nucleic acid complex at excitation and emission wavelengths of 520 and 590 nm, respectively. After 2 min of incubation with 5 μM ethidium bromide (*arrows*), an MTS reagent or solvent alone (dimethyl sulfoxide-ethyl acetate [3:1, vol/vol]) was added to 2 ml of cell suspension. Additions were as follows: curve 1, 3.2 μl of solvent; curve 2, 4 μM 5-MTS (3.2 μl of a 2.5 mM stock solution); curve 3, 4 μM MTS-2-MTS (3.2 μl of a 2.5 mM stock solution). CCCP (40 μM final concentration) was added to cells with pS(CLB)$_3$ as a control to show the effect of total inactivation of AcrB pump. *A.U.* arbitrary unit. From [7]

However, these approaches are still not ideal because the results may be affected not only by the degradation of the linked trimer, but also by the possible truncation of the trimer due to premature halting in transcription or translation, as described above. An ideal approach would use cells producing a functional linked trimer, and to inactivate only one of the monomeric units rapidly, so as to enable us to observe the effect of this alteration in "real time" through the continued observation of the efflux process. This can be achieved by introduction of two Cys residues at opposite sides of the wide-open cleft in the periplasmic domain of AcrB, and cross-linking them by the fast-acting MTS cross-linker.

1. Construct pS(CLB)$_3$ plasmids in which one protomer has Cys residues replacing F666 and Q830 residues, as described in Subheading 3.2. These two residues are far away from each other (>9 Å) in Access and Binding protomers, but become close (4.4 Å) in the Extrusion protomer. This will result in the generation of pS(FQ-B$_2$), pS(B-FQ-B), and pS(B$_2$-FQ), where FQ denotes the protomer containing these Cys residues.
2. Use, as the host BL21YBDR, which contains an additional mutation *dsbA1*::*kan* to prevent the premature formation of disulfide bonds and inactivation of the linked AcrB. Inoculate LB broth with ampicillin with a single colony of the freshly transformed BL21YBDR with plasmids described above, and also pS(CLB)$_3$ (as a control) and grow the strains overnight without shaking at 30 °C. Dilute these cultures in fresh LB medium with ampicillin (10 ml), so that the initial OD660 of 0.08 will be obtained for all strains, and grow the cultures with shaking at 37 °C until they reach an OD660 of 0.8–0.9, without IPTG induction. Harvest cells by centrifugation at 5000 × *g* for 10 min at room temperature, wash once with 50 mM sodium phosphate buffer (pH 7.0) containing 0.1 M NaCl and 0.1% (vol/vol) glycerol, and resuspended in the same buffer to produce OD660 of about 0.2.
3. Monitor the entry of ethidium bromide (final concentration: 5 μM) into cells (which is counterbalanced by an active efflux by AcrB) by following the cellular fluorescence with a Shimadzu RF-5301PC spectrofluorometer. Set the excitation and emission wavelengths at 520 and 590 nm, respectively, with slit widths at 5 nm for excitation and 10 nm for emission. After 2 min of preincubation with ethidium bromide, add 2.5 mM 1,2-ethanediyl bismethanethiosulfonate (MTS-2-MTS) or 2.5 mM pentyl methanethiosulfonate (5-MTS) (Toronto Research Chemicals, Toronto, Ontario, Canada) (*see* **Notes 1** and **2**) to the spectrofluorometer cells with constant magnetic stirring, so that the final concentration will become 4 μM. (These reagents are freshly dissolved in dimethyl sulfoxide-ethyl acetate (3:1, vol/vol).)

This will produce the results similar to those shown in Fig. 4. Ethidium entry into cells containing pS(CLB)$_3$ continues to be slow even when thiol agents were added, because cysteineless AcrB is not affected by them and continues to pump out ethidium (Fig. 4a). However, in cells expressing the linked trimer with a monomeric unit containing F666C and Q830C substitutions, the MTS-2-MTS crosslinker (curves 3) produced instant inactivation of the efflux activity, as seen by an instantaneous increase in the net entry rate of ethidium, regardless of the position of the monomeric unit within the linked trimer (Fig. 4b–d). Furthermore, the increased influx rate of ethidium will be similar to the rate found in CCCP-treated, pS(CLB)$_3$-containing cells (Fig. 3a), where the AcrB pump is completely inactivated by the uncoupling of the proton-motive force. That the inactivation in curves 3 (Fig. 4b–d) is due to the cross-linking of the two Cys residues, not to the effect of modification of individual Cys residues, is seen by the observation that a noncrosslinker 5-MTS, which produces a similar alkyl-S modification to sulfhydryl groups of Cys [26], has little effect on the rate of ethidium entry (curves 2).

4 Notes

1. MTS-2-MTS is soluble in DMSO, but 5-MTS is not. Although 5MTS is quite soluble in ethyl acetate, this solvent alone tends to affect the influx rate of ethidium. Thus to minimize this effect and to obtain reasonable solubility for both MTS-2-MTS and 5-MTS, we use the 3:1 (vol/vol) mixture of DMSO and ethyl acetate.
2. MTS reagents are sensitive to moisture and are hydrolyzed in water over a period of time. The reagents should be stored in a desiccator at −20 °C and warmed up to room temperature before opening of the vial. We made up the solutions in organic solvent immediately prior to use and the solutions, kept on ice, were used within ~1.5 h.
3. The procedures for plasmid construction may be unnecessarily complex, in view of the progress in the field in the intervening 6 years. We also tried to utilize intermediate constructs in our earlier attempt to produce pUC19-based giant plasmid. Nevertheless, we give here the way we have successfully created the plasmid and reported in reference [7].
4. Plasmid pU19$^{\Delta H}$-linkXSmH, used as the template, was generated by first introducing the cysteineless *acrB* sequence (amplified by PCR by using primers MetBFw and CdHisXRv with pSCLBH as the template) into pUC19 cut with BamHI and XbaI, and then cutting this intermediate plasmid with XbaI and HindIII, followed by the insertion of the linker sequence PCR-amplified with primers linkXFw and linkSmHRv with pSCLBH as the template.
5. *Escherichia coli* BL21 strains (especially BL21(DE3), which is just a lysogen of BL21) are often described as carrying mutations in *ompT* and *lon*. OmpT is indeed missing in the genome sequence of BL21(DE3) (Genbank NC_012971.2); however OmpT protease is thought to be involved in the degradation of

misfolded proteins in the periplasm, and it may not play a large role in the degradation of the linked trimer with its "unnatural" sequences in the cytosol. Lon protease is thought to play a major role in the degradation of misfolded proteins in the cytosol [25]. However, in the BL21(DE3) genome there is no evidence that its *lon* gene is defective. Although there are four SNPs in comparison with the K-12 (strain MG1655) sequence, most of them occur in the third letter of the codon, and none changes the amino acid coded. Thus we are not able to supply an easy explanation for the stability of the linked trimer in BL21YBR, but BL21 comes from a completely different lineage and must contain many features not present in K-12.

Acknowledgements

This work was supported by a grant from the U.S. Public Health Service (AI-009644). I thank Y. Takatsuka for supplying the details and for suggestions.

References

1. Li X-Z, Plesiat P, Nikaido H (2015) The challenge of efflux-mediated antibiotic resistance in Gram-negative bacteria. Clin Microbiol Rev 28:337–418
2. Murakami S, Nakashima R, Yamashita E et al (2002) Crystal structure of bacterial multidrug efflux transporter AcrB. Nature 419:587–593
3. Murakami S, Nakashima R, Yamashita E et al (2006) Crystal structures of a multidrug transporter reveal a functionally rotating mechanism. Nature 443:173–179
4. Seeger MA, Schiefner A, Eicher T et al (2006) Structural asymmetry of AcrB trimer suggests a peristaltic pump mechanism. Science 313:1295–1298
5. Sennhauser G, Amstutz P, Briand C et al (2007) Drug export pathway of multidrug exporter AcrB revealed by DARPin inhibitors. PLoS Biol e7:5
6. Cha HJ, Pos KM (2014) Cooperative transport mechanism and proton-coupling in the multidrug efflux transporter complex AcrAB-TolC. In: Kramer R, Ziegler C (eds) Membrane transport mechanisms: 3D structure and beyond. Springer, Heidelberg, pp 207–232
7. Takatsuka Y, Nikaido H (2009) Covalently linked trimer of the AcrB multidrug efflux pump provides support for the functional rotating mechanism. J Bacteriol 191:1729–1737
8. Kinana AD, Vargiu AV, Nikaido H (2013) Some ligands enhance the efflux of other ligands by the Escherichia coli multidrug pump AcrB. Biochemistry 52:8342–8351
9. Kim HS, Nikaido H (2012) Different functions of MdtB and MdtC subunits in the heterotrimeric efflux transporter MdtB(2)C complex of *Escherichia coli*. Biochemistry 51:4188–4197
10. Sambrook J, Fritsch EF, Maniatis T (1989) Molecular cloning: a laboratory manual, 2nd edn. Cold Spring Harbor Laboratory Press, Cold Spring Harbor, NY
11. Rietsch A, Belin D, Martin N et al (1996) An in vivo pathway for disulfide bond isomerization in Escherichia coli. Proc Natl Acad Sci U S A 93:13048–13053
12. Okusu H, Ma D, Nikaido H (1996) AcrAB efflux pump plays a major role in the antibiotic resistance phenotype of Escherichia coli multiple-antibiotic-resistance (Mar) mutants. J Bacteriol 178:306–308
13. Takatsuka Y, Nikaido H (2007) Site-directed disulfide cross-linking shows that cleft flexibility in the periplasmic domain is needed for the multidrug efflux pump AcrB of *Escherichia coli*. J Bacteriol 189:8677–8684

14. Churchward G, Belin D, Nagamine Y (1984) A pSC101-derived plasmid which shows no sequence homology to other commonly used cloning vectors. Gene 31:165–171
15. Ma D, Cook DN, Alberti M et al (1993) Molecular cloning and characterization of *acrA* and *acrE* genes of *Escherichia coli*. J Bacteriol 175:6299–6313
16. Takatsuka Y, Nikaido H (2006) Threonine-978 in the transmembrane segment of the multidrug efflux pump AcrB of *Escherichia coli* is crucial for drug transport as a probable component of the proton relay network. J Bacteriol 188:7284–7289
17. Schagger H, von Jagow G (1991) Blue native electrophoresis for isolation of membrane protein complexes in enzymatically active form. Anal Biochem 199:223–231
18. Laemmli UK (1970) Cleavage of structural proteins during the assembly of the head of bacteriophage T4. Nature 227:680–685
19. Zgurskaya HI, Nikaido H (1999) Bypassing the periplasm: reconstitution of the AcrAB multidrug efflux pump of *Escherichia coli*. Proc Natl Acad Sci U S A 96:7190–7195
20. Takatsuka Y, Nikaido H (2010) Site-directed disulfide cross-linking to probe conformational changes of a transporter during its functional cycle: *Escherichia coli* AcrB multidrug exporter as an example. Methods Mol Biol 634:343–354
21. Su CC, Li M, Gu R et al (2006) Conformation of the AcrB multidrug efflux pump in mutants of the putative proton relay pathway. J Bacteriol 188:7290–7296
22. Nikaido H (1994) Isolation of outer membranes. Methods Enzymol 235:225–234
23. Elkins CA, Nikaido H (2002) Substrate specificity of the RND-type multidrug efflux pumps AcrB and AcrD of *Escherichia coli* is determined predominantly by two large periplasmic loops. J Bacteriol 184:6490–6498
24. Bryson V, Szybalski W (1952) Microbial selection. Science 116:45–51
25. Goldberg AL (2003) Protein degradation and protection against misfolded or damaged proteins. Nature 426:895–899
26. Kenyon GL, Bruice TW (1977) Novel sulfhydryl reagents. Methods Enzymol 47:407–430

Chapter 10

Determining Ligand Path Through a Major Drug Transporter, AcrB, in *Escherichia coli*

Fasahath Husain and Hiroshi Nikaido

Abstract

An experimental approach to detect the path a substrate takes through a complex membrane protein is described with emphasis on technical approach and theoretical considerations. The protocols for bacterial culture preparation, membrane protein purification, fluorescent assay standardization, data collection, and data analysis are provided. Useful software tools are recommended.

Key words Substrate path, Membrane transporter, Membrane channel, Fluorescent assay, Bodipy FL Maleimide, Membrane protein purification

1 Introduction

Loss-of-function mutations [1–3], crystallographic studies [4–6] and lately computational molecular dynamics [7, 8] have been the predominant methods of determining the ligand binding domains of a multidrug efflux transporter protein, such as AcrB of *Escherichia coli*. The loss-of-function mutations have for decades provided insight into protein-ligand interaction in other proteins, but the major limitation is that most alterations in AcrB and its homologs do not confer a noticeable phenotype and occasionally it is hard to interpret whether the loss-of-function was due to general structural defect or compromised ligand-protein interaction. In spite of complexity of crystallography studies (especially with membrane proteins), the ligand-bound crystal structures have provided a remarkable insight into ligand-protein interaction. The crystallographic studies too are limited in scope because the ligands bind to specific sites leaving the other transitionary interactions unknown. Using the crystal structures, molecular dynamics approach has managed to fill many of those gaps but still are limited in robustness and interpretation without direct biochemical evidence is difficult.

Akihito Yamaguchi and Kunihiko Nishino (eds.), *Bacterial Multidrug Exporters: Methods and Protocols*, Methods in Molecular Biology, vol. 1700, https://doi.org/10.1007/978-1-4939-7454-2_10, © Springer Science+Business Media, LLC 2018

The Bodipy-FL-Maleimide assay described here [9, 10] complements the crystallographic approach, loss-of-mutations studies and molecular dynamics. Using this assay, we managed to provide a more comprehensive picture of the path a drug takes through inner membrane protein, AcrB. AcrB is a major antibiotic efflux pump in *Escherichia coli* that confers resistance to myriad classes of drugs. We use AcrB as model because of its wide substrate spectrum and clinical importance. The Bodipy-FL-Maleimide assay involves modifying of the target residues in AcrB to a cysteine side-chain and allowing a cysteine-reactive fluorescent compound (Bodipy-FL-Maleimide) to covalently react with cysteine side-chain. The irreversible reaction is observed and quantified.

The target residues were identified using published crystal structures, loss-of-function mutations and molecular dynamics. Site-directed-mutagenesis was utilized to modify the native side-chains of protein to cysteine side-chain. The live cells in a controlled assay were allowed to react with Bodipy-FL-maleimide. The fluorescent stain-treated protein was purified and resolved on SDS-PAGE gel. Using phosphorimager, coomassie blue staining, and ImageJ software the fluorescent staining of AcrB was quantified.

2 Materials

All solutions were prepared with ultrapure pure water and analytical-grade reagents. All the kits and reagents are commercially available.

2.1 Bacterial Strains for Bodipy-FL-Maleimide Assay

RAM-1334 (*Escherichia coli* MC4100 Δ*ara* Δ*acrB*), a gift from Dr. Rajeev Misra (Arizona State University, Tempe, AZ), was used for the assays.

2.2 Bacterial Strains for Site-Directed Mutagenesis

XL10-Gold ultracompetent cells were provided with QuikChange II XL Site-Directed Mutagenesis Kit (Agilent, Santa Clara, CA, Catalog # 200521).

2.3 Vectors

The vector, pSportI-*acrB*, could be requested from Dr. Hiroshi Nikaido (University of California, Berkeley, CA). This vector expresses AcrB that lacks the two native cysteines (*see* **Note 1**) and has an additional tag containing six Histidine residues at C-terminus of AcrB. The product of this construct is referred to as CL-AcrB$_{\text{His}}$ [11].

2.4 Identification of Target Residues

PyMol (https://www.pymol.org/) [12], and the PyMol plug-in Caver (http://caver.cz/) [13] was used.

2.5 Vector for Protein Modification

QuikChange II XL Site-Directed Mutagenesis Kit (Agilent, Santa Clara, CA, Catalog # 200521), was used for the all the site directed mutagenesis.

2.6 Growth of Bacterial Strain

1. Luria–Bertani (LB) broth, 1 L: 10 g Tryptone, 5 g Bacto-Yeast extract, and 10 g NaCl, prepared in 950 mL. The pH is adjusted to 7.3 with 0.1 N NaOH and made up to 1 L with water. Sterilize by autoclaving and store at room temperature.
2. Ampicillin (Amp) solution, 50 mg/mL: Prepare Ampicillin solution by adding 500 mg of Ampicillin to 7 mL water and dissolve. Make up the solution to 10 mL with water. Sterilize by filtering through 0.45 μm syringe filter. Prepare aliquots of 1 mL and store in −20 °C.
3. LB-Amp medium: Just before inoculation add 5 μL of Amp to 5 mL of LB.
4. 37 °C shaker incubator, shaking at 200 rpm.

2.7 Preparation of Cells for Bodipy-FL-Maleimide Assay

1. Phosphate buffer: 50 mM potassium phosphate, 0.5 mM $MgCl_2$, pH 7.0.
2. Centrifuge: Microcentrifuge 5418 (Eppendorf, Hamburg, Germany, Catalog # 5418 000.017) set at 13,000 × *g*, unless specified.
3. Spectrophotometer: Shimadzu 1240.

2.8 Staining of Cells for Bodipy-FL-Maleimide Assay

1. Bodipy-FL-Maleimide 6 mM: 6 mM Bodipy-FL-Maleimide (Thermo Scientific, San Diego, CA, Catalog # B-10250) is prepared in DMSO solvent. 50 μL aliquots are stored at −20 °C.
2. Glucose solution 40%: 10 mL solution of 40% glucose solution in water is prepared, filter-sterilized, and stored at room temperature.
3. Bodipy-FL-Maleimide staining solution: 6 μM Bodipy-FL-Maleimide, 0.4% glucose, 50 mM potassium phosphate, 0.5 mM $MgCl_2$, pH 7.0. Prepared fresh just before use by adding Bodipy-FL-Maleimide and glucose in phosphate buffer.
4. Wash solution: 0.4% glucose, 50 mM potassium phosphate, 0.5 mM $MgCl_2$, pH 7.0. Prepared fresh just before use by adding glucose in phosphate buffer.
5. Vortexer: VWR vortexer mini 230 V (VWR, Radnor, PA, Catalog # 58816-123).

2.9 Protein Purification for Bodipy-FL-Maleimide Assay

1. Sonicator: Q500 Sonicator (Qsonica, Newton, CT, Catalog #Q500-110).
2. Centrifuge: Microcentrifuge 5418 (Eppendorf, Hamburg, Germany, Catalog # 5418 000.017) set at 13,000 × *g* unless specified.

3. Ultracentrifuge: Ultracentrifuge (Thermo Scientific, San Diego, CA, Catalog # 46960) with fixed angle S50-A rotor (Thermo Scientific, San Diego, CA, Catalog # 45540), set at 95,000 × *g* for 90 min and at 4 °C, unless specified.
4. 1% DDM phosphate buffer: 1% *n*-dodecyl-β-D-maltoside (DDM) (Sigma-Aldrich, St. Louis, MO, Catalog # D4641), 50 mM potassium phosphate, 0.5 mM $MgCl_2$, pH 7.0. Prepared fresh just before use by adding the DDM in the phosphate buffer.
5. 0.02% DDM wash buffer: 0.02% *n*-dodecyl-β-D-maltoside, 500 mM Imidazole (Sigma-Aldrich, St. Louis, MO, Catalog # I5513), 50 mM potassium phosphate, 0.5 mM $MgCl_2$, pH 7.0. Prepared fresh just before use by adding DDM and imidazole in the phosphate buffer.
6. 500 mM imidazole elution buffer: 0.02% *n*-dodecyl-β-D-maltoside, 50 mM potassium phosphate, 0.5 mM $MgCl_2$, pH 7.0. Prepared fresh just before use by adding the DDM in the phosphate buffer.
7. Rocker: Compact digital rocker (Thermo Scientific, San Diego, CA, Catalog # 88880019) set 50 rpm at 4 °C.

2.10 SDS-PAGE for Protein Resolution

11% SDS page gel was prepared as previously described [14].

2.11 Fluorescent Imaging

Phosphorimager: Typhoon FLA 9500 (GE Healthcare Life Sciences, Marlborough, MA, Catalog # 28-9969-43), the filters were set for excitation at 488 nm and emission at 510 nm.

2.12 Coomassie Staining and Drying

1. Coomassie blue staining solution: 1% Coomassie brilliant blue R250, 40% methanol, 10% glacial acetic acid. Store at room temperature.
2. Destaining solution: 40% methanol, 10% glacial acetic acid.
3. Store at room temperature.

2.13 Analysis and Comparison of AcrB Staining

Image J, available for free download from National Institute of Health (http://imagej.nih.gov/ij/) [15] was used for analysis.

3 Methods

3.1 Vector Modification

3.1.1 Identification of Target Residues

1. Previous research studies were used as reference to identify residues in AcrB that are predicted to play a role in ligand binding. These residues were considered positive control.
2. With this point of reference, PyMol was used to recognize the residues in closest proximity to the residues that are predicted to be positive in the assay (Fig. 1).
3. Caver, a plug-in for PyMol, was used to identify the cavities by using the positive residues as point of reference.

3.1.2 Modification of Residues

1. QuikChange II XL Site-Directed Mutagenesis Kit was used as per manufacturer's directions to modify the identified residues to Cysteine in each independent vectors (*see* **Note 1**).
2. This step was repeated for the mutants that required two alterations (*see* **Note 2**).

3.2 Growth and Preparation of Bacterial Cells for Bodipy-FL-Maleimide

1. LB with ampicillin (50 μg/mL) was inoculated and grown overnight at 37 °C with vigorous shaking. Include positive and negative with every independent round of assays from this step onward (*see* **Note 1**).
2. The overnight grown cells were washed by pelleting, using table-top centrifuge, and resuspending in equal amount of phosphate buffer, using the vortexer.
3. The wash step was repeated.

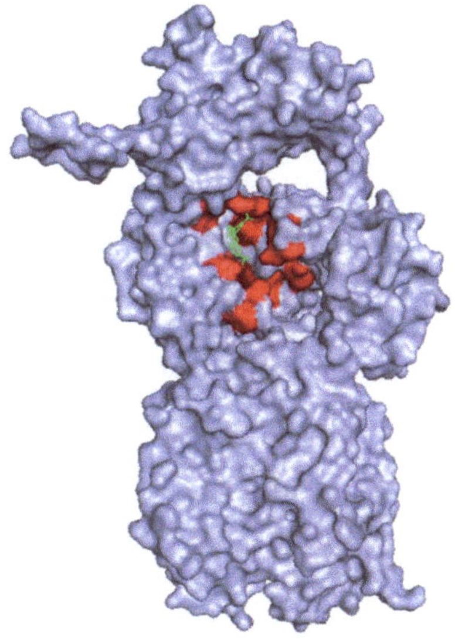

Fig. 1 The ligand and the residues in closest proximity to the ligand are selected as initial targets for modification. The image was processed using PyMol with PDB ID: 2DRD [6]

4. The cell density was adjusted to OD_{660} using the Shimadzu 1240 spectrophotometer (*see* **Note 3**).
5. An aliquot of 5 mL of these cells was used for the assay in the following steps.

3.3 Staining with Bodipy-FL-Maleimide

1. The cells, 5 mL, were treated to a final concentration of 0.4% glucose and 6 mM Bodipy-FL-maleimide (*see* **Note 4**).
2. The mixture was allowed to stand at room temperature with a 2 svortexing step at 30 min.
3. The cells were washed two times with 5 mL phosphate buffer infused with 0.4% glucose.
4. The stained cells were left on ice in preparation for the lysis.

3.4 Lysis, Membrane Separation, Membrane Solubilization, and Protein Purification

3.4.1 Lysis of Cells

1. The tube containing the cells was kept on ice during sonication (*see* **Note 5**).
2. The cells were sonicated with short pulses of 5–10 s with an interval of 20–30 s: always kept on ice through the process.
3. The short pulses were repeated until the solution turned close to translucent.
4. The lysate was centrifuged at 1000 × *g* for 3 min at 4 °C to remove the unlysed cells and the supernatant was used for the membrane isolation.

3.4.2 Isolation and Solubilization of Bacterial Membrane

1. The membrane was pelleted by ultracentrifugation at 95,000 × *g* for 1 h at 4 °C.
2. The membrane pellet was resuspended in 1% DDM phosphate buffer containing 200 mL His-select cobalt affinity gel (for his-tagged protein binding), a total volume of 800 μL.
3. The mix was left on a rocker to solubilize overnight, rocking gently at 4 °C.

3.4.3 Protein Purification

1. In the previous step, during membrane solubilization, the CL-$AcrB_{His}$ was also allowed to bind to His-Select cobalt matrix.
2. The entire 800 μL is applied to Micro-Spin chromatography column and centrifuged at 6000 × *g* for 1 min.
3. The matrix was washed by applying 500 μL 0.02% DDM wash buffer to the cobalt matrix in the column and centrifuged at 6000 × *g* for 1 min. This wash step was repeated one more time.
4. The purified His-tagged AcrB was eluted by applying 100 μL of 500 mM imidazole elution buffer to the matrix and allowed to sit at room temperature for 5 min and centrifuged at 6000 × *g* for 1 min.

3.5 Gel Separation and Image Capture for Analysis

1. 10 μL of the purified protein was resolved in 11% SDS protein gel.
2. The gel was separated from the plate and gently washed with water.
3. An image of gel was captured with Typhoon FLA 9500 with filters set at excitation at 488 nm and emission at 510 nm.
4. The image was printed and the lanes are carefully labeled.
5. The gel was then stained with coomassie stain.
6. The gel was dried using a previously described method.
7. The dried gel was attached to the sheet that contained the picture from Typhoon FLA 9500.
8. A JPEG image of the sheet containing both the gel and image from Typhoon FLA 9500 was created using regular office scanner (HP Officejet 6500). This image was used for the further analysis (Fig. 2).

3.6 Analysis Using ImageJ

1. The scanned image was used for this analysis using ImageJ (Fig. 3).
2. The scanned JPEG image was converted to 8-bit digital output.
3. The background was digitally subtracted using 50 pixel radius.
4. The image was inverted to negative image for quantification.
5. Each gel and assay was always run with a positive control and a negative control.
6. The fluorescent band was normalized against the Coomassie-stained band to correlate with the protein quantity.
7. The relative intensity of Bodipy-FL-Maleimide stained band (normalized for amount) was calculated by comparing each CL-AcrB$_{His}$ variant with the positive control.

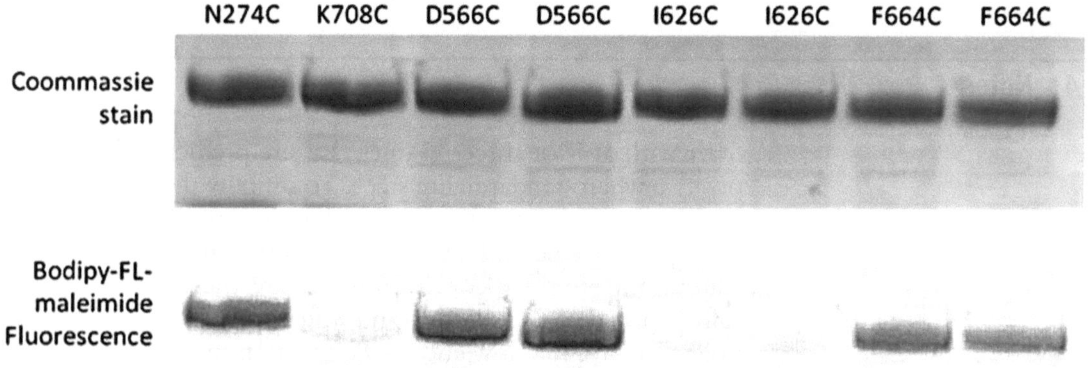

Fig. 2 The images from Spectrofluorometer and the scanned Comassive-stained gel are captured in one image. The AcrB with the N274C residue is the positive control and the K708C residue is the negative control. The positive and the negative controls, that were fully processed along with every new round of assay, were included in every gel

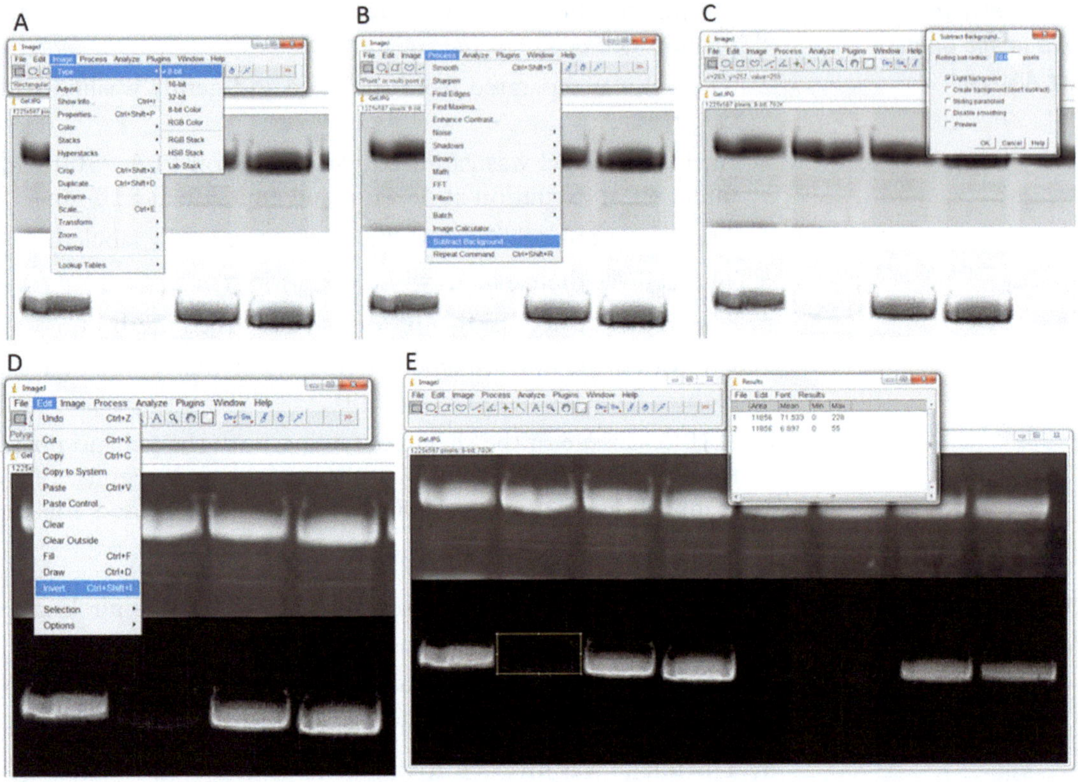

Fig. 3 Processing of the gel image using ImageJ [15]. The image captured in Fig. 2 is processed using the following steps. In step **a**, the image is converted to 8-bit. In steps **b** and **c**, the background is subtracted using 50 pixels radius. In step **d**, the image is inverted. In step **e**, the intensity of all the bands are captured as mean. The area selector once selected is maintained for all the bands. The area selector can be moved using keyboard arrows and Ctrl + M is used to collect the mean. The data from 'results' can be copied to excel for processing

8. Protein variants that had intensity of at least 30% of the positive control were rendered as positive.

4 Notes

1. Start with at least three residues that are predicted to be to be positive when standardizing steps. To simplify the steps use only positive control. Once the predicted positive residues are stained, start using predicted negative controls to standardize dye concentration. Negative control residues are the residues that are fully exposed to the solution and will react with the reactive dye unless the dye is at low enough concentration. The dye concentration has to be low enough that it does not react spontaneously with any exposed cysteine side-chain unless it binds specifically to a domain that consists of the positive residue.

2. It is important to remove the native cysteines. The native cysteines could get specifically stained and may confound the results.
3. The concentration of dye, the cell density, and cell permeability are variables. Depending on the stain background all of these need to be standardized.
4. Bodipy-FL-Maleimide may not be a substrate for all the proteins of interest. Try several dyes for developing an assay.
5. The cells should remain cold during sonication. Any significant increase in temperature will damage the protein.

Acknowledgments

I am thankful to Dr. Etsuko Sugawara for suggestions and Dr. Yumiko Takatsuka and Dr. Rajeev Misra for materials. This work was carried out in the laboratory of Dr. Hiroshi Nikaido at University of California, supported by AI-09644 grant from U.S. Public Health Service.

References

1. Bohnert JA, Schuster S, Seeger MA et al (2008) Site-directed mutagenesis reveals putative substrate binding residues in the *Escherichia coli* RND efflux pump AcrB. J Bacteriol 190:8225–8229
2. Wang Z, Zhong M, Lu W et al (2015) Repressive mutations restore function-loss caused by the disruption of trimerization in *Escherichia coli* multidrug transporter AcrB. Front Microbiol 6:4
3. Soparkar K, Kinana AD, Weeks JW et al (2015) Reversal of the drug binding pocket defects of the AcrB multidrug efflux pump protein of *Escherichia coli*. J Bacteriol 197:3255–3264
4. Seeger MA (2006) Structural asymmetry of AcrB trimer suggests a peristaltic pump mechanism. Science 313:1295–1298
5. Kim LHH, Murakami S (2013) Crystal structure of AcrB complexed with linezolid at 3.5 Å resolution. J Struct Funct Genom 14:71–75
6. Murakami S, Nakashima R, Yamashita E et al (2006) Crystal structures of a multidrug transporter reveal a functionally rotating mechanism. Nature 443:173–179
7. Vargiu AV, Nikaido H (2012) Multidrug binding properties of the AcrB efflux pump characterized by molecular dynamics simulations. Proc Natl Acad Sci U S A 109:20637–20642
8. Ruggerone P, Vargiu AV, Collu F et al (2013) Molecular dynamics computer simulations of multidrug RND efflux pumps. Comput Struct Biotechnol J 5:e201302008
9. Husain F, Nikaido H (2010) Substrate path in the AcrB multidrug efflux pump of *Escherichia coli*. Mol Microbiol 78:320–330
10. Husain F, Bikhchandani M, Nikaido H (2011) Vestibules are part of the substrate path in the multidrug efflux transporter AcrB of *Escherichia coli*. J Bacteriol 193:5847–5849
11. Takatsuka Y, Nikaido H (2007) Site-directed disulfide cross-linking shows that cleft flexibility in the periplasmic domain is needed for the multidrug efflux pump AcrB of *Escherichia coli*. J Bacteriol 189:8677–8684
12. DeLano, WL (2002) The PyMOL user's manual. DeLano Scientific, San Carlos, CA, 452
13. Chovancova E, Pavelka A, Benes P et al (2012) CAVER 3.0: a tool for the analysis of transport pathways in dynamic protein structures. PLoS Comput Biol 8:e1002708
14. Laemmli UK (1970) Cleavage of structural proteins during the assembly of the head of bacteriophage T4. Nature 227:680–685
15. Schneider CA, Rasband WS, Eliceiri KW (2012) NIH Image to ImageJ: 25 years of image analysis. Nat Methods 9:671–675

Part III

Computational Analysis of Bacterial Multidrug Exporters

Chapter 11

Molecular Modeling of Multidrug Properties of Resistance Nodulation Division (RND) Transporters

Pierpaolo Cacciotto, Venkata K. Ramaswamy, Giuliano Malloci, Paolo Ruggerone, and Attilio V. Vargiu

Abstract

Efflux pumps of the resistance nodulation division (RND) superfamily are among the major contributors to intrinsic and acquired multidrug resistance in Gram-negative bacteria. Structural information on AcrAB-TolC and MexAB-OprM, major efflux pumps of *Escherichia coli* and *Pseudomonas aeruginosa* respectively, boosted intensive research aimed at understanding the molecular mechanisms ruling the active extrusion processes. In particular, several studies were devoted to the understanding of the determinants behind the extraordinary broad specificity of the RND transporters AcrB and MexB. In this chapter, we discuss the ever-growing role computational methods have been playing in deciphering key structural and dynamical features of these transporters and of their interaction with substrates and inhibitors. We further discuss and illustrate examples from our lab of how molecular docking, homology modeling, all-atom molecular dynamics simulations and in silico free energy estimations can all together give precious insights into the processes of recognition and extrusion of substrates, as well as on the possible inhibition strategies.

Key words Efflux pumps, Gram-negative bacteria, Membrane proteins, RND transporters, MD simulations, Molecular docking, Homology modeling, Free energy calculations, MM/GBSA

1 Introduction

The tolerance of bacteria to several antibiotics from different classes (a phenomenon known as multidrug resistance) has become one of the most serious threats for public health [1] and the scientific community is nowadays aware that fighting antibacterial resistance is an urgent and not deferrable task [2–5]. Gram-negative bacteria are of particular concern as they represent most of the clinical strains endowed with multi, extreme, or total drug resistance (MDR, XDR, or TDR respectively) for which there are no promising antibiotics in the pipeline [6, 7]. Among the various mechanisms related to intrinsic and acquired MDR in

Pierpaolo Cacciotto and Venkata K. Ramaswamy contributed equally to this work.

Akihito Yamaguchi and Kunihiko Nishino (eds.), *Bacterial Multidrug Exporters: Methods and Protocols*, Methods in Molecular Biology, vol. 1700, https://doi.org/10.1007/978-1-4939-7454-2_11, © Springer Science+Business Media, LLC 2018

Gram-negative pathogens, a particularly important one involves multidrug efflux pumps belonging to the Resistance-Nodulation-cell Division (RND) superfamily [8–17].

These pumps are very complex machineries formed by at least three units whose assembly span the whole periplasmic space [18–28], and they are composed by: (1) a secondary RND antiporter embedded in the inner membrane [15, 29–34], which is the engine of the pump and is responsible of drug recognition and selectivity [35–40]; (2) a channel embedded in the outer membrane (Outer Membrane Factor) [41–45], through which toxic molecules reach the outer medium; (3) a number of several membrane fusion proteins (MFP), linking the two former components and setting up a trans-envelope efflux system [22, 24–27, 46–50]. In addition, a new protein (AcrZ) has been discovered [51] as fourth component of the AcrAB-TolC efflux pump [25, 51]. However, such fourth component has not been identified in other efflux pumps yet. Therefore, in the following general scheme of RND efflux pumps (Fig. 1) the presence of the fourth component will be neglected.

The last 15 years have seen a rush by structural biologists to provide information on key properties of both the single components of RND efflux pumps, as well as on their assembly. This competition has generated a large number of structures of the main efflux pumps of *Escherichia coli* and *Pseudomonas aeruginosa*, AcrAB(Z)-TolC, and MexAB-OprM respectively, being AcrB and MexB the transporters, AcrA and MexA the MFPs, and TolC and OprM the OMFs [22, 25, 26, 28, 30–34, 42–44, 47–50, 52–58].

The availability of structural information, together with the development of new robust simulation techniques [59–71] and the recent increase in computational power [72–75] has boosted several computational studies aiming at elucidating mechanistic aspects of extrusion process by these pumps [15, 17, 76]. Due to the large size of these molecular complexes, most in silico studies have been still limited to investigations on the single components [15, 39, 76–96]. As the engine of these pumps is the RND transporter, it is not surprising that a large fraction of studies focused on the functional dynamics of these proteins, as well as on their detailed interactions with substrates and inhibitors. The most studied transporter is by far AcrB of *E. coli*, able to recognize an extremely large number of unrelated compounds, from organic solvents to high molecular-mass antibiotics and detergents, thus being a paradigm of multidrug transporter [15, 17, 97].

In this chapter we describe the computational procedures adopted in our lab to study AcrB or other RND transporters. We discuss the main tools/protocols to model the properties of these giant proteins, namely the homology modeling, the molecular

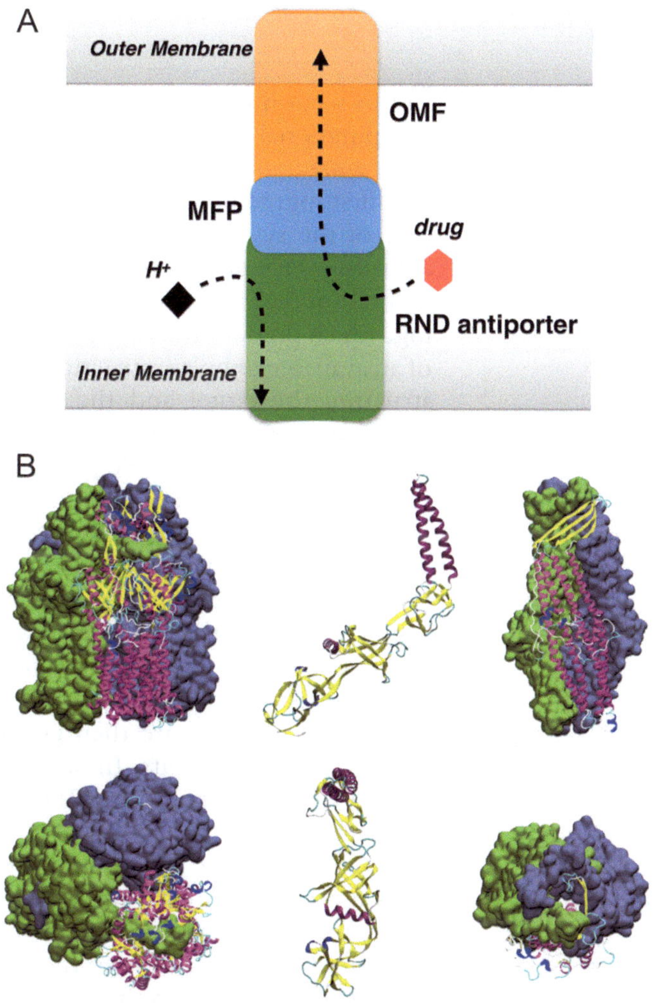

Fig. 1 (**A**) Functioning scheme of a typical RND efflux pump. The three components are arranged in a complex that spans the entire periplasm, connecting the inner and the outer membranes. A proton flux from the periplasm to the cytoplasm activates the RND transporter by inducing conformational changes that allow the complex to capture substrates and extrude them to the outside. (**B**) Side (*upper row*) and top (*lower row*) views of the components of a typical RND efflux pump, AcrAB-TolC from E. coli: AcrB (*left column*), AcrA (*middle column*), and TolC (*right column*)

docking of ligands, the insertion of the complex into the membrane, the equilibration procedure of the resulting system and the subsequent molecular dynamics (MD) simulations with estimation of the free energy of binding.

2 Theory

In this section we briefly discuss the theoretical background of the computational methods used to study the RND transporters.

2.1 Homology Modeling

The first ingredient of any MD simulation is a proper initial structure of the system under study. One of the main computational approaches to model proteins structures not available from experimental sources is the so-called homology (or comparative) modeling, which predicts the 3D structure of a target protein on the basis of an available template structure [98–102]. To predict a reasonable structure the target and the template should have a significant amino acid sequence similarity.

A widely used homology modeling program is MODELLER [102, 103], which generates the target protein structure by computing a set of positional restraints from the template 3D structure. It is assumed that the distances between corresponding residues in the target and the template are similar. This procedure is further supplemented by the addition of stereochemical restraints on bond lengths, bond angles, dihedral angles, and nonbonded atom–atom contacts, usually obtained from a molecular mechanics force field [104, 105]. Models are then generated by minimizing the violations of all the restraints. In Subheading 4 we will describe how to use MODELLER to build a 3D model for an RND transporter.

2.2 Molecular Docking

Molecular docking is a widely used computational method that aims at predicting the bound conformation and the binding affinity of a complex from the unbound structure of its constituents. Here, we restrict the description to the interaction between a small molecule (ligand) and a larger macromolecule (receptor). This subject is particularly relevant due to its applications in medicinal chemistry, such as computer-aided-drug-design, virtual screening of chemical libraries to hit identification and lead optimization [61, 106, 107]. The original concept of molecular docking is that of lock (the receptor protein) and key (the ligand). However, experimental observations of ligand-receptor binding reveal, in most cases, conformational changes of the native unbound structures. The historical lock-and-key model has therefore evolved to the so-called induced-fit model and the available docking software packages predict protein–ligand association taking sometimes into account flexibility (see the compilation of available docking software and their main features in Ref. [108]). In particular, a full-flexible docking in which both ligand and protein are flexible should be the best option but the high computational cost prevents its massive application. The preferred approach is therefore a semi-flexible docking in which the protein is kept fixed in space (or only some torsional angles in the active site are set as rotatable [65]) and different conformations of the ligand are generated on the fly during the docking process. An alternative and reliable approach is to rigidly dock ensemble of conformations of a ligand on ensemble of receptor conformations [109, 110]. In this approach the conformational analyses of ligand and receptor are made externally, for example they are generated from MD simulations. To compute relative affinities of putative poses of ligands docking programs use the so-called scoring function, which approximates the free energy of binding by means of a "master equation" summing up the several terms contributing to the binding.

Molecular docking has proven to be a powerful tool to investigate the mechanism of recognition of different compounds by the MDR transporters AcrB and MexB [80–82, 111]. In this chapter we describe the use of AutoDock VINA, which is among the most used programs for molecular docking [106, 108]. It is based on the definition of a rectangular grid of points where the molecular potential of the receptor is evaluated and sensed by the ligand during the docking process, and allows to set rotatable bonds for ligand and receptor (although flexible backbone or induced-fit are not taken into account explicitly).

2.3 All-atom MD Simulations

Computational approaches like MD simulations are nowadays commonly used in biomedical research to complement experimental results and to investigate dynamical properties that might be difficult to study in laboratories [61–64, 68, 71, 112]. Although a detailed description of any complex atomicistic system should take into account its quantum mechanical properties, it is sometimes useful and relatively safe to neglect them either to speed up the simulation or to study large systems using classical mechanics instead.

Despite this simplification, in order to properly study a dynamical system such as a protein, several interactions must still be taken into account and force fields are the "containers" of such information. Usually, force fields used for biomolecular simulations are made up by adding pair-wise interaction terms, commonly divided into bonded (covalent bond-stretching, angle-bending, proper and improper dihedrals), and nonbonded (electrostatic—modeled by Coulomb law—and van der Waals—usually approximated by a 6–12 Lennard-Jones potential) as in Eq. 1.

$$E = \overbrace{E_{bond} + E_{angle} + E_{dihe} + E_{improper}}^{E_b} + \overbrace{E_{elec} + E_{vdW}}^{E_{nb}} \quad (1)$$

This functional form is nowadays adopted by most of the force fields employed in biomolecular simulations [113–117], often taking implicitly into account changes in polarization of the electron density with additional terms [118].

2.3.1 Force Fields for Small General Molecules

While most force fields used today in biomolecular simulations are generally considered as highly reliable [117], the parametrization of generic molecules (drugs, dyes, etc.) remains often a nontrivial task [119], despite the efforts in developing (semi-)automatic parametrization tools (*see*, e.g., [120–122]). Obtaining reliable force fields for general molecules requires the combination of several tools and expertise, from chemical characterization [123] to classical [124] and/or quantum calculations [125] at different levels, as well as chemical, physical, and biological intuition. In our lab, we have experience with the AMBER/GAFF force fields for proteins, nucleic acids and general molecules [113, 115–117, 126–131] and we assume that the reader already has a basic knowledge of the theoretical background on force fields and on how they are implemented into an MD simulation program. Recently, as part of the activity of the TRANSLOCATION consortium within the Innovative Medicines Initiative antimicrobial resistance program, New Drugs for Bad Bugs [132], we have undertaken a long-term project with the goal of building a large database of antimicrobial compounds [130, 131]. The database is freely accessible at http://www.dsf.unica.it/translocation/db/ and contains, for each molecule, all-atom parameters compatible with the AMBER/GAFF force field [115], as well as physicochemical descriptors extracted from quantum-mechanical calculations and classical μs-long MD trajectories.

2.3.2 Force Fields for Phospholipid Membranes

The phospholipid membrane is a crucial component of any molecular simulation of transporters working at the interface between cellular compartments, such as the RND proteins. The hydrophobic effect drives the assembly of phospholipids in a bilayer conformation that represents: (1) a matrix in which membrane proteins or ion channels can be embedded [133], and (2) a barrier regulating the transport to the inside or the outside the cell. The different roles of a membrane in the cell are investigated experimentally by means of X-ray and neutron scattering, IR/Raman, NMR spectroscopy, and computationally by means of MD simulations allowing a description of membrane behavior at atomic-level resolution, although coarse-grained or united-atoms models are also available [126, 128, 134–138]. As for any simulation, the reliability of computational studies dealing with membrane-like environments strongly depends on the accuracy of the force field used. However, while force fields for proteins and nucleic acids have been developed and optimized since long [139–143], obtaining accurate description of membrane dynamics in silico is a relatively recent achievement [144–150]. Among the different force fields developed to mimic phospholipid bilayer dynamics, in this chapter we will refer to Lipid14 [127], which is included in the AMBER15 package [124].

While tutorials about the equilibration of the membrane and automatic protein insertion are available from other sources [151–153], in Subheading 4 we provide an example of by-hand protein insertion into the membrane, taking into account the presence of lipids inside the protein.

2.4 Postprocessing of MD Trajectories

In this section we briefly describe the theory behind the main analyses performed on the trajectories generated by MD simulations of membrane proteins and their complexes.

2.4.1 Analysis of Membrane Structure and Dynamics

The structural properties of the lipid bilayer extracted from the MD simulations provide information about the correctness of the force-field parameters. Experimental values of lipid bilayer structure usually suffer a certain degree of uncertainty [154] but they can be used as reference to validate the procedure adopted during the system set up [147, 155–158]. The first structural indicator that can be easily checked is the surface area occupied by each lipid, or area per lipid (ApL). Other indicators of the membrane stability are the volume per lipid, which can be compared with more accurate measures, and the membrane thickness. Another property that can be checked is the electron density profile, which can be used to calculate scattering form factors [159].

We labeled as *easy* the computation of ApL referring to the case of simulations in which a complete lipid bilayer is involved, which can be performed with several available software [160, 161] or hand-made scripts. The idea behind such calculations is to use the dimensions of the simulation box [127, 160, 162]. However, the insertion of a membrane protein into the lipid bilayer makes the calculation of the ApL (and of the other properties as well) more difficult since the area (or the volume) occupied by the protein must be taken into account. To our knowledge only a few software, like Grid MAT-MD [161], can calculate the ApL of a protein-membrane system, while other approaches can be used to estimate the area occupied by the protein [163–165]. Thus, while the reader should calculate such properties as first check in membrane protein simulations, we strongly suggest to double check the results at every stage. In fact, bad values of ApL (and other properties) suggest either the presence of an issue with the system or a possible error in taking into account the presence of the protein.

2.4.2 Free Energy of Binding: The MM/GBSA Method

The binding free energy of a compound or ligand [B] to the protein or receptor [A] is a thermodynamic quantity describing the affinity between A and B, thus related to the ratio between the concentration [AB] of the complex and the product of those of the partners [A] and [B] [70, 166] as shown in Eq. 2.

$$K_0 = {}^{[AB]}/_{[A][B]} \tag{2}$$

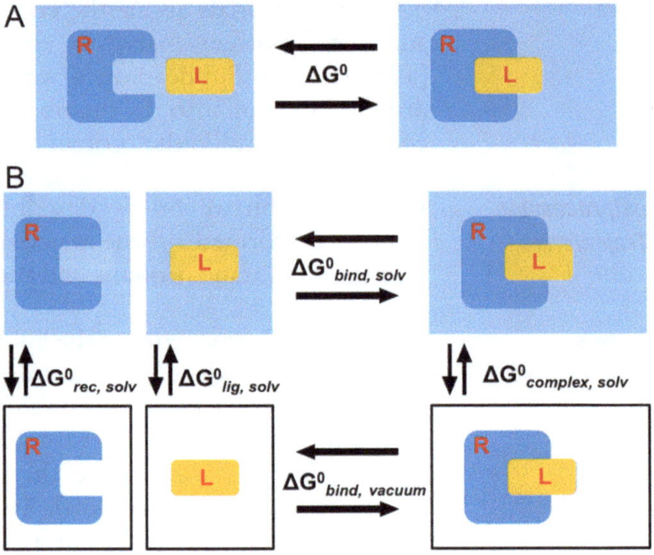

Fig. 2 (**A**) Scheme of the ideal mechanism to calculate the binding free energy for a receptor (R)–ligand (L) system in the solvent, and (**B**) scheme of the cycle used in MM-G(P)BSA

Free energies of binding can be calculated with a plethora of methods [167–171], including some computationally cheap ones characterized by an implicit treatment of solvent, namely the Molecular Mechanics-Generalized Born Surface Area (MM-GBSA) and the MM-Poisson-Boltzmann Surface Area (MM-PBSA) methods [172–175], which have been used extensively by us in order to estimate the affinity of substrates and inhibitors to RND transporters AcrB and MexB [81, 82].

The idea of such methods is to calculate the free energy difference between two states, which we can assume as representatives of the bound ([AB]) and unbound ([A] + [B]) states of the system, by decomposing the total free energy of binding into gas-phase energies, solvation free energies, and entropic contributions [173] (Fig. 2).

In the following we restrict the description of free energy calculations to concepts only, since the reader can easily find detailed descriptions of the methods in the literature [172–175]. In the MM-G(P)BSA framework the free energy of binding for each compound is evaluated as:

$$\Delta G_{\mathrm{E}} = G_{\mathrm{com}} - (G_{\mathrm{rec}} + G_{\mathrm{lig}}) \tag{3}$$

where G_{com}, G_{rec}, and G_{lig} are the absolute free energies of complex, receptor and ligand respectively, averaged over the trajectory at equilibrium of the complex (single trajectory approach). Each term in the equation above can be decomposed as:

$$\Delta G = \Delta E_{\mathrm{MM}} + \Delta G_{\mathrm{solv}} - T\Delta S_{\mathrm{conf}} \quad (4)$$

where the free energy is the sum of the molecular mechanics energy difference ΔE_{MM}, the solvation-free energy difference ΔG_{solv} and the solute conformational entropy difference ΔS_{conf}. The term ΔE_{MM} contains the molecular mechanics energy contribution of bonded (E_{bond}, E_{angle}, E_{torsion}) and nonbonded terms (E_{vdW}, E_{elec}, with no cutoff) estimated from the force field and, thus, can be calculated as:

$$\Delta E_{\mathrm{MM}} = \Delta E_{\mathrm{bond}} + \Delta E_{\mathrm{angle}} + \Delta E_{\mathrm{torsion}} + \Delta E_{\mathrm{vdW}} + \Delta E_{\mathrm{elec}} \quad (5)$$

For the second term of Eq. 4 the solvation-free energy is modeled as sum of an electrostatic [evaluated with MM-G(P)BSA approach] and a nonpolar contribution (proportional to the difference in the solvent-exposed surface area).

$$\Delta G_{\mathrm{solv}} = \Delta G_{\mathrm{solv,p}} + \Delta G_{\mathrm{solv,np}} \quad (6)$$

The last term in Eq. 4 is the solute entropy contribution, which is composed by two terms: the rototranslational contribution, calculated through classical statistical mechanics, and the vibrational term, which can be estimated through normal-mode analysis [176].

The MM-GBSA approach also provides an alternative to the alanine scanning approach [177] by means of a per-residue decomposition of the contributions to the binding free energy. In this case, the energy contribution of a single residue is calculated by summing its interactions over all residues in the system. In Subheading 4 the reader can find an example of the binding free energy computation including the per-residue decomposition.

3 Materials

The protocols described in the following Subheading 4 require the combined use of different software programs briefly described below.

MODELLER. It is an open-source program for homology or comparative modeling of protein three-dimensional structures [98, 100, 103]. The program also provides additional tools such as modeling of loops [178], optimization of various models of protein structure with respect to an objective function, clustering, comparison of protein structures, etc. MODELLER is available for download at www.salilab.org/modeller/.

MolProbity. It is a web service for structure-validation and refinement of proteins, nucleic acids, and complexes. It is available at http://molprobity.biochem.duke.edu.

SAVES. This web service offers a collection of programs to check and validate protein structures, during and after model refinement. It is available at http://services.mbi.ucla.edu/SAVES/.

AutoDock VINA. It is an open-source software designed to perform molecular docking runs [65]. It is available at http://vina.scripps.edu.

AMBER15. AMBER is a commercial suite of programs, designed for MD simulations. The AMBER15 suite can be purchased at http://ambermd.org.

VMD. VMD [179] is an open-source program designed for modeling and visualization, offering a large number of analysis tools, which can be extended by users with the *Tk/Tcl* console. In particular, VMD offers the possibility to build membranes (and to insert protein into them by using the *Tk/Tcl* console), to solvate the system and to study the properties of the lipid bilayer such as membrane thickness, lipid tilt, and area per lipid [152, 153]. VMD is available at http://www.ks.uiuc.edu/Research/vmd.

4 Methods

In this section, we provide examples of the methods and techniques used in our lab to study membrane proteins that belong to the RND superfamily. We assume that the reader has access to the programs described in Subheading 3, and is familiar with the UNIX environment and BASH shell scripting.

4.1 Homology Modeling of Protein Structures

If the structure of a protein has not been resolved yet by experimental means (e.g., for the RND transporters AcrD and MexY), homology modeling is a method allowing to build a 3D model in silico on the basis of the structure of a homologous protein (template). In the following, we describe the steps to be done (*see* Fig. 3) in order to create such homology model structure of a target protein using the software MODELLER [98].

1. *Sequence download*: The very first step in homology modeling is to collect, from any database (e.g., UniProtKB at http://www.uniprot.org/help/uniprotkb) the amino acid sequences needed for the process, i.e., those of the target and of the template(s). While the target sequence can be identified very easily, the selection of a suitable template is not immediate. Programs like the Basic Local Alignment Search Tool (BLAST) [180–182] can be extremely useful to identify the protein with

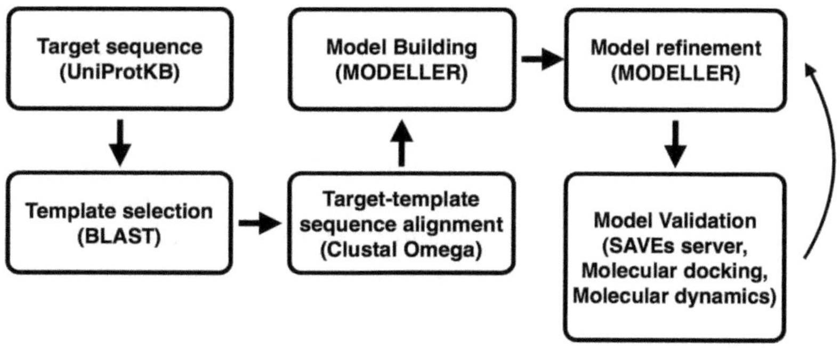

Fig. 3 Typical workflow of a homology modeling run: from target and template sequence selection to alignment, homology modeling, and finally model refinement

the highest sequence identity, maximum query coverage, lowest number of gaps or lowest *E*-value (the number of hits *expected by chance*).

2. *Model building*: Once one (or more) template protein has been identified, its 3D structure can be downloaded in PDB format and homology modeling can be performed. In the following, we provide a simple example on how to build 25 different models for a generic system. More advanced methods can be found in the official manual (https://salilab.org/modeller/manual). The first step is to align the target and the template sequences. To do so, we create an alignment file (alignment.ali) in which we define the sequence and the PDB structure to be used as template.

```
>P1;target_code
sequence:target_name:::::::0.00: 0.00
FASTA sequence of the target
*
>P1;template_code
structureX:template_name::::::0:0
*
```

Subsequently, we run the actual homology modeling that will produce the 3D structure of the target protein by using spatial restraints from the template structure (*see* **Note 1**). The reader can easily complete the scheme properly, according to the system under study, by adding the sequence in FASTA format of the target protein and the name of the template (*see* **Note 2**). The example file structure described above is used to build a homology model for a protein using only one template. However, if two or more useful templates are available, the alignment.ali file can be modified in order to use all these structures during the modeling.

Once the alignment file is ready, there is only one more step to do: build the input file of MODELLER. Several input files are available with different options. Here, we present the basic input, which we call model-example.py, based on model-default.py of the official MODELLER package. The first lines in the input file typically refer to important information and usually they should not be edited. The next lines can be easily changed following the example, in which 25 models of the target (identified by the target_name in alignment.ali) protein will be produced using the template_name.pdb as reference for the 3D structure. Note that the values of the *knowns* and *sequence* variables in model-example.py must be the same as in the alignment file, otherwise MODELLER would not be able to find the information needed for the run. The models are then automatically scored according to DOPE (Discrete Optimized Protein Energy) score [183].

```
...
a = automodel(env,
alnfile = 'alignment.ali',
knowns = 'template_name',
sequence = 'target_name',
assess_methods=assess.DOPE)
a.starting_model= 1
a.ending_model = 25
a.auto_align()
a.make()
```

Finally, the homology modeling can be performed with the command:

```
$ mod9vXX model-example.py
```

where XX identifies the MODELLER version installed.

3. *Model selection and refinement*: Once the homology modeling is finished, 25 files with names like target_name.B999900XX.pdb have been generated each corresponding to a 3D structure of the target. At this stage it is always a good option to visually inspect the structures with a program like VMD (or PyMOL, *see* **Note 3**). In this example we choose the top five models according to the lowest DOPE [183] score and use them in the validation process. During this step we can use web servers like MolProbity and SAVEs to check the correctness of backbone angles, side chains flips, rotamers, steric clashes, etc. However,

depending on the system, the reader might also need to use the refining tool provided by MODELLER itself, such as the loop refinement, and short in-vacuo minimization steps to optimize the structures (*see* **Note 4**).

4.2 Force Field Parameters for Substrates and Inhibitors of RND Transporters

The 3D structure of the ligand of interest can be downloaded from one of the many databases available (https://pubchem.ncbi.nlm.nih.gov; http://www.chemspider.com; http://www.drugbank.ca). For example, the 3D structure data file (SDF format) of a large number of antimicrobial compounds is available in the PubChem database [184]. If not available, the 3D conformation can be generated by sketching the 2D structure and then building the corresponding 3D one using Open Babel [185] or the ChemAxon's Marvin suite of programs [123]. The latter programs can then be used to calculate the protonation/charge state most populated at physiological pH = 7.4, and the corresponding microspecies distribution.

The structure of the major species determined in the step above can be used as an input to perform quantum-chemical calculations. We use the Density Functional Theory level [186, 187] as implemented in the Gaussian09 package [125]. In particular, we adopt the widely used hybrid B3LYP functional, a combination of exact exchange with local and gradient-corrected exchange and correlation terms [188, 189], in conjunction with the 6-31G** Gaussian basis-set [190]. The ground-state optimized structure is obtained employing the Polarizable Continuum Model [191] as to mimic the effect of water solvent particularly to avoid formation of possibly spurious intra-molecular H-bonds. To confirm the geometry obtained to be a global minimum on the potential energy surface a full vibrational analysis is finally executed.

Atomic partial charges are generated by fitting the molecular electrostatic potential calculated on the optimized geometry by means of B3LYP/6-31G** single-point energy calculations in vacuum. Under the constraint of reproducing the overall electric dipole moment of the molecule, we use the so-called Merz-Singh-Kollman Scheme [192] to construct a grid of points around the molecule. Atomic partial charges can then be generated through the two-step restrained electrostatic potential (RESP) method [193] implemented in the Antechamber module of the AMBER package [194].

The General AMBER Force Field (GAFF) parameters [115] determined as detailed above are then used to perform all-atom MD simulations in the presence of explicit water and ions (0.1 M *KCl* solution) using the AMBER15 package [124]. For the purpose of the present Chapter, we will skip the description of how to run simulations of a small molecule in explicit solvent, remanding the reader to well-established online tutorials (*see*, e.g., http://ambermd.org/tutorials/basic/tutorial4b).

4.3 Molecular Docking

In this session we describe our general docking protocol in which ensemble of conformations of a ligand are docked on ensemble of receptor conformations. This section is not meant to demonstrate the use of AutoDock VINA but aims just to illustrate the direct usage of this program as potential tool to dock antibiotics/inhibitors onto RND efflux transporters.

To run a docking experiment from scratch one first needs to prepare the starting structures properly. In particular, missing hydrogens must be added to both the receptor and the ligand. For the receptor it is possible to use among the others the PDB2PQR Server [195] (http://nbcr-222.ucsd.edu/pdb2pqr_2.0.0/); for the ligand, the MarvinSketch ChemAxon program [123] or the PRODRG server [196] can be used. All the waters, solvent molecules and noninteracting ions must be removed from experimental structures, which can be easily done manually or using some visualization tool (VMD, PyMOL, UCSF Chimera, etc.). Finally, all of the available receptor configurations must be aligned. We therefore assume that different configurations of both receptor (rec_i.pdb, $i = 1, 2, \ldots, R$) and ligand (lig_j.pdb, $j = 1,2,\ldots,L$) are available in a standard pdb file format. These files must then be converted to the input file format required by AutoDock VINA, namely pdbqt, that can be seen as an extension of the pdb file format. The whole procedure can be done either using the graphical user interface of AutoDockTools (http://mgltools.scripps.edu/downloads) or by command-line input:

```
$ prepare_receptor4.py -r receptor.pdb -o receptor.pdbqt
$ prepare_ligand4.py -l ligand.pdb -o ligand.pdbqt
```

The pdbqt files are similar to the original pdb files, but contain an additional column reporting the Gasteiger type charges for each atom, and the information on rotatable bonds. Since we will perform a rigid docking using ligand configurations generated externally (e.g., by MD simulations) we will inactivate all active torsions with the -Z option:

```
$ prepare_ligand4.py -Z -l ligand.pdb -o ligand.pdbqt
```

All of the available options of the Python scripts prepare_ligand4.py and prepare_receptor4.py are listed by simply typing these commands without any argument. Before running AutoDock VINA one needs to prepare the docking configuration file, which contains all the parameters needed to perform the actual docking calculation. The available options can be listed simply by typing:

```
$ vina or vina -help
```

A configuration file config.txt can then be edited accordingly, and the program can be run by typing:

```
$ vina -config config.txt
```

A typical configuration file for a docking search in a cubic box 40 Å wide and centered at the origin looks like the following:

```
receptor = rec.pdbqt
ligand = lig.pdbqt
out = docked.pdbqt
log = log.txt
center_x = 0.0
center_y = 0.0
center_z = 0.0
size_x = 40.0
size_y = 40.0
size_z = 40.0
cpu = 12
exhaustiveness = 128
```

Note that receptor file, ligand file and search space (*center_k*, *size_k*, $k = x,y,z$, expressed in Å) are mandatory fields, while *cpu* (number of CPUs to use), *exhaustiveness* (that refers to the *exhaustiveness* of the global search—roughly proportional to the running time—and should be set to the highest value compatible with the hardware and time available), *out* and *log* are optional fields. The output models, i.e., the poses generated during the docking runs, are written in the docked.pdbqt file and the corresponding binding affinities (in *kcal/mol*) are reported in the *log* file log.txt. The maximum number of binding modes to generate can be changed with the *num_modes* keyword (default is 9).

To analyze the results one can load the docked.pdbqt file into PyMOL or UCSF chimera (by instructing the program to open the pdbqt file as a pdb one). In VMD one can load multiple single files obtained through the tool *vina_split available within the VINA package*:

```
$ vina_split -input docked.pdbqt
```

Alternatively, one can convert the docked.pdbqt file into a pdb file. In a Linux-Mac environment the simplest way is to type in the console:

```
$ grep -v BRANCH docked.pdbqt | grep -v ROOT | grep -v TORS >
docked.pdb
```

Overall, assuming the config.txt file in the local directory and the receptor (rec_i.pdb, $i = 1,2,\ldots,R$) and ligand (lig_j.pdb, $j = 1,2,\ldots,L$) files in the rec and lig directories respectively, the following shell script runs the complete ensemble-docking experiment:

```
N_replicas=6
for i in $(ls rec/rec*.pdb)
do
filerec=$(basename ${i})
dir=$(echo $file | awk 'BEGIN{FS="."}{print $1}')
mkdir $dir ; cd $dir
ln -s ../rec/$filerec rec.pdb
prepare_receptor4.py -r rec.pdb -o rec.pdbqt
rm rec.pdb
for j in $(ls ../lig/*.pdb)
do
filelig=$(basename ${j})
lig=$(echo $file | awk 'BEGIN{FS="."}{print $1}')
mkdir $lig ; cd $lig
ln -s ../../lig/$filelig lig.pdb
prepare_ligand4.py -Z -l .lig.pdb -o lig.pdbqt
rm lig.pdb
for k in $(seq 1 1 $N_replicas)
do
mkdir $k ; cd $k
ln -s ../lig.pdbqt ln -s ../rec.pdbqt
ln -s ../../../config.txt
vina -config config.txt rm lig.pdbqt rec.pdbqt config.txt
grep -v -e BRANCH -e ROOT -e TORS docked.pdbqt > docked.pdb
cd ..
done
cd ..
done
cd ..
done
```

Due to the stochastic nature of the algorithm implemented in the program, it is a good practice to set the exhaustiveness parameter as large as possible and/or to repeat a single run a given number of times set by the *$N_replicas* variable.

4.4 MD Simulations

In this section we provide example protocols that can be used to study proteins partially embedded into phospholipid bilayers (such as RND transporters). It is worth pointing out that the reader should always adapt such examples according to his/her needs, since part of the protocols and input strongly depend on the system under study and on the version of the MD simulation packages (as the protocol could change depending on the implementation of some specific features within the MD codes).

4.4.1 Embedding the Protein in a Phospholipid Membrane Model

Selection of a membrane model. The role of the phospholipid bilayer is crucial for the study of proteins that are fully or partially embedded into a membrane [197–200] although, under certain assumptions (see next paragraphs), we could safely neglect the contribution to the protein dynamics from the membrane and/or the protein domains embedded therein [79, 81, 82]. Clearly, using such a reduced models is not always possible and the reader should consider this option very carefully.

Here we describe how to prepare the full protein-membrane system following two different methods. One way is to use the CharmmGUI web server [201, 202] (http://www.charmm-gui.org), from which a complete system including the protein, the membrane and the solvent (water and ions) can be built and downloaded. In this case, the file must be adapted to the AMBER file format since the lipid format is different in the two cases (*see* **Note 5**). Another way is offered by VMD through the Membrane builder plugin [152] (http://sourceforge.net/p/membplugin/wiki/Home) to build a lipid bilayer suitable for the system under study. While the first method will provide the full system with a pre-equilibrated membrane, the second will produce only the membrane, which should be properly equilibrated before protein insertion. A protocol for the membrane equilibration can be found in the official tutorial of the AMBER Lipid14 force field [127] (http://ambermd.org/tutorials/advanced/tutorial16).

Insertion of the protein into the membrane. The insertion of a membrane protein into a lipid bilayer can be performed with different methods. An automatic and easy way is provided by the LAMBADA/InflateGro2 packages [151]. However, the insertion of phospholipids inside proteins, as it occurs within the central cavity of RND transporters, may reveal itself to be a tricky stage also when using (semi)automated protocols. Here we describe an example procedure to insert by hand the protein into a lipid bilayer. Assuming the structure of a pre-equilibrated membrane patch of the proper size is available (file membrane_eq.pdb, see the previous section and *see* **Note 6**) together with a structure of the protein (file solute.pdb, either determined by experimental means or by homology modeling as described in the Subheading 2.1) the first step is to visualize both the protein structure and the membrane with VMD:

```
$ vmd -m solute.pdb membrane_eq.pdb
```

Then, alignment of both the protein and the membrane to a common axis (typically the *z* axis) and subsequent insertion of the protein into the lipid bilayer is required. This process can be performed by hand, if the reader already knows which part of the protein is embedded into the membrane, or using tools that automatically accommodate the protein in the membrane by optimizing the hydrophobic patch in the transmembrane (TM) domain and the lipid tails [151, 203–205]. Assuming the reader aims to insert the protein in the membrane by hand, the next step is to superimpose the two structures in order to get the TM component of the protein fully buried in the membrane. Of course, this simple structural alignment will produce several overlaps and steric clashes. However, we can neglect this issue at this stage since we are just in the process of building the starting structure of the system, therefore some physical inconsistencies are unavoidable.

Once the superposition is done and the user is reasonably sure that the TM region of the protein is optimally buried in the membrane, it is possible to save the new coordinates of the membrane and the protein into two temporary files (e.g., tmp_solute.pdb and tmp_membrane.pdb) and, subsequently, merge them into a unique PDB file (e.g., tmp_solute_membrane.pdb) and visualize this final structure with VMD. The user should immediately see the physical inconsistencies described above, which we describe now how to deal with. Before working on the structure with the *Tk/Tcl* Console, it would be better to do some visualization with the Graphics Representation tool in Graphics menu.

By typing:

```
noh and (same residue as within 10 of protein) and (not protein)
```

we delete all the phospholipid molecules having at least one atom within 10 Å of the protein. This cutoff is deliberately large and the reader should thus reduce it to the optimal value for the system under study. During this step the reader should check not only the lipids surrounding the protein but, depending on the system, also the lipids located in the central cavity (*see* Fig. 4). This is the case for AcrB, in which a certain number of lipids are supposed to reside inside a central cavity lined by the TM domains of the three monomers constituting the protein [30]. Therefore, the adjustment of the cutoff should not remove such elements. Once the phospholipids overlapping with the protein have been removed we can save the new structure as solute_membrane.pdb with the *Tk/Tcl* Console (*see* **Note 7**).

This new file still contains physical inconsistencies due to a vacuum layer between the protein and the membrane (with the possible exception of the lipid inside the protein in the transmembrane domain), which will be removed following the procedure outlined in the next section.

Structural relaxation of the system. Once the TM part of the protein has been inserted into the phospholipid bilayer, the next

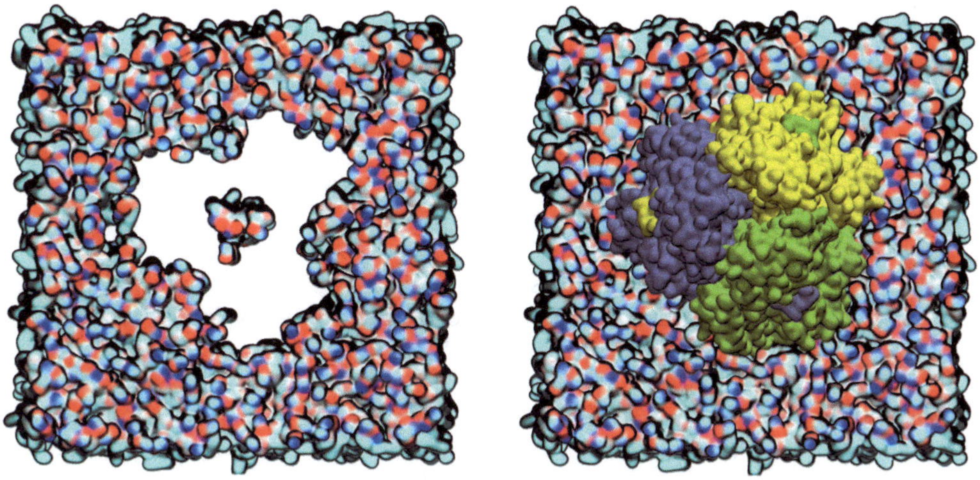

Fig. 4 Protein insertion into a pre-equilibrated membrane, by removing the lipids that are within a certain distance of protein and then inserting the protein [in this example AcrB, PDB code 2J8S, was used together with a bilayer of 1-Palmitoyl-2-oleoyl-sn-glycero-3-phosphoethanolamine (POPE)]. The protein and the phospholipid bilayer are both shown with the QuickSurf option of VMD; the three monomers of AcrB are colored *yellow*, *green*, and *iceblue*, while the bilayer is colored according to atom names

step is to perform a structural relaxation of the system in vacuo before adding water and ions (*see* **Note 8**). This allows optimizing the interactions at the interface between the protein and the membrane, so that possible large voids therein could be removed. The protocol we generally follow at this stage is to run different consecutive steps of structural optimization of the system, applying different restraints to the protein and to the phospholipids. The rationale is to keep the protein frozen in its original conformation, while allowing phospholipids to stick onto the protein surface. Therefore, protein Cα atoms are kept fixed (using the *ibelly* flag in *sander/pmemd*), while partial restraints are applied on the atoms of the membrane, usually only to the z component of the position of P atoms. This avoids large movements of the lipid heads along the z-axis and perhaps the disruption of the membrane (indeed, the stabilizing hydrophilic environment—water solution—is absent on both leaflets), while allowing movements within the xy plane. Unfortunately, at the time of writing, AMBER15 did not include a single axis restraint yet, and thus the reader should edit the code or seek for another program that includes such kind of restraints (*see* **Note 9**).

Assuming the protein has been already optimized in vacuo, the modified CPU version of *pmemd* can be used (indeed the GPU version does not support the option *ibelly* $\neq 0$ yet) to allow the lipid heads to fill the empty space between the membrane and the protein. An input example for such steps is provided in the following. It is worthwhile stressing that the number of steps in the example below is just representative, since the speed at which the lipids fill the empty space is strongly system-dependent.

```
#Filling the empty space with z-restraints on lipids head
&cntrl
imin = 1, maxcyc = 10000, ncyc = 5000,
ntb = 1, ntp = 0, ntf = 1, ntc = 1, igb = 0, cut = 12,
ntpr = 100, ntwx = 250, ntwr = 250, ig = -1,
ntr=1, restraint_wt = 10.0, restraintmask = '@P31',
ibelly=1, bellymask = '!@CA'/
```

During this stage the user is strongly suggested to check the evolution of membrane properties, in order to identify possible issues as described in Subheading 2.4.1.

4.4.2 Reduced Model of Protein

As the recognition of substrates by the giant RND transporters is thought to occur predominantly in the large periplasmic loops of the protein [36], it could be convenient to use a reduced model of the protein, generated by removing both the trans-membrane domain of the protein and the lipid bilayer [81]. This step can be easily achieved by editing the PDB file and by removing all the residues belonging to the trans-membrane domain. Our group already published a few studies using such a reduced model of AcrB [39, 79, 81, 82] for which extensive validation was performed [81], and we will not detail this step further. Nonetheless, considering the solute and the solvent (see next section), the truncation of AcrB allows greatly reducing the size of the system (from ~450,000 to ~132,000 atoms).

4.4.3 Adding Solvent to the System

Either the user prefers adopting a full model of the RND transporter embedded in phospholipid bilayer or a truncated one, the next step is to insert this "solute" in a solvent (water and ions), and finally to generate the initial structure and the topology files needed to perform MD simulations. We generally perform simulations in 0.1 M *KCl* water solution [81], using a TIP3P model of water [206], together with the parameters refined by Joung and Cheatham for monovalent ions [207]. Assume that we have generated the AMBER *prep* and *frcmod* files for the ligand (named ligand.prep and ligand.frcmod respectively, the first containing the indication "LIG" for the residue name of the ligand; see section above), and we have a structure file containing the proper geometry of the protein, the ligand and the membrane (in case we do not use a truncated model of the protein), which we name solute.pdb. The following bash script is an example of how to generate, starting from the above files, all the needed files automatically, and in particular the files solv_system.parm7, solv_system.rst7 and solv_system.pdb which will be used to perform MD simulations with *pmemd*, supposing that 35 *Cl-* ions are needed to reach a 0.1 M concentration of *KCl* in water (the rough number of water molecules in the box should be known). Before loading the system into the program *leap* (a tool of the AmberTools package [208]), its dimensions must be determined in order to set the proper periodic boundary conditions while generating the simulation box (vide infra). This can be done through VMD by typing the command:

```
$ vmd -dispdev none solute.pdb -e measure_box_and_center.tcl
```

where measure_box_and_center.tcl is a *Tcl* script measuring dimensions and center of the system and writing them into a file called box_dimension.dat:

```
set out [open ./box_dimension.dat w]; set all [atomselect top all]
puts $out "# BOX SIZE X Y Z ; center coords"
set x_min [lindex [lindex [measure minmax $all] 0] 0]
set y_min [lindex [lindex [measure minmax $all] 0] 1]
...
set x_max [lindex [lindex [measure minmax $all] 1] 0]
...
set x_box [expr $x_max - $x_min]
...
set center [measure center $all]
puts $out "$x_box $y_box $z_box $center"
close $out
quit
```

In the following example, the force fields ff14SB, GAFF and Lipid14 [115, 127, 129], will be used with the whole model of the protein by typing:

```
$ source ./solvate_leap.sh LIG ligand
```

where the solvate_leap.sh script contains the following instructions:

```
drug=$1; drugfiles=$2; nions1=35; offset=20; cutxy=10
if [ -f "leap.log" ]; then
rm leap.log
fi
#step1: check charge of complex before adding counterions and ions
cat<<EOF>charge.leap loadamberprep ${drugfiles}.prep
loadamberparams ${drugfiles}.frcmod
com=loadpdb solute.pdb
charge com
saveoff com solute.off
saveamberparm com solute.parm7 solute.rst7
quit
EOF
${AMBERHOME}/bin/tleap -s -f \
${AMBERHOME}/dat/leap/cmd/leaprc.ff14SB -f
${AMBERHOME}/dat/leap/cmd/leaprc.gaff -f \
${AMBERHOME}/dat/leap/cmd/leaprc.lipid14 -f charge.leap
```

```
ambpdb -p solute.parm7 < solute.rst7 > solute.pdb
#step2: only with membrane-containing system, change the box in
solute
boxfile="box_dimension.dat"
x=$(tail -n 1 ${boxfile} | awk '{printf"%.6f", $2}')
y=$(tail -n 1 ${boxfile} | awk '{printf"%.6f", $3}')
z=$(tail -n 1 ${boxfile} | awk '{printf"%.6f", $4}')
xp=$(tail -n 1 ${boxfile} | awk '{printf"%.6f", $2-'${cutxy}'}')
yp=$(tail -n 1 ${boxfile} | awk '{printf"%.6f", $3-'${cutxy}'}')
dz=$(echo "(2*${z}*(${yp}*(${x}-${xp})+${xp}*(${y}-\ ${yp})))/
(${xp}*${yp})" | bc -l)
zp=$(tail -n 1 ${boxfile} | awk '{printf"%.6f",'${dz}'+'${off-
set}'}')
cat<<EOF>changebox.leap
loadoff solute.off
set com box {${xp} ${yp} ${zp}}
saveoff com solute.off
quit
EOF
${AMBERHOME}/bin/tleap -s -f\
${AMBERHOME}dat/leap/cmd/leaprc.ff14SB -f\
${AMBERHOME}dat/leap/cmd/leaprc.gaff -f\
${AMBERHOME}dat/leap/cmd/leaprc.lipid14 -f charge.leap
#step3: create topology and structure files
charge=$(grep "Total unperturbed charge" leap.log | \
awk '{printf"%5d",$4}')
charge_abs=$(echo "sqrt($charge*$charge)" | bc)
nions2=$(($charge_abs+$nions1))
if [ $charge -lt 0 ]
then cat<<EOF>solvate_TIP3P.leap
loadamberparams ${AMBERHOME}dat/leap/parm/frcmod.ionsjc_tip3p
loadamberprep ${drugfiles}.prep
loadamberparams ${drugfiles}.frcmod
loadoff solute.off
com_solv=copy com
solvatebox com_solv TIP3PBOX 0
# Use "solvateoct com_solv TIP3PBOX $(209)" with the reduced model.
This will generate a truncated octahedral box.
addionsrand com_solv K+ ${nions2} Cl- ${nions1}
saveoff com_solv solv_system.off
saveamberparm com_solv solv_system.parm7 solv_system.rst7
quit
EOF
${AMBERHOME}/bin/tleap -s -f\ ${AMBERHOME}dat/leap/cmd/leaprc.
ff14SB -f\ ${AMBERHOME}dat/leap/cmd/leaprc.gaff -f\ ${AMBERHOME}
dat/leap/cmd/leaprc.lipid14 -f solvate_TIP3P.leap
else
```

```
cat<<EOF>solvate_TIP3P.leap
loadamberparams ${AMBERHOME}dat/leap/parm/frcmod.ionsjc_tip3p
loadamberprep ${drugfiles}.prep
loadamberparams ${drugfiles}.frcmod
loadoff solute.off
com_solv=copy com
solvatebox com_solv TIP3PBOX 0
# Use "solvateoct com_solv TIP3PBOX $(209)" with the reduced model.
addionsrand com_solv K+ ${nions1} Cl- ${nions2}
saveoff com_solv solv_system.off
saveamberparm com_solv solv_system.parm7 solv_system.rst7
quit
EOF
${AMBERHOME}/bin/tleap -s -f\
${AMBERHOME}dat/leap/cmd/leaprc.ff14SB -f\
${AMBERHOME}dat/leap/cmd/leaprc.gaff -f\
${AMBERHOME}dat/leap/cmd/leaprc.lipid14 -f solvate_TIP3P.leap
fi
ambpdb -p solv_system.parm7 < solv_system.rst7 > solv_system.pdb
```

As can be seen from **step 2** in the script above, the box information in the file solute.off has been modified in order to shrink the dimensions along the *xy* plane while recovering the "lost" volume on the *z* direction (Fig. 5). This step is necessary because the

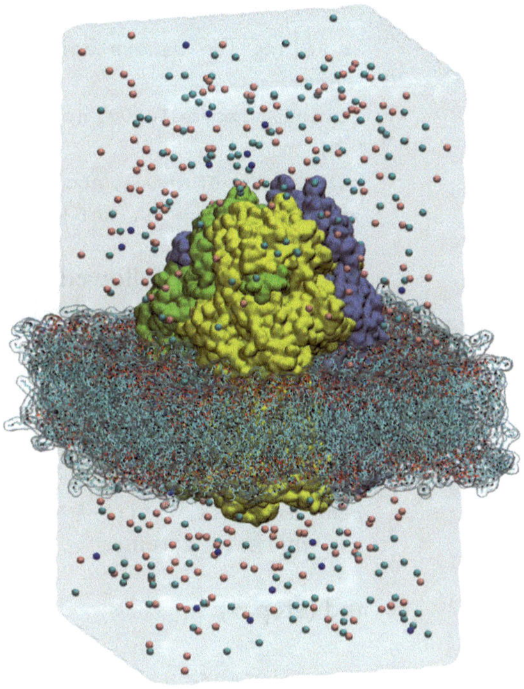

Fig. 5 Example of solvated system in which the water molecules close to the borders of the lipid bilayer have been removed.

solvatebox command within *tleap* fills with water the box as it is found within the off file. Thus, if the box is enclosing membrane in the *xy* plane, several water molecules will be placed at the interface between borders of the membrane in adjacent periodic cells, which is undesirable as the time needed to reach equilibrium from this conformation can be very large (*see* **Note 10**). An output example of the above mentioned procedure is shown in Fig. 5.

4.4.4 Running MD Simulations

Once a reliable topology file and an initial structure are available for the system of interest, we are ready to perform MD simulations to study its behavior. In the following we will describe only the protocol followed to perform unbiased MD simulations with AMBER [124], but the procedure can be easily adapted to other programs, such as NAMD [210] or GROMACS [211].

Structural optimization and system equilibration. First, it is necessary to allow a relaxation of possible steric clashes between the solvent and the solute. This is generally done by applying soft restraints on all heavy atoms of the solute, thus leaving the solvent free to rearrange. In the following example input file of the programs *pmemd* or *sander* (the main MD engines within the AMBER15 package), which we name relax1.in, 10,000 cycles of structural optimization are performed (1000 via the steepest descent algorithm, the remaining through the conjugate gradient one) while applying restraints with a force constant of 1 kcal/mol/Å^2 (**step 1**):

```
# Relaxation waters and ions
&cntrl
imin=1, maxcyc=10000, ncyc=1000, cut=12.0, ntr=1, restraint_wt =
1.0,
ntb=0, ntp=0, ntf=1, ntc=1, ntpr=1000, ntwr=1000,
restraintmask = "!(:WAT | :K+ | :Cl-) & !@H="/
```

This step is followed by two additional ones, whose *pmemd* inputs are relax2.in and relax3.in. Restraints are applied only on backbone and Cα atoms of the protein respectively, and on all the nonhydrogenous atoms of the ligand and of the membrane. In this case, the last line of the input file for *pmemd* must be changed as (**step 2**):

```
restraintmask = "(:1-#NRESPROT@CA, C, O, N) | (:#NRESFIRSTLI-
PID-#NRESLIG & !@H=)
```

and (**step 3**):

```
restraintmask = "(:1-#NRESPROT@CA) | (:#NRESFIRSTLIPID-
#NRESLIG & !@H=)"
```

where #NRESPROT is the resid number of the last residue in the protein, and #NRESFIRSTLIPID and #NRESLIG are the resid numbers of the first phospholipid and of the ligand respectively (*see* **Note 11**).

Finally, in order to optimize the specific interactions established between each ligand and its putative binding site before switching on the thermostat and the barostat in the heating and box equilibration phases, we perform a further optimization (**step 4**) allowing only residues of the protein within a given cutoff from the ligand (usually 7 Å) to fully relax [81]. Residues within this selection can be easily identified through *Tcl* scripting within VMD, after loading the file solv_system.pdb (assuming that residues identifying phospholipids are called POP) as follows:

```
set cut 7
set selfree [atomselect top "protein and same residue as within
$(cut) of (not (protein or water or resname POP "Cl.*" "K.
*"))"]
set selfreelist [lsort -integer -uniq [$selfree get resid]]
set out [open "resid_free.dat" w]
puts $out $selfreelist
close $out
quit
```

After rearranging the file resid_free.dat to make the string shorter than 256 characters (mask strings are currently limited to this number of characters), we can substitute the right line into the file relax1.in and get something like the following (relax4.in):

```
restraintmask = "!(:667,719-722,731-735,756-760,852-861,866-
870,904-907,960-972,978-982) & @CA"
```

Relaxation of the system can be finally performed by running the various steps described above (either with *pmemd* or *sander*):

```
$ pmemd -O -i relax1.in -o relax1.out -p solv_system.parm7 -c
solv_system.rst7 -ref solv_system.rst7 -r relax1.rst7
$ pmemd -O -i relax2.in -o relax2.out -p solv_system.parm7 -c
relax1.rst7 -ref relax1.rst7 -r relax2.rst7
$ pmemd -O -i relax3.in -o relax3.out -p solv_system.parm7 -c
relax2.rst7 -ref relax2.rst7 -r relax3.rst7
$ pmemd -O -i relax4.in -o relax4.out -p solv_system.parm7 -c
relax3.rst7 -ref relax3.rst7 -r relax4.rst7
```

Once the system has been relaxed, it is heated to the desired temperature. In this example (**step 5**, file heat.in), we perform heating by linearly increasing the temperature of the system from 0 to 310 K in 2 ns of simulation time and using the NTP ensemble with anisotropic pressure coupling (*see* http://ambermd.org/tutorials/advanced/tutorial and **Note 12**), while keeping the Cα atoms of the protein and the P atoms of the membrane restrained (this is strongly suggested at this stage, since the wrong initial density of water—usually lower than 1 g/cm^3—usually leads to the appearance of artificial vacuum bubbles at the corners of the box):

```
#Heating to 310K in the NTP ensemble and with restraints on
protein and membrane
&cntrl
imin = 0, irest = 0, ntx = 1, ntb = 1, cut = 12.0, ig = -1, ntf
= 2, ntc = 2,
ntp = 2, taup = 5.0, ntt = 3, gamma_ln = 1.0, nstlim =
1000000,
ntpr = 10000, ntwx = 10000, ntwr = 10000, ntwe = 10000,
dt = 0.002, ioutfm=1, ntr = 1, nmropt=1
restraint_wt = 10.0, restraintmask = '@CA | @P31',
/
&wt type='TEMP0', istep1=0, istep2=1000000, value1=0.0,
value2=310.0
/
```

While is almost guaranteed that this step will bring the system to the right temperature, it is quite likely that the structure of the solvent and of the membrane need to be further equilibrated before productive MD can be performed (this is particularly true for medium/large-sized systems such as those considered here). Therefore, starting from the final conformation generated at **step 5**, we perform a few further steps of equilibration, by gradually relieving the restraints from both the protein and the membrane.

Namely, in **step 6** (file equil1.in) a 1 ns long plain MD run is performed where the restraints are softened (note the change of the *irest* and *ntx* parameters, needed to read also the velocities from the output of the previous step):

```
irest = 1, ntx = 5,
restraint_wt = 1.0, restraintmask = '@CA | @P31'
```

Then, restraints are removed from the membrane (**step 7**, 2 ns, file equil2.in):

```
restraint_wt = 1.0, restraintmask = '@CA'
```

and finally from every atom in the system (**step 8**, 10 ns, file equil3.in). Clearly, the user should visually inspect the output of all the steps above for convergence and structural integrity of all the macromolecules in the system. Eventually, if convergence is not reached in some stage, the user should change the timings of the corresponding run according to his/her needs. Equilibration of the system can be finally performed by running consecutively **steps 5–8**:

```
$ pmemd -O -i heat.in -o heat.out -p solv_system.parm7 -c
relax4.rst7 -ref relax4.rst7 -r heat.rst7
...
```

Productive MD. Once the system is reasonably equilibrated and structural parameters of the protein/ligand complex and of the membrane seem to be converged, the user can finally run productive dynamics (**step 9**, file md.in). Basically the input is the same as for **step 8**, with changes in the parameters *taup* from 5.0 to 1.0 and *gamma_ln* from 1.0 to 5.0. *See* **Note 13** about the use of a 4 fs timestep in AMBER15 and **Notes 14** and **15** about productive dynamics of protein/ligand complexes.

Notes on the MPI and GPU accelerated codes. The AMBER packages comes with MPI support for both *pmemd* and *sander* (*pmemd.MPI* and *sander.MPI* binaries) and with GPU support for pmemd (*pmemd.cuda* and *pmemd.cuda.MPI*). If the parallel version has been installed, the user could consider using the mpirun command:

```
$ mpirun -np #PROCS pmemd.MPI (or sander.MPI, pmemd.cuda.MPI)
...
```

If the GPU/MPI version of *pmemd* is used, the user should specify the GPUs to be used:

```
$ export CUDA_VISIBLE_DEVICES=ID1, ID2, ID3, ...
$ mpirun -np #GPUs pmemd.cuda.MPI ...
```

Here the first line sets the GPUs to be used, so that 0, 2, 3 will tell *pmemd.cuda.MPI* to run on the GPUs number 0, 2, and 3 (*see* **Note 16**). If the user decides to use the GPU code, it is necessary to increase the *skinnb* value during the equilibration phase with periodic boundary condition and to restart the MD simulation after a shorter time, typically 0.5 ns. This is due to the fact that box dimensions are changing, but for performance reasons, the GPU code does not recalculate the nonbond list cells during a simulation, and thus the code could halt with an error related to the *skinnb* parameter. Thus, in **steps 6–8**, one should add the following line at the end of each input file:

```
/&ewald skinnb=5, ! Increase skinnb to avoid skinnb errors/
```

and split all of the single steps in shorter ones (0.5 ns each accordingly to the Lipid14 tutorial). Once the system is equilibrated the box size fluctuations should be small and so this should not be an issue during production.

4.4.5 Post-processing Analysis

In this section we describe how to process the results of the simulations focusing on the computation of the free energies of binding of ligands to RND transporters using the MM/GBSA method. More basic analyses can be easily performed with the plethora of tools available, e.g., with the *ptraj* or *cpptraj* tools of AMBER (see, e.g., the Tutorial at this webpage: http://ambermd.org/tutorials/advanced/tutorial16).

4.4.6 MM/GBSA

The MM/GBSA method has been implemented in several ways within the AMBER15 package (*see* Subheading 2 and the AMBER manual for a detailed description of the method [124]). The simplest way in our opinion is to use the *MMPBSA.py* python script [212], which we describe here in the framework of the "single trajectory" approach. Three topology files are required, one for the entire complex, one for the ligand and one for the receptor (with the addition of the topology for the solvate complex, if the MD simulation is performed in explicit water). Clearly the above-mentioned topologies must be cross-compatible, i.e., they must have the same charges for the same atoms, they must use the same force field and they must have the same Bondi radii (PBRadii) [213–216]. There are several ways of setting these radii (e.g., through *leap*, *parmed*, or *ante-MMPBSA.py* tools of AMBER). The following one-line command gives an example of how to build the topologies for the complex, the ligand and the receptor with *ante-MMPBSA.py*, setting the PBradii to mbondi3 [214, 215], which must be used in conjunction with the Generalized Born model developed by Mongan et al. [216] (set through the flag *igb = 8* in the input of *MMPBSA.py*, vide infra):

```
$ ante-MMPBSA.py -p solv_system.parm7 -c com_mbondi3.parm7
-r rec_mbondi3.parm7 -l lig_mbondi3.parm7
-s ":#FIRST-NON-SOLUTE-RESIDUE-10000000"
-m ":#FIRST-REC-RESIDUE-#LAST-REC-RESIDUE"
```

It is important to set to the proper values to be used in conjunction with the Generalized Born model adopted in the calculation, and the above scripts can be modified to adapt the Bondi radii to the all the GB models available in AMBER. After these preliminary steps, the actual MM/GBSA calculation can be performed following the simple input example reported below (mmpbsa.in):

```
#MMPBSA GB calculations
&general
startframe=1, interval=1, use_sander=1/
&gb
igb=8, saltcon=0.1/
&decomp
idecomp=2, dec_verbose=3, csv_format=0, print_res="X,Y-
Z,..."/
```

which is run through the command:

```
$ MMPBSA.py -O -i mmpbsa.in -o mmpbsa.out -cp com_mbondi3.
parm7
-rp rec_mbondi3.parm7 -lp lig_mbondi3.parm7 -y trajectory
```

Here the file trajectory is the trajectory of the solute (that is containing the protein, the ligand and possibly the membrane atoms).

In the input file, the flag *saltcon* indicates the salt concentration of the system in *mol/l*. The section *&decomp* indicates that the free energy of binding will be decomposed on a per-residue basis, with residues whose contribution to the binding will be indicated by the flag *print_res* = "X,Y-Z,...". Also an MPI version of the code, named *MMPBSA.py.MPI*, is available. The reader is encouraged to check in the manual the advanced options to be used for a more detailed computation.

5 Notes

1. Since MODELLER uses spatial restraints from the template structure, the target and the template sequences should have the same length. While this is not strictly required, in case of target sequence longer than the template one, the homology modeling might lead to randomly modeled sections.
2. The empty space ":::" in the alignment.ali file needs to be completed according to the reader needs. As example, we can define the initial/final residue to be considered in the template structure, the chain(s), and other properties:

```
structureX:template_pdb_name:FIRST:A:LAST:A:::0:0
```

3. To visualize all of the structures at once, simply use the command:

```
$ vmd -m template_name.pdb target_name.BXX{1..25}.pdb
```

4. The minimization in vacuo can be performed with the *sander* program in the AMBER15 package. Examples of minimization steps are provided in Subheading 4.4.1.
5. The reader can use any available script to rename the lipids or write his/her own script. However, an easy way to adapt the format is to use the *charmmlipid2amber.py* script, provided together with the AmberTools package.
6. The reader should set the membrane size large enough so that the images of the protein, introduced with the boundary conditions, would not interact with the protein in the original box.

A reasonable safe distance between the protein images is around 30 Å and thus the membrane size in the xy plane would be 30 Å plus the approximate diameter of the protein.

7. Open the *Tk* Console in the Extension menu of VMD and type:

```
set membrane_hole [atomselect top "noh and (same residue
as within 10 of protein) and not protein]
$membrane_hole writepdb memb_hole.pdb
set protein [atomselect top "noh and protein"]
$protein writepdb protein.pdb
```

Once the two files are generated we can merge them:

```
$ cat protein.pdb memb_hole.pdb > system.pdb
```

Now, open the file system.pdb and make sure that the file format is fine. In particular, check if more than one "END" is present. If so, remove the extra "END".

8. This step is necessary to reduce the optimization time. In fact, water molecules accidentally inserted in the empty space between the protein and the lipid bilayer would be trapped into the membrane and their release would require an increase of computational time.

9. As first step create a new AMBER15 folder in which we will change the code.

```
$ cp $path/AMBER15 $path/AMBER15_z
```

Subsequently, we edit the following file:

```
$ emacs $path/AMBER15_z/src/pmemd/src/constraints.F90
```

and simply change line 249 and 250 as follows:

```
wx = 0.d0
wy = 0.d0
```

Finally recompile the code. If everything goes well, the reader will be able to use a modified CPU version of *pmemd* with restraints that act only in the *z* direction.

10. The user is strongly suggested to visually inspect for the presence of water molecules within the membrane, and change accordingly the values of the *cutxy* parameter in the script solvate_leap.sh so as to remove all the unwanted waters. Another way of proceeding is to simply use the dimensions

x and *y* of the box as they come from the measure_box_and_center.tcl script, adding the offset along the *z* direction—*solvatebox com_solv TIP3PBOX {0 0 ${offset}}* in *leap*. Then, unwanted water molecules can be simply removed, e.g., by loading the file solv_system.pdb into VMD, calculate the average *z* position of P atoms in the inner and outer leaflet of the membrane and then remove all waters whose value of the *z* component out of the $[-z_{in}; z_{out}]$ interval. This edited file can be then saved as pdb and loaded back in *leap*, from where a new solv_system.off file can be generated. This file needs in turn to be edited in order to insert the right box information—which is not saved properly by *leap*. Finally, the modified solv_system.off file is loaded again into *leap* and the topology and coordinate files are saved.

11. We suppose here that the topology file has been generated so that the listing of residues proceeds sequentially from the protein to the phospholipids to the ligand.
12. In cases where the barostat is thought to induce instability in the system, the user can consider to run this step in the NVT ensemble, and to continue then with the next **step 6**.
13. The AMBER15 version allows the users to use a 4 fs time step, for which it is also needed to reweight all the masses in the system, according to the AMBER15 manual. This last step can be performed through the *HMassRepartition* command within the tool *parmed.py* included in the AMBER15 suite.
14. Running MD simulations for more than the best docking pose - possibly up to the top five ones—is encouraged, as it is known that the scoring functions in docking are approximated [217].
15. Since there is no absolute method to assess the convergence of MD simulations, it is a good idea to perform, for each pose, several production runs using different initial velocities [218]. This is of course time consuming, and it should also be noted that convergence in MD simulations of even smaller proteins than those investigated here has proven to be very difficult [219], thus it is a good practice to extend the simulation time of each trajectory as much as the user can afford.
16. The tool NVIDIA system management interface (nvidia-smi) can be used to check the state of each NVIDIA GPU card.

Acknowledgment

The research leading to the results discussed here was partly conducted as part of the Translocation Consortium (http://www.translocation.eu/) and has received support from the Innovative Medicines Joint Undertaking under Grant Agreement no. 115525, resources, which are composed of financial contribution from the European Union's Seventh Framework Programme (FP7/2007-2013) and EFPIA companies in kind contribution. VKR is a Marie Skłodowska-Curie fellow within the "Translocation" Network, project no. 607694.

References

1. Antimicrobial resistance: global report on surveillance (2014) World Health Organization
2. Livermore DM (2004) The need for new antibiotics. Clin Microbiol Infect 10:1–9
3. Bush K, Courvalin P, Dantas G, Davies J, Eisenstein B, Huovinen P, Jacoby GA, Kishony R, Kreiswirth BN, Kutter E, Lerner SA, Levy S, Lewis K, Lomovskaya O, Miller JH, Mobashery S, Piddock LJV, Projan S, Thomas CM, Tomasz A, Tulkens PM, Walsh TR, Watson JD, Witkowski J, Witte W, Wright G, Yeh P, Zgurskaya HI (2011) Tackling antibiotic resistance. Nat Rev Microbiol 9:894–896
4. Bassetti M, Merelli M, Temperoni C, Astilean A (2013) New antibiotics for bad bugs: where are we? Ann Clin Microbiol Antimicrob 12:22
5. Rex JH, Goldberger M, Eisenstein BI, Harney C (2014) The evolution of the regulatory framework for antibacterial agents. Ann N Y Acad Sci 1323:11–21
6. Taubes G (2008) The bacteria fight back. Science 321:356–361
7. Pitout JDD (2010) The latest threat in the war on antimicrobial resistance. Lancet Infect Dis 10(9):578
8. Poole K, Krebes K, Mcnally C, Neshat S (1993) Multiple antibiotic-resistance in pseudomonas-aeruginosa – evidence for involvement of an efflux operon. J Bacteriol 175:7363–7372
9. Li XM, Zolli-Juran M, Cechetto JD, Daigle DM, Wright GD, Brown ED (2004) Multicopy suppressors for novel antibacterial compounds reveal targets and drug efflux susceptibility. Chem Biol 11:1423–1430
10. Piddock LJV (2006) Clinically relevant chromosomally encoded multidrug resistance efflux pumps in bacteria. Clin Microbiol Rev 19:382–402
11. Nikaido H (2009) Multidrug resistance in bacteria. Annu Rev Biochem 78:119–146
12. Poole K (2011) Pseudomonas aeruginosa: resistance to the max. Front Microbiol 2:65. https://doi.org/10.3389/fmicb.2011.00065.
13. Nikaido H, Pagès J-M (2012) Broad-specificity efflux pumps and their role in multidrug resistance of Gram-negative bacteria. FEMS Microbiol Rev 36:340–363
14. Schweizer HP (2012) Understanding efflux in Gram-negative bacteria: opportunities for drug discovery. Expert Opin Drug Discov 7:633–642
15. Ruggerone P, Murakami S, Pos KM, Vargiu AV (2013) RND efflux pumps: structural information translated into function and inhibition mechanisms. Curr Top Med Chem 13:3079–3100
16. Blair JMA, Richmond GE, Piddock LJV (2014) Multidrug efflux pumps in Gram-negative bacteria and their role in antibiotic resistance. Future Microbiol 9:1165–1177
17. Li X-Z, Plésiat P, Nikaido H (2015) The challenge of efflux-mediated antibiotic resistance in Gram-negative bacteria. Clin Microbiol Rev 28:337–418
18. Dinh T, Paulsen IT, Saier MH (1994) A family of extracytoplasmic proteins that allow transport of large molecules across the outer membranes of Gram-negative bacteria. J Bacteriol 176:3825–3831
19. Nikaido H (1996) Multidrug efflux pumps of gram-negative bacteria. J Bacteriol 178:5853–5859
20. Tikhonova EB, Zgurskaya HI (2004) AcrA, AcrB, and TolC of Escherichia coli form a stable intermembrane multidrug efflux complex. J Biol Chem 279:32116–32124

21. Lobedanz S, Bokma E, Symmons MF, Koronakis E, Hughes C, Koronakis V (2007) A periplasmic coiled-coil interface underlying TolC recruitment and the assembly of bacterial drug eff lux pumps. Proc Natl Acad Sci U S A 104:4612–4617
22. Symmons MF, Bokma E, Koronakis E, Hughes C, Koronakis V (2009) The assembled structure of a complete tripartite bacterial multidrug efflux pump. Proc Natl Acad Sci U S A 106:7173–7178
23. Pos KM (2009) Trinity revealed: stoichiometric complex assembly of a bacterial multidrug efflux pump. Proc Natl Acad Sci U S A 106:6893–6894
24. Su CC, Long F, Zimmermann MT, Rajashankar KR, Jernigan RL, Yu EW (2011) Crystal structure of the CusBA heavy-metal efflux complex of Escherichia coli. Nature 470:558–562
25. Du D, Wang Z, James NR, Voss JE, Klimont E, Ohene-Agyei T, Venter H, Chiu W, Luisi BF (2014) Structure of the AcrAB-TolC multidrug efflux pump. Nature 509:512–515
26. Jin-Sik K, Hyeongseop J, Saemee S, Hye-Yeon K, Kangseok L, Jaekyung H, And Nam-Chul H (2015) Structure of the tripartite multidrug efflux pump AcrAB-TolC suggests an alternative assembly mode. Mol Cells 38:180–186
27. Zgurskaya HI, Weeks JW, Ntreh AT, Nickels LM, Wolloscheck D (2015) Mechanism of coupling drug transport reactions located in two different membranes. Front Microbiol 6:100
28. Du D, Van Veen HW, Luisi BF (2015) Assembly and operation of bacterial tripartite multidrug efflux pumps. Trends Microbiol 23:311–319
29. Zgurskaya HI, Nikaido H (1999) Bypassing the periplasm: reconstitution of the AcrAB multidrug efflux pump of Escherichia coli. Proc Natl Acad Sci U S A 96:7190–7195
30. Murakami S, Nakashima R, Yamashita E, Yamaguchi A (2002) Crystal structure of bacterial multidrug efflux transporter AcrB. Nature 419:587–593
31. Murakami S, Nakashima R, Yamashita E, Matsumoto T, Yamaguchi A (2006) Crystal structures of a multidrug transporter reveal a functionally rotating mechanism. Nature 443:173–179
32. Seeger MA, Schiefner A, Eicher T, Verrey F, Diederichs K, Pos KM (2006) Structural asymmetry of AcrB trimer suggests a peristaltic pump mechanism. Science 313:1295–1298
33. Sennhauser G, Amstutz P, Briand C, Storchenegger O, Grutter MG (2007) Drug export pathway of multidrug exporter AcrB revealed by DARPin inhibitors. PLoS Biol 5:106–113
34. Sennhauser G, Bukowska MA, Briand C, Grutter MG (2009) Crystal structure of the multidrug exporter MexB from Pseudomonas aeruginosa. J Mol Biol 389:134–145
35. Mazzariol A, Cornaglia G, Nikaido H (2000) Contributions of the AmpC beta-lactamase and the AcrAB multidrug efflux system in intrinsic resistance of Escherichia coli K-12 to beta-lactams. Antimicrob Agents Chemother 44:1387–1390
36. Elkins CA, Nikaido H (2002) Substrate specificity of the RND-type multidrug efflux pumps AcrB and AcrD of Escherichia coli is determined predominately by two large periplasmic loops. J Bacteriol 184:6490–6498
37. Baucheron S, Imberechts H, Chaslus-Dancla-E, Cloeckaert A (2002) The AcrB multidrug transporter plays a major role in high-level fluoroquinolone resistance in salmonella enterica serovar typhimurium phage type DT204. Microb Drug Resist 8:281–289
38. Middlemiss JK, Poole K (2004) Differential impact of MexB mutations on substrate selectivity of the MexAB-OprM multidrug efflux pump of Pseudomonas aeruginosa. J Bacteriol 186:1258–1269
39. Kinana AD, Vargiu AV, Nikaido H (2013) Some ligands enhance the efflux of other ligands by the Escherichia coli multidrug pump AcrB. Biochemistry 52:8342–8351
40. Kobayashi N, Tamura N, Van Veen HW, Yamaguchi A, Murakami S (2014) β-Lactam selectivity of multidrug transporters AcrB and AcrD resides in the proximal binding pocket. J Biol Chem 289:10680–10690
41. Koronakis V, Sharff A, Koronakis E, Luisi B, Hughes C (2000) Crystal structure of the bacterial membrane protein TolC central to multidrug efflux and protein export. Nature 405:914–919
42. Higgins MK, Eswaran J, Edwards P, Schertler GFX, Hughes C, Koronakis V (2004) Structure of the ligand-blocked periplasmic entrance of the bacterial multidrug efflux protein TolC. J Mol Biol 342:697–702
43. Phan G, Benabdelhak H, Lascombe MB, Benas P, Rety S, Picard M, Ducruix A, Etchebest C, Broutin I (2010) Structural and dynamical insights into the opening

mechanism of P. aeruginosa OprM channel. Structure 18:507–517

44. Akama H, Kanemaki M, Yoshimura M, Tsukihara T, Kashiwagi T, Yoneyama H, Narita S, Nakagawa A, Nakae T (2004) Crystal structure of the drug discharge outer membrane protein, OprM, of Pseudomonas aeruginosa – dual modes of membrane anchoring and occluded cavity end. J Biol Chem 279:52816–52819
45. Bavro VN, Pietras Z, Furnham N, Perez-Cano L, Fernandez-Recio J, Pei XY, Misra R, Luisi B (2008) Assembly and channel opening in a bacterial drug efflux machine. Mol Cell 30:114–121
46. Poole K (2001) Multidrug resistance in Gram-negative bacteria. Curr Opin Microbiol 4:500–508
47. Akama H, Matsuura T, Kashiwagi S, Yoneyama H, Narita SI, Tsukihara T, Nakagawa A, Nakae T (2004) Crystal structure of the membrane fusion protein, MexA, of the multidrug transporter in Pseudomonas aeruginosa. J Biol Chem 279:25939–25942
48. Higgins MK, Bokma E, Koronakis E, Hughes C, Koronakis V (2004) Structure of the periplasmic component of a bacterial drug efflux pump. Proc Natl Acad Sci U S A 101:9994–9999
49. Mikolosko J, Bobyk K, Zgurskaya HI, Ghosh P (2006) Conformational flexibility in the multidrug efflux system protein AcrA. Structure 14:577–587
50. Xu Y, Lee M, Moeller A, Song S, Yoon B-Y, Kim H-M, Jun S-Y, Lee K, Ha N-C (2011) Funnel-like hexameric assembly of the periplasmic adapter protein in the tripartite multidrug efflux pump in Gram-negative bacteria. J Biol Chem 286:17910–17920
51. Hobbs EC, Yin X, Paul BJ, Astarita JL, Storz G (2012) Conserved small protein associates with the multidrug efflux pump AcrB and differentially affects antibiotic resistance. Proc Natl Acad Sci U S A 109:16696–16701
52. Elkins CA, Nikaido H (2003) Chimeric analysis of AcrA function reveals the importance of its c-terminal domain in its interaction with the AcrB multidrug efflux pump. J Bacteriol 185:5349–5356
53. Elkins CA, Nikaido H (2003) 3D structure of AcrB: the archetypal multidrug efflux transporter of Escherichia coli likely captures substrates from periplasm. Drug Resist Updat 6:9–13
54. Trépout S, Taveau J-C, Benabdelhak H, Granier T, Ducruix A, Frangakis AS, Lambert O (2010) Structure of reconstituted bacterial membrane efflux pump by cryo-electron tomography. BBA-Biomembranes 1798:1953–1960
55. Nakashima R, Sakurai K, Yamasaki S, Nishino K, Yamaguchi A (2011) Structures of the multidrug exporter AcrB reveal a proximal multisite drug-binding pocket. Nature 480:565–569
56. Pei X-Y, Hinchliffe P, Symmons MF, Koronakis E, Benz R, Hughes C, Koronakis V (2011) Structures of sequential open states in a symmetrical opening transition of the TolC exit duct. Proc Natl Acad Sci U S A 108:2112–2117
57. Xu Y, Moeller A, Jun S-Y, Le M, Yoon B-Y, Kim J-S, Lee K, Ha N-C (2012) Assembly and channel opening of outer membrane protein in tripartite drug efflux pumps of Gram-negative bacteria. J Biol Chem 287:11740–11750
58. Hinchliffe P, Symmons MF, Hughes C, Koronakis V (2013) Structure and operation of bacterial tripartite pumps. Annu Rev Microbiol 67(67):221–242
59. Friesner RA, Banks JL, Murphy RB, Halgren TA, Klicic JJ, Mainz DT, Repasky MP, Knoll EH, Shelley M, Perry JK, Shaw DE, Francis P, Shenkin PS (2004) Glide: a new approach for rapid, accurate docking and scoring. 1. Method and assessment of docking accuracy. J Med Chem 47:1739–1749
60. Nissink JWM, Murray C, Hartshorn M, Verdonk ML, Cole JC, Taylor R (2002) A new test set for validating predictions of protein-ligand interaction. Proteins 49:457–471
61. Jorgensen WL (2004) The many roles of computation in drug discovery. Science 303:1813–1818
62. Van Gunsteren WF, Bakowies D, Baron R, Chandrasekhar I, Christen M, Daura X, Gee P, Geerke DP, Glattli A, Hunenberger PH, Kastenholz MA, Ostenbrink C, Schenk M, Trzesniak D, Van Der Vegt NFA, Yu HB (2006) Biomolecular modeling: goals, problems, perspectives. Angew Chem Int Ed 45:4064–4092
63. Dodson GG, Lane DP, Verma CS (2008) Molecular simulations of protein dynamics: new windows on mechanisms in biology. EMBO Rep 9:144–150
64. Lee EH, Hsin J, Sotomayor M, Comellas G, Schulten K (2009) Discovery through the computational microscope. Structure 17:1295–1306
65. Trott O, Olson AJ (2010) Autodock vina: improving the speed and accuracy of docking with a new scoring function, efficient

optimization, and multithreading. J Comput Chem 31:455–461
66. De Vries SJ, Zacharias M (2012) ATTRACT-EM: a new method for the computational assembly of large molecular machines using cryo-EM maps. PLoS One 7:e49733
67. Karplus M, Lavery R (2014) Significance of molecular dynamics simulations for life sciences. Isr J Chem 54:1042–1051
68. Dror RO, Dirks RM, Grossman JP, Xu HF, Shaw DE (2012) Biomolecular simulation: a computational microscope for molecular biology. Annu Rev Biophys 41:429–452
69. Ruiz-Carmona S, Alvarez-Garcia D, Foloppe N, Garmendia-Doval AB, Juhos S, Schmidtke P, Barril X, Hubbard RE, Morley SD (2014) rDock: a fast, versatile and open source program for docking ligands to proteins and nucleic acids. PLoS Comput Biol 10:e1003571
70. Gilson MK, Zhou HX (2007) Calculation of protein-ligand binding affinities. Annu Rev Biophys Biomol Struct 36:21–42
71. Mortier J, Rakers C, Bermudez M, Murgueitio MS, Riniker S, Wolber G (2015) The impact of molecular dynamics on drug design: applications for the characterization of ligand-macromolecule complexes. Drug Discov Today 20:686–702
72. Shaw DE, Deneroff MM, Dror RO, Kuskin JS, Larson RH, Salmon JK, Young C, Batson B, Bowers KJ, Chao JC, Eastwood MP, Gagliardo J, Grossman JP, Ho CR, Ierardi DJ, Kolossvary I, Klepeis JL, Layman T, Mcleavey C, Moraes MA, Mueller R, Priest EC, Shan YB, Spengler J, Theobald M, Towles B, Wang SC (2008) Anton, a special-purpose machine for molecular dynamics simulation. Commun ACM 51:91–97
73. Anderson JA, Lorenz CD, Travesset A (2008) General purpose molecular dynamics simulations fully implemented on graphics processing units. J Comput Phys 227:5342–5359
74. Harvey MJ, Giupponi G, Fabritiis GD (2009) ACEMD: accelerating biomolecular dynamics in the microsecond time scale. J Chem Theory Comput 5:1632–1639
75. Le Grand S, Gotz AW, Walker RC (2013) SPFP: speed without compromise—a mixed precision model for GPU accelerated molecular dynamics simulations. Comput Phys Commun 184:374–380
76. Ruggerone P, Vargiu AV, Collu F, Fischer N, Kandt C (2013) Molecular dynamics computer simulations of multidrug RND efflux pumps. Comput Struct Biotechnol J 5: e201302008
77. Schulz R, Vargiu AV, Collu F, Kleinekathofer U, Ruggerone P (2010) Functional rotation of the transporter AcrB: insights into drug extrusion from simulations. PLoS Comput Biol 6:e1000806
78. Schulz R, Vargiu AV, Ruggerone P, Kleinekathofer U (2011) Role of water during the extrusion of substrates by the efflux transporter AcrB. J Phys Chem B 115:8278–8287
79. Vargiu AV, Collu F, Schulz R, Pos KM, Zacharias M, KleinekathöFer U, Ruggerone P (2011) Effect of the F610A mutation on substrate extrusion in the AcrB transporter: explanation and rationale by molecular dynamics simulations. J Am Chem Soc 133:10704–10707
80. Collu F, Vargiu AV, Dreier J, Cascella M, Ruggerone P (2012) Recognition of imipenem and meropenem by the RND-transporter MexB studied by computer simulations. J Am Chem Soc 134:19146–19158
81. Vargiu AV, Nikaido H (2012) Multidrug binding properties of the AcrB efflux pump characterized by molecular dynamics simulations. Proc Natl Acad Sci U S A 109:20637–20642
82. Vargiu AV, Ruggerone P, Opperman TJ, Nguyen ST, Nikaido H (2014) Molecular mechanism of MBX2319 inhibition of Escherichia coli AcrB multidrug efflux pump and comparison with other inhibitors. Antimicrob Agents Chemother 58:6224–6234
83. Blair JMA, Bavro VN, Ricci V, Modi N, Cacciotto P, Kleinekathoefer U, Ruggerone P, Vargiu AV, Baylay AJ, Smith HE, Brandon Y, Galloway D, Piddock LJV (2015) AcrB drug-binding pocket substitution confers clinically relevant resistance and altered substrate specificity. Proc Natl Acad Sci U S A 112:3511–3516
84. Schulz R, Vargiu AV, Ruggerone P, Kleinekathoefer U (2015) Computational study of correlated domain motions in the AcrB efflux transporter. Biomed Res Int 2015:487298
85. Kinana AD, Vargiu AV, May T, Nikaido H (2016) Aminoacyl β-naphthylamides as substrates and modulators of AcrB multidrug efflux pump. Proc Natl Acad Sci U S A 113:1405–1410
86. Sjuts H, Vargiu AV, Kwasny SM, Nguyen ST, Kim H-S, Ding X, Ornik AR, Ruggerone P, Bowlin TL, Nikaido H, Pos KM, Opperman TJ (2016) Molecular basis for inhibition of AcrB multidrug efflux pump by novel and powerful pyranopyridine derivatives. Proc Natl Acad Sci U S A 113:3509–3514

87. Fischer N, Kandt C (2011) Three ways in, one way out: water dynamics in the transmembrane domains of the inner membrane translocase AcrB. Proteins 79:2871–2885
88. Fischer N, Kandt C (2013) Porter domain opening and closing motions in the multidrug efflux transporter AcrB. BBA-Biomembranes 1828:632–641
89. Fischer N, Raunest M, Schmidt TH, Koch DC, Kandt C (2014) Efflux pump-mediated antibiotics resistance: insights from computational structural biology. Interdiscip Sci 6:1–12
90. Koch DC, Raunest M, Harder T, Kandt C (2013) Unilateral access regulation: ground state dynamics of the Pseudomonas aeruginosa outer membrane efflux duct OprM. Biochemistry 52:178–187
91. Raunest M, Kandt C (2012) Locked on one side only: ground state dynamics of the outer membrane efflux duct TolC. Biochemistry 51:1719–1729
92. Yamane T, Murakami S, Ikeguchi M (2013) Functional rotation induced by alternating protonation states in the multidrug transporter AcrB: all-atom molecular dynamics simulations. Biochemistry 52:7648–7658
93. Wang B, Weng J, Wang W (2015) Substrate binding accelerates the conformational transitions and substrate dissociation in multidrug efflux transporter AcrB. Front Microbiol 6:302
94. Vaccaro L, Koronakis V, Sansom MSP (2006) Flexibility in a drug transport accessory protein: molecular dynamics simulations of MexA. Biophys J 91:558–564
95. Vaccaro L, Scott KA, Sansom MSP (2008) Gating at both ends and breathing in the middle: conformational dynamics of TolC. Biophys J 95:5681–5691
96. Wang B, Weng J, Fan K, Wang W (2012) Interdomain flexibility and pH-induced conformational changes of AcrA revealed by molecular dynamics simulations. J Phys Chem B 116:3411–3420
97. Yamaguchi A, Nakashima R, Sakurai K (2015) Structural basis of RND-type multidrug exporters. Front Microbiol 6:327
98. Sali A, Blundell TL (1993) Comparative protein modelling by satisfaction of spatial restraints. J Mol Biol 234:779–815
99. Bower MJ, Cohen FE, Dunbrack RL (1997) Prediction of protein side-chain rotamers from a backbone-dependent rotamer library: a new homology modeling tool. J Mol Biol 267:1268–1282
100. Marti-Renom MA, Stuart AC, Fiser A, Sanchez R, Melo F, Sali A (2000) Comparative protein structure modeling of genes and genomes. Annu Rev Biophys Biomol Struct 29:291–325
101. Schwede T, Kopp J, Guex N, Peitsch MC (2003) SWISS-MODEL: an automated protein homology-modeling server. Nucleic Acids Res 31:3381–3385
102. Šali A, Potterton L, Yuan F, Van Vlijmen H, Karplus M (1995) Evaluation of comparative protein modeling by MODELLER. Proteins 23:318–326
103. Eswar N, Webb B, Marti-Renom MA, Madhusudhan MS, Eramian D, Shen M-Y, Pieper U, Sali A (2006) Comparative protein structure modeling using modeller. Curr Protoc Bioinformatics 15:5.6:5.6.1–5.6.30
104. Sali A, Overington JP (1994) Derivation of rules for comparative protein modeling from a database of protein structure alignments. Protein Sci 3:1582–1596
105. Webb B, Sali A (2014) Comparative protein structure modeling using MODELLER. Curr Protoc Bioinformatics 47:5.6:5.6.1–5.6.32
106. Sousa SF, Fernandes PA, Ramos MJ (2006) Protein-ligand docking: current status and future challenges. Proteins 65:15–26
107. Rodrigues JP, Karaca E, Bonvin AM (2015) Information-driven structural modelling of protein-protein interactions. Methods Mol Biol 1215:399–424
108. Chen YC (2015) Beware of docking! Trends Pharmacol Sci 36:78–95
109. Amaro RE, Baron R, Mccammon JA (2008) An improved relaxed complex scheme for receptor flexibility in computer-aided drug design. J Comput Aided Mol Des 22:693–705
110. Huang S-Y, Zou X (2007) Ensemble docking of multiple protein structures: considering protein structural variations in molecular docking. Proteins 66:399–421
111. Takatsuka Y, Chen C, Nikaido H (2010) Mechanism of recognition of compounds of diverse structures by the multidrug efflux pump AcrB of Escherichia coli. Proc Natl Acad Sci U S A 107:6559–6565
112. Karplus M, Mccammon JA (2002) Molecular dynamics simulations of biomolecules. Nat Struct Biol 9:646–652
113. Van Gunsteren WF, Billeter SR, Eising AA, Hunenberger PH, Kruger P, Mark AE, Scott WRP, Tironi IG (1996) Biomolecular simulation: the GROMOS96 manual and user guide. Hochschulverlag AG an der ETH, Zurich

114. Jorgensen WL, Tirado-Rives J (1988) The OPLS force field for proteins. Energy minimizations for crystals of cyclic peptides and crambin. J Am Chem Soc 110:1657–1666
115. Wang J, Wolf RM, Caldwell JW, Kollman PA, Case DA (2004) Development and testing of a general amber force field. J Comput Chem 25:1157–1174
116. Brooks BR, Brooks CL 3rd, Mackerell AD Jr, Nilsson L, Petrella RJ, Roux B, Won Y, Archontis G, Bartels C, Boresch S, Caflisch A, Caves L, Cui Q, Dinner AR, Feig M, Fischer S, Gao J, Hodoscek M, Im W, Kuczera K, Lazaridis T, Ma J, Ovchinnikov V, Paci E, Pastor RW, Post CB, Pu JZ, Schaefer M, Tidor B, Venable RM, Woodcock HL, Wu X, Yang W, York DM, Karplus M (2009) CHARMM: the biomolecular simulation program. J Comput Chem 30:1545–1614
117. Ponder JW, Case DA (2003) Force fields for protein simulations. Adv Protein Chem 66:27–85
118. Halgren TA, Damm W (2001) Polarizable force fields. Curr Opin Struct Biol 11:236–242
119. Graen T, Hoefling M, Grubmuller H (2014) AMBER-DYES: characterization of charge fluctuations and force field parameterization of fluorescent dyes for molecular dynamics simulations. J Chem Theory Comput 10:5505–5512
120. Malde AK, Zuo L, Breeze M, Stroet M, Poger D, Nair PC, Oostenbrink C, Mark AE (2011) An automated force field topology builder (ATB) and repository: version 1.0. J Chem Theory Comput 7:4026–4037
121. Vanquelef E, Simon S, Marquant G, Garcia E, Klimerak G, Delepine JC, Cieplak P, Dupradeau FY (2011) R.E.D. Server: a web service for deriving RESP and ESP charges and building force field libraries for new molecules and molecular fragments. Nucleic Acids Res 39: W511–W517
122. Mayne CG, Saam J, Schulten K, Tajkhorshid E, Gumbart JC (2013) Rapid parameterization of small molecules using the force field toolkit. J Comput Chem 34:2757–2770
123. Marvin, Marvin 14.8.25.0 (2012) http://www.chemaxon.com
124. Case DA, Berryman JT, Betz RM, Cerutti DS, Cheatham Iii TE, Darden TA, Duke RE, Giese TJ, Gohlke H, Goetz AW, Homeyer N, Izadi S, Janowski P, Kaus J, Kovalenko A, Lee TS, Legrand S, Li P, Luchko T, Luo R, Madej B, Merz KM, Monard G, Needham P, Nguyen H, Nguyen HT, Omelyan I, Onufriev A, Roe DR, Roitberg A, Salomon-Ferrer R, Simmerling CL, Smith W, Swails J, Walker RC, Wang J, Wolf RM, Wu X, York DM, Kollman PA (2015) AMBER 2015. University of California, San Francisco
125. Frisch MJ, Trucks GW, Schlegel HB, Scuseria GE, Robb MA, Cheeseman JR, Scalmani G, Barone V, Mennucci B, Petersson GA, Nakatsuji H, Caricato M, Li X, Hratchian HP, Izmaylov AF, Bloino J, Zheng G, Sonnenberg JL, Hada M, Ehara M, Toyota K, Fukuda R, Hasegawa J, Ishida M, Nakajima T, Honda Y, Kitao O, Nakai H, Vreven T Jr, Montgomery JA, Peralta JE, Ogliaro F, Bearpark M, Heyd JJ, Brothers E, Kudin KN, Staroverov VN, Kobayashi R, Normand J, Raghavachari K, Rendell A, Burant JC, Iyengar SS, Tomasi J, Cossi M, Rega N, Millam JM, Klene M, Knox JE, Cross JB, Bakken V, Adamo C, Jaramillo J, Gomperts R, Stratmann RE, Yazyev O, Austin AJ, Cammi R, Pomelli C, Ochterski JW, Martin RL, Morokuma K, Zakrzewski VG, Voth GA, Salvador P, Dannenberg JJ, Dapprich S, Daniels AD, Farkas Ö, Foresman JB, Ortiz JV, Cioslowski J, Fox DJ (2009) Gaussian 09, Revision A.1. Gaussian, Inc., Wallingford, CT
126. Klauda JB, Venable RM, Freites JA, O'connor JW, Tobias DJ, Mondragon-Ramirez C, Vorobyov I, Mackerell AD Jr, Pastor RW (2010) Update of the CHARMM all-atom additive force field for lipids: validation on six lipid types. J Phys Chem B 114:7830–7843
127. Dickson CJ, Madej BD, Skjevik AA, Betz RM, Teigen K, Gould IR, Walker RC (2014) Lipid14: the amber lipid force field. J Chem Theory Comput 10:865–879
128. Poger D, Van Gunsteren WF, Mark AE (2010) A new force field for simulating phosphatidylcholine bilayers. J Comput Chem 31:1117–1125
129. Hornak V, Abel R, Okur A, Strockbine B, Roitberg A, Simmerling C (2006) Comparison of multiple amber force fields and development of improved protein backbone parameters. Proteins 65:712–725
130. Malloci G, Vargiu A, Serra G, Bosin A, Ruggerone P, Ceccarelli M (2015) A database of force-field parameters, dynamics, and properties of antimicrobial compounds. Molecules 20:13997

131. Malloci G, Serra G, Bosin A, Vargiu AV (2016) Extracting conformational ensembles of small molecules from molecular dynamics simulations: ampicillin as a test case. Computation 4:5
132. Stavenger RA, Winterhalter M (2014) TRANSLOCATION project: how to get good drugs into bad bugs. Sci Transl Med 6:228ed7
133. Van Meer G, Voelker DR, Feigenson GW (2008) Membrane lipids: where they are and how they behave. Nat Rev Mol Cell Biol 9:112–124
134. Tieleman DP, Marrink SJ, Berendsen HJ (1997) A computer perspective of membranes: molecular dynamics studies of lipid bilayer systems. Biochim Biophys Acta 1331:235–270
135. Berger O, Edholm O, Jahnig F (1997) Molecular dynamics simulations of a fluid bilayer of dipalmitoylphosphatidylcholine at full hydration, constant pressure, and constant temperature. Biophys J 72:2002–2013
136. Marrink SJ, Risselada HJ, Yefimov S, Tieleman DP, De Vries AH (2007) The MARTINI force field: coarse grained model for biomolecular simulations. J Phys Chem B 111:7812–7824
137. Orsi M, Essex JW (2011) The ELBA force field for coarse-grain modeling of lipid membranes. PLoS One 6:e28637
138. Marrink SJ, Tieleman DP (2013) Perspective on the Martini model. Chem Soc Rev 42:6801–6822
139. Cheatham TE 3rd, Case DA (2013) Twenty-five years of nucleic acid simulations. Biopolymers 99:969–977
140. Vargiu AV, Magistrato A (2014) Atomistic-level portrayal of drug-DNA interplay: a history of courtships and meetings revealed by molecular simulations. ChemMedChem 9:1966–1981
141. Perez A, Marchan I, Svozil D, Sponer J, Cheatham TE 3rd, Laughton CA, Orozco M (2007) Refinement of the AMBER force field for nucleic acids: improving the description of {alpha}/{gamma} conformers. Biophys J 92:3817–3829
142. Soares TA, Hünenberger PH, Kastenholz MA, Kräutler V, Lenz T, Lins RD, Oostenbrink C, Van Gunsteren WF (2005) An improved nucleic acid parameter set for the GROMOS force field. J Comput Chem 26:725–737
143. Mackerell AD, Bashford D, Bellott M, Dunbrack RL, Evanseck JD, Field MJ, Fischer S, Gao J, Guo H, Ha S, Joseph-Mccarthy D, Kuchnir L, Kuczera K, Lau FT, Mattos C, Michnick S, Ngo T, Nguyen DT, Prodhom B, Reiher WE, Roux B, Schlenkrich M, Smith JC, Stote R, Straub J, Watanabe M, Wiorkiewicz-Kuczera J, Yin D, Karplus M (1998) All-atom empirical potential for molecular modeling and dynamics studies of proteins. J Phys Chem B 102:3586–3616
144. Siu SWI, Vácha R, Jungwirth P, Böckmann RA (2008) Biomolecular simulations of membranes : physical properties from different force fields. J Chem Phys 128:125103
145. Marrink SJ, De Vries AH, Tieleman DP (2009) Lipids on the move: simulations of membrane pores, domains, stalks and curves. Biochim Biophys Acta 1788:149–168
146. Lyubartsev AP, Rabinovich AL (2011) Recent development in computer simulations of lipid bilayers. Soft Matter 7:25–39
147. Leftin A, Brown MF (2011) An NMR database for simulations of membrane dynamics. Biochim Biophys Acta 1808:818–839
148. Bennett WF, Tieleman DP (2013) Computer simulations of lipid membrane domains. Biochim Biophys Acta 1828:1765–1776
149. Chavent M, Reddy T, Goose J, Dahl AC, Stone JE, Jobard B, Sansom MS (2014) Methodologies for the analysis of instantaneous lipid diffusion in MD simulations of large membrane systems. Faraday Discuss 169:455–475
150. Javanainen M (2014) Universal method for embedding proteins into complex lipid bilayers for molecular dynamics simulations. J Chem Theory Comput 10:2577–2582
151. Schmidt TH, Kandt C (2012) LAMBADA and InflateGRO2: efficient membrane alignment and insertion of membrane proteins for molecular dynamics simulations. J Chem Inf Model 52:2657–2669
152. Balabin IA (2010) Membrane Plug-in
153. Guixa-Gonzalez R, Rodriguez-Espigares I, Ramirez-Anguita JM, Carrio-Gaspar P, Martinez-Seara H, Giorgino T, Selent J (2014) MEMBPLUGIN: studying membrane complexity in VMD. Bioinformatics 30:1478–1480
154. Nagle JF, Tristram-Nagle S (2000) Lipid bilayer structure. Curr Opin Struct Biol 10:474–480
155. Lis LJ, Mcalister M, Fuller N, Rand RP, Parsegian VA (1982) Interactions between neutral phospholipid bilayer membranes. Biophys J 37:657–665

156. Rand RP, Parsegian VA (1989) Hydration forces between phospholipid bilayers. Biochem Biophys Acta 988:351–376
157. Zimmerberg J (1987) Molecular mechanisms of membrane fusion: steps during phospholipid and exocytotic membrane fusion. Biosci Rep 7:251–268
158. Rand RP. Structural parameters of aqueous phospholipid mixtures. Available from: https://brocku.ca/researchers/peter_rand/lipid/
159. Kucerka N, Katsaras J, Nagle JF (2010) Comparing membrane simulations to scattering experiments: introducing the SIMtoEXP software. J Membr Biol 235:43–50
160. Skjevik AA, Madej BD, Walker RC, Teigen K (2012) LIPID11: a modular framework for LIPID simulations using amber. J Phys Chem B 116(36):11124
161. Allen WJ, Lemkul JA, Bevan DR (2009) GridMAT-MD: a grid-based membrane analysis tool for use with molecular dynamics. J Comput Chem 30:1952–1958
162. Dickson CJ, Rosso L, Betz RM, Walker RC, Gould IR (2012) GAFFlipid: a general amber force field for the accurate molecular dynamics simulation of phospholipid. Soft Matter 8:9617–9627
163. Shinoda W, Okazaki S (1998) A Voronoi analysis of lipid area fluctuation in a bilayer. J Chem Phys 109:1517–1521
164. Mori T, Ogushi F, Sugita Y (2012) Analysis of lipid surface area in protein-membrane systems combining Voronoi tessellation and Monte Carlo integration methods. J Comput Chem 33:286–293
165. Gapsys V, De Groot BL, Briones R (2013) Computational analysis of local membrane properties. J Comput Aided Mol Des 27:845–858
166. Gilson MK, Given JA, Bush BL, Mccammon JA (1997) The statistical-thermodynamic basis for computation of binding affinities: a critical review. Biophys J 72:1047–1069
167. Woo HJ, Roux B (2005) Calculation of absolute protein-ligand binding free energy from computer simulations. Proc Natl Acad Sci U S A 102:6825–6830
168. Jiao D, Golubkov PA, Darden TA, Ren P (2008) Calculation of protein-ligand binding free energy by using a polarizable potential. Proc Natl Acad Sci U S A 105:6290–6295
169. Mitomo D, Fukunishi Y, Higo J, Nakamura H (2009) Calculation of protein-ligand binding free energy using smooth reaction path generation (SRPG) method: a comparison of the explicit water model, gb/sa model and docking score function. Genome Inform 23:85–97
170. Steinbrecher T, Labahn A (2010) Towards accurate free energy calculations in ligand protein-binding studies. Curr Med Chem 17:767–785
171. Rathore RS, Sumakanth M, Reddy MS, Reddanna P, Rao AA, Erion MD, Reddy MR (2013) Advances in binding free energies calculations: QM/MM-based free energy perturbation method for drug design. Curr Pharm Des 19:4674–4686
172. Srinivasan J, Cheatham Iii TE, Cieplak P, Kollman PA, Case DA (1998) Continuum solvent studies of the stability of DNA, RNA, and phosphoramidate-DNA helices. J Am Chem Soc 120:9401–9409
173. Kollman PA, Massova I, Reyes C, Kuhn B, Huo SH, Chong L, Lee M, Lee T, Duan Y, Wang W, Donini O, Cieplak P, Srinivasan J, Case DA, Cheatham TE (2000) Calculating structures and free energies of complex molecules: combining molecular mechanics and continuum models. Acc Chem Res 33:889–897
174. Fogolari F, Brigo A, Molinari H (2003) Protocol for MM/PBSA molecular dynamics simulations of proteins. Biophys J 85:159–166
175. Hou T, Wang J, Li Y, Wang W (2010) Assessing the performance of the MM/PBSA and MM/GBSA methods. 1. The accuracy of binding free energy calculations based on molecular dynamics simulations. J Chem Inf Model 51:69–82
176. Kongsted J, Ryde U (2009) An improved method to predict the entropy term with the MM/PBSA approach. J Comput Aided Mol Des 23:63–71
177. Massova I, Kollman PA (1999) Computational alanine scanning to probe protein-protein interactions: a novel approach to evaluate binding free energies. J Am Chem Soc 121:8133–8143
178. Fiser A, Do RKG, Sali A (2000) Modeling of loops in protein structures. Protein Sci 9:1753–1773
179. Humphrey W, Dalke A, Schulten K (1996) VMD: visual molecular dynamics. J Mol Graph 14:33–38
180. Altschul SF, Gish W, Miller W, Myers EW, Lipman DJ (1990) Basic local alignment search tool. J Mol Biol 215:403–410
181. Gish W, States DJ (1993) Identification of protein coding regions by database similarity search. Nat Genet 3:266–272

182. Madden TL, Tatusov RL, Zhang J (1996) Applications of network BLAST server. Methods Enzymol 266:131–141
183. Shen MY, Sali A (2006) Statistical potential for assessment and prediction of protein structures. Protein Sci 15:2507–2524
184. Bolton EE, Wang Y, Thiessen PA, Bryant SH (2008) PubChem: integrated platform of small molecules and biological activities. In: Wheeler RA, Spellmeyer DC (eds) Annual reports in computational chemistry. Elsevier, Oxford, pp 217–241
185. O'boyle NM, Banck M, James CA, Morley C, Vandermeersch T, Hutchison GR (2011) Open babel: an open chemical toolbox. J Cheminform 3:33
186. Hohenberg P, Kohn W (1964) Inhomogeneous electron gas. Phys Rev 136: B864–B871
187. Kohn W, Sham L (1965) Self-consistent equations including exchange and correlation effects. Phys Rev 140:A1133–A1138
188. Becke AD (1993) Density-functional thermochemistry. 3. The role of exact exchange. J Chem Phys 98:5648–5652
189. Kim K, Jordan KD (1994) Comparison of density-functional and Mp2 calculations on the water monomer and dimer. J Phys Chem 98:10089–10094
190. Pople JA (1999) Quantum chemical models (Nobel lecture). Angew Chem Int Ed 38:1894–1902
191. Tomasi J, Mennucci B, Cammi R (2005) Quantum mechanical continuum solvation models. Chem Rev 105:2999–3093
192. Singh UC, Kollman PA (1984) An approach to computing electrostatic charges for molecules. J Comput Chem 5:129–145
193. Laio A, Vandevondele J, Rothlisberger U (2002) D-RESP : dynamically generated electrostatic potential derived charges from quantum mechanics/molecular mechanics simulations. J Phys Chem B 106:7300–7307
194. Wang J, Wang W, Kollman PA, Case DA (2006) Automatic atom type and bond type perception in molecular mechanical calculations. J Mol Graph Model 25:247–260
195. Dolinsky TJ, Nielsen JE, Mccammon JA, Baker NA (2004) PDB2PQR: an automated pipeline for the setup of Poisson-Boltzmann electrostatics calculations. Nucleic Acids Res 32:W665–W667
196. Schuttelkopf AW, Van Aalten DM (2004) PRODRG: a tool for high-throughput crystallography of protein-ligand complexes. Acta Crystallogr D Biol Crystallogr 60:1355–1363
197. Kandasamy SK, Larson RG (2006) Molecular dynamics simulations of model transmembrane peptides in lipid bilayers: a systematic investigation of hydrophobic mismatch. Biophys J 90:2326–2343
198. Lindahl E, Sansom MSP (2008) Membrane proteins : molecular dynamics simulations. Curr Opin Struct Biol 18:425–431
199. Gurtovenko AA, Vattulainen I (2009) Calculation of the electrostatic potential of lipid bilayers from molecular dynamics simulations: methodological issues. J Chem Phys 130:215107
200. Stansfeld PJ, Sansom MS (2011) Molecular simulation approaches to membrane proteins. Structure 19:1562–1572
201. Jo S, Kim T, Iyer VG, Im W (2008) CHARMM-GUI: a web-based graphical user interface for CHARMM. J Comput Chem 29:1859–1865
202. Wu EL, Cheng X, Jo S, Rui H, Song KC, Davila-Contreras EM, Qi Y, Lee J, Monje-Galvan V, Venable RM, Klauda JB, Im W (2014) CHARMM-GUI membrane builder toward realistic biological membrane simulations. J Comput Chem 35:1997–2004
203. Lomize MA, Pogozheva ID, Joo H, Mosberg HI, Lomize AL (2012) OPM database and PPM web server: resources for positioning of proteins in membranes. Nucleic Acids Res 40: D370–D376
204. Kimmett T, Smith N, Witham S, Petukh M, Sarkar S, Alexov E (2014) ProBLM web server: protein and membrane placement and orientation package. Comput Math Methods Med 2014:838259
205. Kufareva I, Lenoir M, Dancea F, Sridhar P, Raush E, Bissig C, Gruenberg J, Abagyan R, Overduin M (2014) Discovery of novel membrane binding structures and functions. Biochem Cell Biol 92:555–563
206. Jorgensen WL, Chandrasekhar J, Madura JD, Impey RW, Klein ML (1983) Comparison of simple potential functions for simulating liquid water. J Chem Phys 79:926–935
207. Joung IS, Cheatham TE (2008) Determination of alkali and halide monovalent ion parameters for use in explicitly solvated biomolecular simulations. J Phys Chem B 112:9020–9041
208. Pearlman DA, Case DA, Caldwell JW, Ross WS, Cheatham Iii TE, Debolt S, Ferguson D, Seibel G, Kollman P (1995) AMBER, a package of computer programs for applying molecular mechanics, normal mode analysis, molecular dynamics and free energy calculations to simulate the structural and energetic properties of molecules. Comput Phys Commun 91:1–41

209. Wimberly BT, Guymon R, Mccutcheon JP, White SW, Ramakrishnan V (1999) A detailed view of a ribosomal active site: the structure of the L11-RNA complex. Cell 97:491–502
210. Phillips JC, Braun R, Wang W, Gumbart J, Tajkhorshid E, Villa E, Chipot C, Skeel RD, Kale L, Schulten K (2005) Scalable molecular dynamics with NAMD. J Comput Chem 26:1781–1802
211. Pronk S, Pall S, Schulz R, Larsson P, Bjelkmar P, Apostolov R, Shirts MR, Smith JC, Kasson PM, Van Der Spoel D, Hess B, Lindahl E (2013) GROMACS 4.5: a high-throughput and highly parallel open source molecular simulation toolkit. Bioinformatics 29:845–854
212. Miller BR, Mcgee TD, Swails JM, Homeyer N, Gohlke H, Roitberg AE (2012) MMPBSA.py: an efficient program for end-state free energy calculations. J Chem Theory Comput 8:3314–3321
213. Tsui V, Case DA (2000) Molecular dynamics simulations of nucleic acids with a generalized born solvation model. J Am Chem Soc 122:2489–2498
214. Bondi A (1964) van der Waals Volumes and Radii. J Phys Chem 68:441–451
215. Onufriev A, Bashford D, Case DA (2004) Exploring protein native states and large-scale conformational changes with a modified generalized born model. Proteins 55:383–394
216. Mongan J, Simmerling C, Mccammon JA, Case DA, Onufriev A (2007) Generalized Born model with a simple, robust molecular volume correction. J Chem Theory Comput 3:156–169
217. Morris GM, Huey R, Lindstrom W, Sanner MF, Belew RK, Goodsell DS, Olson AJ (2009) AutoDock4 and AutoDockTools4: automated docking with selective receptor flexibility. J Comput Chem 30:2785–2791
218. Grossfield A, Zuckerman DM (2009) Quantifying uncertainty and sampling quality in biomolecular simulations. Annu Rep Comput Chem 5:23–48
219. Grossfield A, Feller SE, Pitman MC (2007) Convergence of molecular dynamics simulations of membrane proteins. Proteins 40:31–40

Chapter 12

A Transcriptomic Approach to Identify Novel Drug Efflux Pumps in Bacteria

Liping Li, Sasha G. Tetu, Ian T. Paulsen, and Karl A. Hassan

Abstract

The core genomes of most bacterial species include a large number of genes encoding putative efflux pumps. The functional roles of most of these pumps are unknown, however, they are often under tight regulatory control and expressed in response to their substrates. Therefore, one way to identify pumps that function in antimicrobial resistance is to examine the transcriptional responses of efflux pump genes to antimicrobial shock. By conducting complete transcriptomic experiments following antimicrobial shock treatments, it may be possible to identify novel drug efflux pumps encoded in bacterial genomes. In this chapter we describe a complete workflow for conducting transcriptomic analyses by RNA sequencing, to determine transcriptional changes in bacteria responding to antimicrobials.

Key words Multidrug efflux, Drug resistance, Transcriptomics, RNA-Seq, Gene expression

1 Introduction

Efflux pumps are encoded by all bacterial species analyzed by whole genome sequencing to date (www.membranetransport.org). Most bacterial genomes contain large numbers of genes encoding putative efflux pumps, and in some cases these genes account for more than 2% of the predicted protein-coding potential of the genome [1]. Different pumps are able to export a broad range of substrates that may include metabolic waste products, signaling molecules, antimicrobial secondary metabolites, siderophores, and/or drugs. As such, efflux pumps participate in diverse housekeeping and protective functions linked to different physiological states of the cell [2]. Efflux pumps that function in antimicrobial resistance have received significant attention from a human health perspective. Many research groups have sought to identify efflux pumps with the capacity to transport clinically relevant antimicrobials to better understand mechanisms of drug resistance.

Several approaches have been devised to identify new efflux pumps. Many of the first genes encoding drug efflux pumps to be

Akihito Yamaguchi and Kunihiko Nishino (eds.), *Bacterial Multidrug Exporters: Methods and Protocols*, Methods in Molecular Biology, vol. 1700, https://doi.org/10.1007/978-1-4939-7454-2_12, © Springer Science+Business Media, LLC 2018

identified in bacteria were found on mobile genetic elements such as plasmids and transposons. These elements increased drug resistance levels when introduced into susceptible strains [3, 4]. Historically, new chromosomally encoded pumps were typically identified using plasmid clone libraries containing random genome fragments from resistant strains. The introduction of these plasmid libraries into a susceptible strain followed by plating onto selective media enabled the identification of plasmids that increased host resistance, including those carrying a drug efflux pump gene [5–7]. With the advent of whole genome sequencing it became common to specifically target genes with a possible role in drug efflux for functional characterization. These genes, which encoded proteins resembling known drug efflux pumps, could be cloned into an expression vector for heterologous characterization in a susceptible strain and/or deleted from the host genome to determine their role in drug resistance and transport [8–11]. In addition to these traditional approaches, we have demonstrated the utility of transcriptomics to identify genes encoding novel efflux pumps in the genomes of bacteria [12, 13]. This approach takes advantage of the regulatory cues for expression of drug efflux pump genes, described below, to identify new candidates from sequenced bacterial genomes.

Efflux pumps can impose a fitness cost on their host when constitutively overexpressed in environments that lack their antimicrobial substrates. Although the mechanism(s) promoting these fitness costs are ill-defined, this phenomenon results in a need for tight regulatory control of efflux pumps, such that they are only expressed when required, e.g., in the presence of an antimicrobial substrate as part of an adaptive resistance response. Therefore, changes in the expression of efflux pump genes in the presence of antimicrobials can be used as an indicator of which pumps are likely to participate in resistance to these compounds. Transcriptomic methods offer simple and effective tools to survey the entire genome for genes that are induced by antimicrobials.

Transcriptomic methods have revolutionized our ability to examine the physiology of bacterial cells, deciphering core features of global stress responses and adaptations. Early transcriptomic based analyses employed whole genome microarrays where oligonucleotide probes were used to measure mRNAs for every known gene in the genome, and tiling microarrays, where the probes covered the entire genome. Contemporary transcriptomics is largely conducted utilizing the Illumina sequencing platform in what is referred to as RNA sequencing (RNA-Seq). Here total cellular RNA is isolated, typically depleted of structural RNA, and then sequenced on an Illumina platform. The resulting short sequence reads are then tiled back to the reference genome to identify regions that were under active transcription when the cells were collected. Sequence read coverage and density for individual genes or genomic regions is used to determine levels of transcription under each tested condition. Relative coverage can be compared between a control cell population and cell populations subjected to an antimicrobial shock treatment to identify genes that were upregulated in response to the treatment. These genes may include novel efflux pumps. In this chapter we describe a complete workflow to conduct RNA-Seq experiments following an antimicrobial shock treatment.

2 Materials

2.1 General Materials

1. Bacterial strain of interest: bacterial glycerol frozen stock or bacterial colonies on an agar plate.
2. Cation-adjusted Muller Hinton (MH) broth and MH agar plates: MH broth powder can be purchased from a commercial supplier, such as Oxoid, and prepared according to their instructions.
3. Thermal shaking incubator.
4. Biological safety cabinet.
5. Filter-sterilized antimicrobial stock solutions: antimicrobials should be prepared at 1000 times working concentration in an appropriate solvent.
6. Spectrophotometer: should be capable of measuring cell densities at 600 nm.
7. Sterile plasticware: 50 ml conical tubes, 50 and 200 ml culture flasks with 0.2 μm filter caps, 1.5 ml microcentrifuge tubes.
8. Nuclease-free plasticware: 1.5 ml microcentrifuge tubes, 0.2 ml thin-walled PCR tubes, 10 μl, 200 μl, and 1000 μl filter tips.
9. RNase*ZAP* (Ambion): a cleaning agent for removing RNases.
10. Microcentrifuge: should be capable of centrifuging 1.5 ml tubes.
11. Refrigerated centrifuge: should be capable of centrifuging 50 ml tubes at 4 °C.
12. Vortex mixer.
13. Thermal cycler, or alternatively a water bath or heating block.
14. RNase-free molecular biology grade water.
15. Fume-hood.
16. NanoDrop Spectrophotometer (Thermo Fisher Scientific), or alternatively a UV spectrophotomer with a small volume cuvette (<50 μl).

2.2 Commercial Kits and Enzymes for RNA Extraction and Purification

1. miRNeasy Mini Kit (Qiagen): a total RNA extraction kit (*see* **Note 1**).
2. RNase-free DNase enzyme: TURBO DNA-*free*™ Kit TURBO™ DNase Treatment and Removal Reagents (Ambion Inc.), DNase I (New England Biolabs), or RNase-Free DNase Set (Qiagen).
3. Ribo-Zero Magnetic kit (bacteria) (Illumina) and a magnetic rack or stand for 1.5 ml tubes.
4. RNeasy™ MinElute® Cleanup Kit (Qiagen).

2.3 Materials for RNA-Seq Sample Preparation and Sequencing

1. TruSeq® Stranded Total RNA Sample Preparation kit (Illumina).
2. HiSeq2500 or NextSeq, 100 or 75 bp paired-end (PE) library sequencing (Illumina).

2.4 Software and Computers for Sequencing Data Analysis

1. An Apple Mac or PC running a Linux operating system.
2. FastQC software (http://www.bioinformatics.babraham.ac.uk/projects/fastqc).
3. EDGE-pro software [14].
4. DESeq2 statistics tool: run using the R package [15].

3 Methods

The following methods are described based on a RNA-Seq experiment for a single bacterial strain with two growth conditions, one with drug treatment and the other without (*see* **Note 2**).

3.1 DNA-Free Total RNA Purification from Bacterial Cell Culture

This section describes methodologies for the extraction of total RNA from a pure culture of bacteria in the laboratory (**steps 1–5** should be performed using aseptic technique).

1. Streak the bacteria on MH agar plates without drug selection and grow overnight at a suitable temperature.
2. Pick an isolated colony and suspend the cells into a sterile 50 ml culture flask or conical tube containing 10 ml MH broth. Likewise, prepare three biological replicates. Grow the cell cultures overnight at a suitable temperature in a thermal incubator/shaker.
3. After ~16–18 h, pipette 300 μl of each overnight cell culture into a sterile 200 ml culture flask containing 30 ml MH broth (1:100 dilutions from overnight culture to fresh subculture). Incubate the subcultures in a thermal shaker and grow cells until mid-exponential phase.

4. Transfer 10 ml of each subculture at mid-exponential phase into each of two new 50 ml sterile culture flasks. Add an appropriate volume of drug solution into one flask (e.g. $0.5 \times$ MIC) and leave the other flask as a no drug control. In total there will be six samples; three biological replicates of drug-treated experimental samples and three biological replicates of drug-free controls. Incubate these six samples in a thermal shaker for 10–30 min (*see* **Note 3**).
5. Transfer the six cultures to 50 ml conical tubes and harvest cells by centrifugation for 5 min at $5000 \times g$ 4 °C. Carefully remove all supernatant (*see* **Note 4**).
6. Wipe all the equipment for the following steps in this section (e.g. pipettes, benches, tube racks, and gloves) with RNase*ZAP* (Ambion) prior to starting the procedure. Use nuclease-free barrier tips and, if possible, a set of dedicated "RNA-work only" pipettes.
7. Disrupt the cells by adding 700 μl of QIAzol reagent (*see* **Note 5**) and extract the total RNA following the miRNeasy Mini Kit Handbook protocol of "Purification of total RNA, including small RNAs, from animal cells" (*see* **Note 6**). Due to the limited nucleic acid binding capacity of the column membrane of the miRNeasy Mini Kit, do not use more than 1×10^{10} bacterial cells. Elute total RNA in 50 μl of RNase-free water in 1.5 ml RNase-free microcentrifuge tubes.
8. Remove any contaminating DNA from the RNA samples by using the TURBO DNA-*free*™ Kit TURBO™ (Ambion) DNase Treatment and Removal Reagents following manufacturer's instructions (*see* **Note 7**). Transfer DNA-free total RNA into fresh nuclease-free 1.5 ml tubes. Measure and record the concentration of the total RNA samples using a Spectrophotometer (*see* **Note 8**). Proceed to Subheading 3.2 immediately, otherwise store the RNA samples at −80 °C.

3.2 Ribosomal RNA Depletion

Ribosomal RNA typically comprises 80–95% of total extracted bacterial RNA [16]. Removing rRNA from total RNA can help to generate a comprehensive transcriptomic profile with reasonable sequencing cost, i.e., avoids excessive sequencing of rRNA and focuses sequencing on messenger RNAs and noncoding RNAs (ncRNA) which may have important regulatory functions. The Epicentre Ribo-Zero™ Magnetic Kit (Bacteria) is recommended for ribosomal RNA depletion in bacteria (*see* **Note 9**). This kit uses 1–5 μg DNA-free RNA, which is dissolved in RNase-free water or TE buffer.

1. Remove rRNA from 5 μg of each RNA sample by following the manufacturer's instructions for the Ribo-Zero™ Magnetic Kit (Bacteria), and purify rRNA-depleted RNA sample through RNeasy™ MinElute® Cleanup Kit.

2. Elute the RNA in 12 μl RNase-free water in nuclease-free 1.5 ml collection tubes supplied in the Qiagen Kit. The final RNA samples are ready for double-stranded cDNA library preparation. If not proceeding immediately store the RNA samples in a −80 °C freezer.

3.3 cDNA Library Preparation and Illumina Sequencing

Commonly applied methods for RNA-Seq require that RNA is first converted into double-stranded (ds) cDNA (*see* **Note 10**), which may then be sequenced using an Illumina system. It is important to employ a sequencing strategy which will result in sufficient sequencing depth to identify differentially expressed genes with statistical significance ($p < 0.001$) (*see* **Note 11**). Another common issue in RNA-Seq experiments is low-quality mapping, which may be due to repetitive sequences in the transcriptome. Paired-end sequencing technology can be used to improve mapping quality.

1. An Illumina cDNA library should be prepared from 0.1–1 μg of depleted RNA following the manufacturer's instructions (*see* **Notes 12** and **13**).
2. The sequence library can be sequenced on an Illumina MiSeq, NextSeq, or HiSeq platform. Each system has different sequencing capacity (*see* **Note 14**).

3.4 Transcriptomic Data Analysis

There will be two sets of data in fastq format generated from PE Illumina sequencing for each sample. Quality control checks should be performed on the raw sequence data to ensure that it is amenable to read mapping. Reads of sufficient quality can be aligned to a reference genome to count the relative transcript copies (count data) of each gene in the sample. There are multiple software packages and platforms suitable for prokaryotic sequence data alignment, including Rockhopper [17], RNA-Rocket [18], and EDGE-pro [14]. The workflow presented here is based on EDGE-pro, which uses the Bowtie aligner for read mapping. Statistical analyses should be conducted to determine the confidence of the gene expression changes between drug and no-drug-treated biological samples. This can be carried out using DESeq2 [19], an R package designed to analyze count data from RNA-Seq and test for differential expression. Both the EDGE-pro and DESeq2 scripts are run from the command line.

3.4.1 Checking Sequence Data Quality by FastQC

1. Download and unzip the FastQC software package from the Babraham Bioinformatics website (http://www.bioinformatics.bbsrc.ac.uk/projects/fastqc/).
2. Run FastQC as an interactive graphical application or through the command line, as per the installation and setup instructions

(http://www.bioinformatics.bbsrc.ac.uk/projects/fastqc/INSTALL.txt).

3. FastQC looks at ten filtering criteria. For detailed information about each filtering criteria or analysis module, please refer to [20] and the documents on this website: http://www.bioinformatics.bbsrc.ac.uk/projects/fastqc/Help/3%20Analysis%20Modules/.
4. RNA-Seq reads differ from gDNA reads, as they will comprise high proportions of duplicate or repeat reads due to multiple transcripts of highly expressed genes. Therefore criteria including "Sequence Duplication Levels" and "Kmer Content" are likely to be marked in the fastqc report as problematic (red cross or orange exclamation mark in summary list) (Fig. 1), even when data quality is acceptable. "Per base sequence content" may also be marked as problematic, due to Illumina systems generating low-quality signal of the first few bases (around 10–15 bp) of each read. This is unlikely to be a concern, as the read alignment tool bowtie 2 [21] used by EDGE-pro tolerates a short section of low-quality bases at either end of reads. It is therefore not generally necessary to perform end trimming of the raw sequence data. If the rest of the criteria pass the FastQC check, the fastq data is likely to be

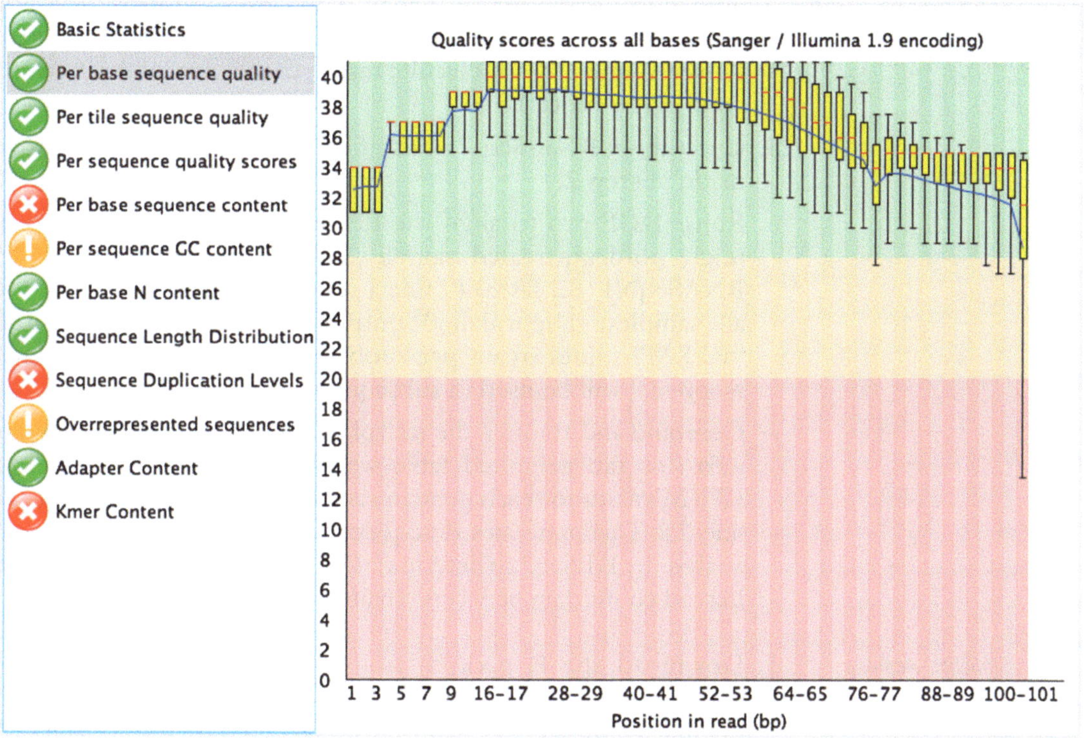

Fig. 1 FastQC screenshot showing a representative analysis of a fastq sequence file generated during an Illumina RNA-Seq workflow. Filtering criteria are listed down the *left side panel*. The *right panel* shows data from the active criteria, "Per base sequence quality" in the example shown

suitable to continue with read alignment. However, should overall sequence quality be low, the reads may be trimmed (*see* **Note 15**).

3.4.2 Gene Expression Level Analysis Using EDGE-pro

1. EDGE-pro can be run on a linux-x86_64 or 64 bit Mac machine (*see* **Note 16**).
2. Create a working directory specifically for EDGE-pro read alignment and DESeq2 differential count analysis.
3. Download the EDGE-pro package from http://ccb.jhu.edu/software/EDGE-pro/ and install following the instructions detailed in the readme file distributed with the software (*see* **Note 17**).
4. In addition to raw or trimmed RNA-Seq read files, EDGE-pro requires the following files as input: an fna file (fasta file containing the bacterial genome sequence), a ptt file (coordinates of coding genes in Genbank format), rnt (coordinates of rRNA and tRNA in Genbank format). Download the appropriate files for your particular reference genome from ncbi (ftp://ftp.ncbi.nih.gov/genomes/Bacteria) or if using a reference genome from another source, ensure that files are correctly formatted for input (refer to NCBI database information for formatting guidelines). Transfer all files into the working directory. For EDGE-pro it is necessary to supply a genome file that represents a single DNA molecule, rather than a mix of scaffolds or contigs (*see* **Note 18**).
5. Copy all six RNA-Seq PE fastq files into the working directory made in **step 2**.
6. Run EDGE-pro to align the PE reads of one of the samples (e.g., no-drug-treated sample 1, RNASeqN1_R1.fastq and RNASeqN1_R2.fastq) (Table 1, command line 1). Repeat for all samples. The rpkm_0 output files, which contain the "RPKM" values (reads per kilobase of gene per million reads mapped), will be used in subsequent steps.
7. Combine and transfer the six rpkm_0 files generated in **step 6** into one tab delimited table which lists the raw counts and RPKM values for each gene in each sample (Table 1, command line 2). This generates deseqFile, which is ready for statistical analysis by DESeq2 to determine if there is a significant difference in read count data between "treated" and "untreated" samples.

3.4.3 Statistical Analysis by DESeq2 in R

1. Download the R package from https://www.r-project.org/ and install following the instructions detailed in the installation manual. Stay in the same working directory and start R (Table 1, command line 3).
2. Download and install DESeq2 in R (Table 1, command lines 4 and 5).

3. Read deseqFile into R (Table 1, command lines 6–8).
4. Transform the raw count data format (deseqFile) into metadata format (Table 1, command lines 9–16), which will have defined biological sample information and features correlating with raw count data. For instance, there are six bacterial samples, representing three biological replicates with two conditions each

Table 1
List of example commands to use in transcriptomic data analysis

Task		Command line	Comments
EDGE-pro alignment	1	$ OMP_NUM_THREADS=n edge.pl –g xxx.fna –p xxx.ptt –r xxx.rnt –t n –u RNASeqN1_R1.fastq –v RNASeqN1_R2.fastq –o RNASeqN1	(1) $ = "working in the working directory", not part of the command line (2) xxx represents the prefix of your reference files (3) THREADS n could be 8, 16, or depending on the linux machine capacity. If THREADS = 8, then the "n" after –t needs to be 8 (4) Please refer to EDGE-pro manual for the features of each parameter (−g, −p, −r, −t, −u, −v, and −o) (5) RNASeqN1 is only an example of file name. You should use your own RNA-Seq data file name (6) The command line is case sensitive
deseqFile generation	2	$ [PATH]edgeToDeseq.perl RNASeqN1.rpkm_0 RNASeqN2.rpkm_0 RNASeqN3.rpkm_0 RNASeqT1.rpkm_0 RNASeqT2.rpkm_0 RNASeqT3.rpkm_0	Specify the path of edgeToDeseq.perl in the command line. RNASeqN1, RNASeqN2 etc. are only examples of file names. You should use your own rpkm_0 file name (*see* comments at command line 1 above)
Start R in the same working directory	3	$ R	
Download DESeq2	4	> source("https://bioconductor.org/biocLite.R")	(1) If the "https" address prefix does not work, try using "http" only (2) ">" = "working in R", not part of the command line
Install DESeq2 in R	5	> biocLite("DESeq2")	
Read deseqFile into R	6 7 8	> "deseqFile" > RNASeqCountTable = read.table("deseqFile", header=TRUE, row.name=1) > head(RNASeqCountTable)	The prefix RNASeq is only an example of file name. You could create your own file name

(continued)

Table 1
(continued)

Task	Command line	Comments
Metadata and DESeq2	9 > RNASeqDesign = data.frame(row.names = colnames(RNASeqCountTable condition = c("untreated", "untreated", "untreated", "treated", "treated", "treated"), libType = c("paired-end", "paired-end", "paired-end", "paired-end", "paired-end", "paired-end")) 10 > RNASeqDesign 11 > pairedSamples = RNASeqDesign $libType == "paired-end" 12 > condition = RNASeqDesign $condition[pairedSamples] 13 > head(RNASeqCountTable) 14 > condition 15 > library("DESeq") 16 > cds = newCountDataSet (RNASeqCountTable, condition)	RNASeq is only an example of file name. You could make your own file name
Normalization of count data	17 > cds = estimateSizeFactors(cds) 18 > sizeFactors(cds)	
Read count dispersion estimation	19 > cds = estimateDispersions(cds)	
Standard comparison between two experimental conditions	20 > res = nbinomTest(cds, "untreated", "treated")	This process could take several hours
Save DESeq2 output to a csv file	21 > write.csv(res, file="output_name.csv")	You can choose your own prefix for the csv file

(drug and no-drug treatment). The respective six sets of PE RNA-Seq raw count data need to be individually defined with conditions (treated and untreated) and library type as PE. This will generate "newCountDataSet" in the appropriate format, enabling DESeq2 to statistically analyze the count data in a meaningful way.

5. Normalize the raw count data based on nondifferentially expressed genes within the respective sample, so that gene expression levels of different biological samples become comparable among each other (Table 1, command lines 17 and 18).

6. Estimate the dispersion (or variance) between the gene expression levels of each individual sample (all biological replicates and drug treatment conditions) and the mean of the normalized counts of each gene (Table 1, command line 19). This is to test that for a given gene an observed difference in read counts is significant due to drug challenge, not random variation (Table 1, command line 20) [15].
7. Finally write the output to a csv file, which can be viewed as a spreadsheet in a program such as Microsoft Excel (command line 21). Please refer to DESeq2 manual for the interpretation of the column features in the table.
8. Open the ptt file of the reference genome as tab-delimited table in Microsoft Excel. Incorporate and correlate its columns of COG and product with the results in the DESeq2 Excel spreadsheet, so that the locus tag in the spreadsheet can be related with biological functions.

Following these data analyses it may be possible to identify genes of particular interest among those that display statistically different levels of transcription between the untreated control and the drug-treated samples. Putative efflux genes or genes encoding hypothetical proteins with predicted transmembrane-spanning segments, which have higher RNA-Seq read-coverage in drug-treated samples than in untreated control samples, are of interest for their putative resistance and transport functions. However, the RNA-Seq approach described in this chapter is only a first step in identifying novel drug efflux pumps. Further phenotypic and biochemical tests are essential to validate any hypotheses drawn from transcriptomic studies.

4 Notes

1. There are various other commercial total RNA extraction kits: PureLink® RNA Mini Kit (Ambion Inc., Applied Biosystems, Life Technologies, USA), UltraClean® Microbial RNA Isolation Kit (MO BIO Laboratories, Inc., USA). Note that some kits may bias against small RNAs which may have important regulatory functions and should ideally be sequenced with other transcripts. If information on the size of RNAs retained during extraction is not provided it is advised to check this with the manufacturer.
2. Read the manufacturer's protocols and associated safety data sheets of each commercial kit mentioned above before starting this experiment. Several reagents used in these kits contain hazardous substances, e.g. QIAzol reagent contains phenol and guanidine thiocyanate.

3. Typically, addition of antimicrobials will slow cell growth. We do not exceed a 30-min treatment to avoid progression of untreated controls into late exponential or stationary phase ahead of the drug-treated experimental samples. Such differences in growth phase could result in broad transcriptional changes that would complicate data interpretation.
4. Remove as much supernatant as possible, as excess residual media could interfere with the extraction procedure and compromise the quality of final product.
5. In our experience Gram-negative bacterial cells will quickly lyse in the QIAzol reagent with gentle agitation. When working with other cell types a more vigorous homogenization method may be required. The cell samples resuspended in QIAzol Lysis Reagent can be stored at −80 °C and should be stable for 1 month. Samples should be placed at −80 °C as soon as possible if not progressing immediately with the procedure.
6. The miRNeasy Mini Kit Handbook can be downloaded from: https://www.qiagen.com/us/resources/download.aspx?id=632801fb-abc5-4e62-b954-ff51f126a34f&lang=en.
7. The Ambion TURBO DNA-*free*™ Kit TURBO™ protocol can be downloaded from: http://tools.thermofisher.com/content/sfs/manuals/cms_055740.pdf.
8. Due to the small sample volume the amount of sample used to determine the concentration and purity should be kept to a minimum. We use 1–2 μl of sample on a Thermo Scientific NanoDrop spectrophotomer. The samples may also be run on a nuclease-free agarose gel to examine quality. At this stage clear bands corresponding to the rRNA should be visible.
9. In our experience, the Epicentre Ribo-Zero™ kit works well for ribosomal RNA depletion in Gram-negative bacteria. However, alternative kits are available, e.g., the MICROBExpress kit (Ambion).
10. Many workflows used for double cDNA library preparation rely on randomly primed cDNA synthesis, e.g., using the TruSeq RNA Sample Preparation v2 kit for Illumina sequencing systems. The data generated using these methods does not provide strand information and does not define transcript ends precisely or elucidate operons with multiple promoters. However, these methods are sufficient to identify differentially expressed protein-coding genes if they are well annotated in the genome sequence and can be useful in the study and discovery of novel drug efflux systems. More recently, with sufficient sequencing depth, strand-specific RNA ligation approaches in cDNA library preparation can ensure full read coverage and precise mapping of both the 5′ and 3′ transcript ends [22]. They enable researchers to reveal the full

transcriptome accurately and define bacterial operon architecture, including all the possible transcripts for protein-coding mRNAs and sRNAs [23], e.g., using the TruSeq® Stranded Total RNA Sample Preparation kit.

11. Coverage can be approximately calculated as [number of reads] × [read length (nt)]/genome size (nt). In general, 500–1000× genome coverage is enough to allow for reliable downstream statistical analyses of gene expression changes following RNA-Seq [24]. Therefore, 20–40 million 100 bp reads would be required to obtain this level of coverage for an average-sized bacterial genome. Most Illumina RNA-Seq work-flows provide for a greater number of sequence reads, e.g., one Illumina NextSeq (medium output kit) flow cell normally produces 260 million paired-end reads. Therefore it is generally possible to multiplex numerous bacterial RNA-Seq samples in a single sequencing run if using this sequencing platform.
12. Due to the specialist instrumentation and expertise required, RNA-Seq should be performed at a sequencing center. Most of these centers will also offer library preparation when provided with rRNA-depleted samples of sufficient quality.
13. We have previously used the TruSeq RNA Sample Preparation v2 kit for Illumina sequencing systems. As mentioned above (*see* **Note 10**) use of the TruSeq Stranded Total RNA Library Prep kit could alternatively be used to provide additional information regarding the transcriptome. Typically these kits require an input of approximately 500 ng of rRNA-depleted RNA sample. If the yield is very low alternative RNA-Seq sample prep methods are available [25]. For instance, Illumina supplies SMARTer® Ultra™ Low Input RNA kit for cDNA synthesis, and Nextera XT DNA library Preparation kit for RNA sequencing library preparation. Please consult with local sequencing suppliers, or the Illumina literature, for guidance.
14. Depending on the nature and scale of each RNA-Seq project, different Illumina sequencing systems may better suit the experimental design and prove to be more economical. To obtain sufficient sequencing depth of an averaged sized bacterial genome (~4 Mbp), MiSeq would be suitable for maximum three samples. However, NextSeq and HiSeq can accommodate many more samples (*see* **Note 11**).
15. Depending on the issue affecting read quality there are several options for read trimming, including trimming to a set read length (i.e. removing the first or last N bases), trimming to remove bases that fall below a certain quality threshold, or trimming to remove reads that fall below a certain overall quality threshold. Several tools for read trimming are freely

available, including FASTQ/A Trimmer (part of the FASTX-Toolkit; http://hannonlab.cshl.edu/fastx_toolkit/), Trim Galore! (http://www.bioinformatics.babraham.ac.uk/projects/trim_galore/), and trimmomatic (http://www.usadellab.org/cms/?page=trimmomatic). Please refer to the developer's notes for installation and usage.

16. The minimum computational requirements to effectively use the EDGE-pro software are detailed in the associated publication [14].
17. The EDGE-pro manual provides a full description of the features and analysis options available with this program (this manual is distributed with the software package).
18. Some genomes will contain multiple replicons (e.g., multiple chromosomes, or a chromosome and one or more plasmids). EDGE-pro can be run separately on each molecule, but the results should be interpreted with care if identical sequences are conserved between molecules.

Acknowledgements

This work was supported by a Macquarie University International Research Excellence Scholarship to L.L., an Australian Research Council Discovery Early Career Research Fellowship to S.G.T. (DE150100009), an Australian National Health and Medical Research Council Project Grant to I.T.P. and K.A.H. (1060895), and a Macquarie University Research Development Grant to K.A.H. (9201401563).

References

1. Kroeger JK, Hassan K, Vörös A et al (2015) Bacillus cereus efflux protein BC3310 – a multidrug transporter of the unknown major facilitator family, UMF-2. Front Microbiol 6:1063
2. Piddock LJ (2006) Multidrug-resistance efflux pumps – not just for resistance. Nat Rev Microbiol 4:629–636
3. McMurry LM, Petrucci RE Jr, Levy SB (1980) Active efflux of tetracycline encoded by four genetically different tetracycline resistance determinants in *Escherichia coli*. Proc Natl Acad Sci U S A 77:3974–3977
4. Tennent JM, Lyon BR, Gillespie MT et al (1985) Cloning and expression of *Staphylococcus aureus* plasmid-mediated quaternary ammonium resistance in *Escherichia coli*. Antimicrob Agents Chemother 27:79–83
5. Ubukata K, Itoh-Yamashita N, Konno M (1989) Cloning and expression of the *norA* gene for fluoroquinolone resistance in *Staphylococcus aureus*. Antimicrob Agents Chemother 33:1535–1539
6. Edgar R, Bibi E (1997) MdfA, an *Escherichia coli* multidrug resistance protein with an extraordinarily broad spectrum of drug recognition. J Bacteriol 179:2274–2280
7. Morita Y, Kodama K, Shiota S et al (1998) NorM, a putative multidrug efflux protein, of *Vibrio parahaemolyticus* and its homolog in *Escherichia coli*. Antimicrob Agents Chemother 42:1778–1782
8. Su XZ, Chen J, Mizushima T et al (2005) AbeM, an H^+-coupled *Acinetobacter baumannii* multidrug efflux pump belonging to the MATE family of transporters. Antimicrob Agents Chemother 49:4362–4364
9. Yamada Y, Hideka K, Shiota S et al (2006) Gene cloning and characterization of SdrM, a

chromosomally-encoded multidrug efflux pump, from *Staphylococcus aureus*. Biol Pharm Bull 29:554–556
10. Hassan KA, Brzoska AJ, Wilson NL et al (2011) Roles of DHA2 family transporters in drug resistance and iron homeostasis in Acinetobacter spp. J Mol Microbiol Biotechnol 20:116–124
11. Hassan KA, Li Q, Henderson PJF et al (2015) Homologs of the *Acinetobacter baumannii* AceI transporter represent a new family of bacterial multidrug efflux systems. MBio 6: e01982–e01914
12. Hassan KA, Jackson SM, Penesyan A et al (2013) Transcriptomic and biochemical analyses identify a family of chlorhexidine efflux proteins. Proc Natl Acad Sci U S A 110:20254–20259
13. Hassan KA, Elbourne LD, Li L et al (2015) An ace up their sleeve: a transcriptomic approach exposes the AceI efflux protein of Acinetobacter baumannii and reveals the drug efflux potential hidden in many microbial pathogens. Front Microbiol 6:333
14. Magoc T, Wood D, Salzberg SL (2013) EDGE-pro: estimated degree of gene expression in prokaryotic genomes. Evol Bioinforma 9:127–136
15. Anders S, Huber W (2010) Differential expression analysis for sequence count data. Genome Biol 11:R106
16. He S, Wurtzel O, Singh K et al (2010) Validation of two ribosomal RNA removal methods for microbial metatranscriptomics. Nat Methods 7:807–812
17. McClure R, Balasubramanian D, Sun Y et al (2013) Computational analysis of bacterial RNA-Seq data. Nucleic Acids Res 41:e140
18. Warren AS, Aurrecoechea C, Brunk B et al (2015) RNA-Rocket: an RNA-Seq analysis resource for infectious disease research. Bioinformatics:1–3
19. Love MI, Huber W, Anders S (2014) Moderated estimation of fold change and dispersion for RNA-seq data with DESeq2. Genome Biol 15:550
20. Reinert K, Langmead B, Weese D et al (2015) Alignment of next-generation sequencing reads. Annu Rev Genomics Hum Genet 16:133–151
21. Langmead B, Salzberg SL (2012) Fast and gapped-read alignment with Bowtie 2. Nat Methods 9:357–359
22. Levin JZ, Yassour M, Adiconis X et al (2010) Comprehensive comparative analysis of stranspecific RNA sequencing methods. Nat Methods 7:709–715
23. Conway T, Creecy JP, Maddox SM et al (2014) Unprecedented high-resolution view of bacterial operon architechture recealed by RNA sequencing. MBio 5:e01442–e01414
24. Haas B, Chin M, Nusbaum C et al (2012) How deep is deep enough for RNA-Seq profiling of bacterial transcriptomes? BMC Genomics 13:734
25. Adiconis X, Borges-Rivera D, Satija R et al (2013) Comparative analysis of RNA sequencing methods for degraded or low-input samples. Nat Methods 10:623–629

Part IV

Biomedical Approach for Bacterial Multidrug Exporters

Part IV

[illegible]

Chapter 13

Regulation of the Expression of Bacterial Multidrug Exporters by Two-Component Signal Transduction Systems

Kunihiko Nishino

Abstract

Bacterial multidrug exporters confer resistance to a wide range of antibiotics, dyes, and biocides. Recent studies have shown that there are many multidrug exporters encoded in bacterial genome. For example, it was experimentally identified that *E. coli* has at least 20 multidrug exporters. Because many of these multidrug exporters have overlapping substrate spectra, it is intriguing that bacteria, with their economically organized genomes, harbor such large sets of multidrug exporter genes. The key to understanding how bacteria utilize these multiple exporters lies in the regulation of exporter expression. Bacteria have developed signaling systems for eliciting a variety of adaptive responses to their environments. These adaptive responses are often mediated by two-component regulatory systems. In this chapter, the method to identify response regulators that affect expression of multidrug exporters is described.

Key words *Escherichia coli*, Expression, Multidrug exporters, Response regulator, Two-component signal transduction system

1 Introduction

Multidrug export is an obstacle to the successful treatment of infectious diseases, and it is mediated by multidrug exporters that recognize and efflux a broad spectrum of chemically dissimilar toxic compounds. Many bacterial genome sequences have been determined, allowing us to identify drug exporter genes encoded in the bacterial genome. In the *E. coli* genome, 37 ORFs are predicted to be drug exporter genes and it was experimentally identified that at least 20 of them contribute to drug resistance to *E. coli* [1]. Because many of these multidrug exporters have overlapping substrate spectra, it is intriguing that bacteria, with their economically organized genomes, harbor such large sets of multidrug export genes. The key to understanding how bacteria utilize these multiple exporters lies in the regulation of exporter expression. Currently available data show that multidrug exporters are often expressed under precise and elaborate transcriptional control [2]. Transcriptional activators

Akihito Yamaguchi and Kunihiko Nishino (eds.), *Bacterial Multidrug Exporters: Methods and Protocols*, Methods in Molecular Biology, vol. 1700, https://doi.org/10.1007/978-1-4939-7454-2_13, © Springer Science+Business Media, LLC 2018

and/or repressors tightly regulate the expression of a number of multidrug exporter genes. The transcriptional regulation of these export systems is achieved by local regulators, which are encoded in the same gene clusters as exporters and by global regulators, which are encoded in other genomic regions and whose regulatory action can also affect functions other than multidrug resistance [2–5].

It is well established that bacterial multidrug exporters confer clinically relevant resistance to antibiotics. Recent studies also showed that these exporters have important physiological roles, such as resistance to natural substances produced by host [6, 7]. It is reasonable to suppose that multidrug exporters are induced not only by antibiotics but also environmental signals existing in where bacteria live. Bacteria can adapt to a wide range of environmental conditions. These adaptive responses are generally mediated by two-component signal transduction systems, which consist of a sensor histidine kinase and its cognate response regulator. Each sensor detects a specific environmental signal, and the histidine residue self-phosphorylates. This phosphate group is then transferred to the specific aspartic acid in the response regulator. The regulator mostly acts as a transcriptional factor to control the expression of genes that have various activities in biological reactions [8, 9]. The entire genomic sequence of *E. coli* allowed us to systematically compile a complete list of genes encoding such two-component signal transduction proteins. It is estimated that *E. coli* has a total of 30 sensors and 34 response regulators [10]. In this article, the effect of the three response regulators, EvgA, BaeR, and CpxR on the expression of multidrug exporters is described [11–17].

2 Materials

2.1 Bacterial Strains and Growth Media

1. *Escherichia coli* MG1655 strain is used for isolation of the genomic DNA and analysis of multidrug exporter expression. *E. coli* TG1 strain is used for cloning the *evgA*, *baeR*, and *cpxR* genes.
2. *Escherichia coli* MG1655Δ*acrB* strain is used for drug susceptibility assay.
3. LB (Lauria-Bertani) Broth, Miller (Becton, Dickinson and Company): Tryptone 10.0 g/L, Yeast Extract 5.0 g/L, Sodium Chloride 10.0 g/L.
4. LB (Lauria-Bertani) Agar, Miller (Becton, Dickinson and Company).
5. Ampicillin is added to the growth media to be concentration of 100 μg/ml when needed.
6. Chloramphenicol is added to the growth media to be concentration of 20 μg/ml when needed (*see* **Note 1**).

2.2 Cloning of Response Regulators

1. Primers *evgA*-F (5′-GCGGAATTCTATTATCTCTCATTTC TCATA-3′) and *evgA*-R (5′-GCGAAGCTTTTAGCCGATTT mTGTTACGTTG-3′) used for the cloning the *evgA* gene into the pUC18 or pHSG398 vector (*see* **Note 2**).
2. Primers *baeR*-F (5′-GCGGAATTCTTGAAGCACATAATGG TCGCA-3′) and *baeR*-R (5′-GCGAAGCTTCTAAACGATG CGGCAGGCGTC-3′) used for the cloning the *baeR* gene into the pUC18 or pHSG398 vector (*see* **Note 2**).
3. Primers *cpxR*-F (5′-GCGGAATTCTTGCTCCCAAAATCTT TCTGT-3′) and *cpxR*-R (5′-GCGAAGCTTTCATGAAGC AGAAACCATCAG-3′) used for the cloning the *cpxR* gene into the pUC18 or pHSG398 vector (*see* **Note 2**). (1, 2, and 3 primers. Introduced restriction sites of *Eco*RI and *Hind*III used for cloning are underlined.)
4. QIAamp DNA Mini Kit (Qiagen) used for purification of the genomic DNA of *E. coli* MG1655.
5. PrimeSTAR Max DNA Polymerase (Takara Bio Inc.) used for the amplification of the *evgA*, *baeR*, and *cpxR* genes by PCR.
6. ProFlex PCR System (Applied Biosystems) used as a thermal cycler for PCR.
7. QIAquick PCR Purification Kit (Qiagen) used for the purification of the amplified DNA.
8. QIAEXII Gel Extraction Kit (Qiagen).
9. TaKaRa DNA Ligation Kit Ver.2.1 (Takara Bio Inc.).
10. QIAprep Spin Miniprep Kit (Qiagen).
11. M13 Primer M4 and M13 Primer RV (TaKaRa Bio Inc.).
12. BigDye Terminator V3.1 (Applied Biosystems).
13. Genetic Analyzer 3100 Avant (Applied Biosystems).

2.3 Drug Susceptibility Assay

1. MicroPlanter MITP-60P (Sakuma Seisakusyo) used for the inoculation of bacterial cell cultures onto LB agar plates containing multiple drugs for measuring Minimum Inhibitory Concentrations (MICs).
2. LB agar plates containing erythromycin, kanamycin, doxorubicin, novobiocin, crystal violet, rhodamine 6G, tetraphenylphosphinium, benzalkonium, sodium dodecyl sulfate, deoxycholate, oxacillin, cefamandole, ceftazidime, aztreonam, amikacin, carbenicillin, cloxacillin, or carmonam at various concentrations used for the measurement of MICs.

2.4 Extraction of RNA

1. RNAprotect Bacteria Reagent (Qiagen) used for the stabilization of RNA before preparation.
2. Lysozyme from chicken egg white (Sigma-Aldrich).

3. TE Buffer, 1×, Molecular Biology Grade, DNase/RNase/Protease None Detected (Promega).
4. SV Total RNA Isolation System (Promega).
5. Nuclease-Free Water (Promega).
6. Collection Tubes, DNase/RNase/Protease None Detected (Ambion).
7. RT-20F, 30 μl Presterilized filter tips, DNA-, DNase, Endotoxin-, RNase-, and ATP-Free (Rainin).
8. RT-200F, 200 μl Presterilized filter tips, DNA-, DNase, Endotoxin-, RNase-, and ATP-Free (Rainin).
9. RT-1000S, 1000 μl Presterilized tips, DNA-, DNase, Endotoxin-, RNase-, and ATP-Free (Rainin).

2.5 Real-Time Quantitative RT-PCR

1. TaqMan Reverse Transcription Reagents (Applied Biosystems).
2. Veriti Thermal Cycler (Applied Biosystems) used for the reverse transcription.
3. MicroAmp, Optical 96-Well Reaction Plate (Applied Biosystems).
4. SYBR Green PCR Master Mix (Applied Biosystems).
5. StepOne Plus, Real-Time PCR System (Applied Biosystems).

2.6 Drug Efflux Assay

1. Hitachi model F-2000 fluorescence spectrophotometer (Hitachi).
2. Doxorubicin, rhodamine 6G and carbonyl cyanide *m*-chlorophenyl hydrazone.

3 Methods

3.1 Cloning of the Response Regulator Genes

1. Under sterile conditions, streak out a 10 μl loop of *E. coli* MG1655 strain onto fresh LB agar plates. Grow over night at 37 °C.
2. On the following day, inoculate 5 ml of LB medium with single colony of the MG1655 strain. Grow the culture overnight at 37 °C with shaking at 200 rpm.
3. Pellet bacteria by centrifugation for 10 min at 5500× *g*.
4. Isolate the genomic DNA from the culture of *E. coli* MG1655 by using QIAamp DNA Mini Kit (Qiagen) (*see* **Note 3**).
5. After the isolation of the genomic DNA, measure the quantity and quality of DNA by the absorption spectrum at a wavelength of 260 and 280 nm (A_{260} and A_{280}).
6. For amplifying the response regulator genes, mix 25 μl PrimeSTAR Max Premix with 15 pmol forward primer, 15 pmol

reverse primer, <200 ng genomic DNA, and sterilized water up to 50 μl. Use primers *evgA*-F and *evgA*-R for *evgA*, *baeR*-F and *baeR*-R for *baeR*, *cpxR*-F, and *cpxR*-R for *cpxA*.

7. PCR with the cycling condition as follows: 35 cycles of 98 °C for 10 s, 55 °C for 15 s, and 72 °C for 10 s.
8. Use gel electrophoresis to check PCR reactions.
9. Purify the amplified DNA by using QIAquick PCR Purification Kit (Qiagen).
10. Digest the amplified DNA fragments, pUC18, and pHSG398 with the restriction enzymes *Eco*RI and *Hind*III.
11. After the gel electrophoresis of the digested DNA fragments, purify DNA with QIAEXII Gel Extraction Kit (Qiagen).
12. Ligate 0.1 pmol DNA fragment containing *evgA*, *baeR*, or *cpxR* with 0.03 pmol digested pUC18 or pHSG398 using TaKaRa DNA Ligation Kit Ver. 2.1 Solution I (Takara Bio Inc.) (*see* **Note 2**).
13. Incubate at 16 °C for 30 min.
14. Transform 10 μl ligated products to 100 μl competent cells of TG1.
15. Streak the transformants onto LB agar plates containing 100 μg/ml ampicillin (for pUC18 plasmids carrying *evgA*, *baeR*, or *cpxR*) or 20 μg/ml chloramphenicol (for pHSG398 plasmids carrying *evgA*, *baeR*, or *cpxR*). Grow overnight at 37 °C.
16. Pickup the colonies and streak onto LB agar plates containing 100 μg/ml ampicillin or 20 μg/ml chloramphenicol again for the selection of single colonies. Grow overnight at 37 °C.
17. The following day, purify plasmid DNA from the bacterial cell cultures using QIAprep Spin Miniprep Kit (Qiagen).
18. Confirm the size of the inserts by gel electrophoresis after digestion with *Eco*RI and *Hind*III. Check the DNA sequences of the inserts by using M13 Primer M4, M13 Primer RV (TaKaRa Bio Inc.), BigDye Terminator V3.1 (Applied Biosystems), and Genetic Analyzer 3100 Avant (Applied Biosystems).
19. Transform the pUC18, pHSG398, pUC18-*evgA*, -*baeR*, or -*cpxR*, and pHSG398-*evgA*, -*baeR*, or -*cpxR* to the competent cells of *E. coli* MG1655 and MG1655Δ*acrB* strain to make the strains for real-time quantitative RT-PCR and drug susceptibility assay.

3.2 Measurement of Minimum Inhibitory Concentrations

1. Under sterile conditions, streak out a 10 μl loop of the *E. coli* wild-type MG1655, MG1655Δ*acrB*, MG1655Δ*acrB*/pUC18, MG1655Δ*acrB*/pHSG398, MG1655Δ*acrB*/pUC18-*evgA*, MG1655Δ*acrB*/pUC18-*baeR*, MG1655Δ*acrB*/pUC18-*cpxR*, MG1655Δ*acrB*/ pHSG398-*evgA*, MG1655Δ*acrB*/ pHSG398-*baeR*, and MG1655Δ*acrB*/pHSG398-*cpxR* strains onto fresh LB agar plates. For the strains harboring the pUC18 vector or plasmids, use LB agar plates containing 100 μg/ml ampicillin. For the strains harboring the pHSG398 vector or plasmids, use LB agar plates containing 20 μg/ml chloramphenicol.
2. On the following day, inoculate 5 ml of LB medium with single colony of the strains above. Grow the culture overnight at 37 °C with shaking at 200 rpm.
3. Prepare antimicrobial solutions and agar plates for agar dilution susceptibility testing. For agar dilution, I recommend following the international guidelines given by the NCCLS (NCCLS document M7-A5 "Methods for dilution antimicrobial susceptibility tests for bacteria that grow aerobically"). Prepare 10 ml LB agar in 90 mm × 15 mm sterile petri dish (BIO-BIK, Ina Optica Co. Ltd). Make agar plates containing various drugs made by the twofold agar dilution technique. Minimum inhibitory concentrations (MICs) of drugs are determined on LB agar plates containing erythromycin (0.5–256 μg/ml), kanamycin (0.5–32 μg/ml), doxorubicin (0.5–256 μg/ml), novobiocin (0.125–256 μg/ml), crystal violet (0.25–32 μg/ml), rhodamine 6G (2–512 μg/ml), tetraphenylphosphinium (2–512 μg/ml), benzalkonium (0.5–256 μg/ml), sodium dodecyl sulfate (16–131,072 μg/ml), deoxycholate (256–32,768 μg/ml), oxacillin (0.125–512 μg/ml), cefamandole (0.0625–8 μg/ml), ceftazidime (0.0625–4 μg/ml), aztreonam (0.0313–4 μg/ml), amikacin (0.125–32 μg/ml), carbenicillin (1–128 μg/ml), cloxacillin (0.5–512 μg/ml), or carmonam (0.0625–4 μg/ml).
4. Dilute the grown bacterial cells into LB broth and inoculate the organisms at a 10^4 cfu/μl onto MIC agar plates with the use of a multipoint inoculator, MicroPlanter MITP-60P (Sakuma Seisakusyo) (Fig. 1).
5. Incubate the inoculated plates at 37 °C for 20 h.
6. Judge the MICs. The MIC is the lowest concentration of drug that inhibits cell growth.

3.3 Extraction of RNA

1. Under sterile conditions, streak out a 10 μl loop of the *E. coli* MG1655/pUC18, MG1655/pUC18-*evgA*, MG1655/ pUC18-*baeR*, MG1655/pUC18-*cpxR* strains onto fresh LB agar plates containing 100 μg/ml ampicillin.

Fig. 1 Inoculation of bacterial cells onto MIC agar plates. A multipoint inoculator, MicroPlanter MITP-60P (Sakuma Seisakusyo) is able to inoculate 1 μl of 60 different bacterial cultures onto an agar plate

2. On the following day, inoculate 5 ml of LB medium containing 100 μg/ml ampicillin with single colony of the strains above. Grow the culture overnight at 37 °C with shaking at 200 rpm.
3. Inoculate bacterial cultures into the LB medium containing 100 μg/ml ampicillin and grow them until an optical density at 600 nm of 0.6.
4. Mix 500 μl of cell cultures with 1 ml RNAprotect Bacteria Reagent (Qiagen) by vortex for 5 s (*see* **Notes 3–6**).
5. Leave the mixture for 5 min at room temperature.
6. Collect samples by centrifugation for 10 min at 6000× *g*.
7. Suspend pellet with 100 μl of TE Buffer containing 0.4 mg/ml lysozyme.
8. Incubate at room temperature for 5 min.
9. Isolate RNA according to the manufacturer's instructions as follows.
10. Add 75 μl of SV RNA Lysis Buffer and then mix by pipetting.
11. Add 350 μl of SV RNA Dilution Buffer then mix by inversion.
12. Add 200 μl of 95% ethanol then mix gently by pipetting.
13. Transfer the mixture to the Spin Column Assembly. Spin-down at 19,000× *g* for 1 min. Empty the solution in the Collection Tube.

14. Add 600 μl of SV RNA Wash Solution. Spin-down at 19,000× *g* for 1 min. Empty the solution in the Collection Tube.
15. Add 50 μl of DNase I incubation mixture directly to the membrane inside the Spin Basket. Incubate at room temperature for 15 min.
16. Add 200 μl of SV DNase Stop Solution to the Spin Basket. Spin-down.
17. Add 600 μl of SV RNA Wash Solution. Spin-down and empty the solution in the Collection Tube.
18. Add 250 μl of SV RNA Wash Solution. Spin-down for 2 min.
19. Prepare 1.5 ml Elution Tube from the packaging. Transfer the Spin Basket from the Collection Tube to the Elution tube.
20. Add 100 μl of Nuclease-Free Water to the membrane. Then, spin-down.
21. Inspect the absence of genomic DNA in RNA samples by agarose electrophoresis gels. Determine the RNA concentrations and their quality by spectrophotometrically at a wavelength of 260 and 280 nm (A_{260} and A_{280}).

3.4 Real-Time Quantitative RT-PCR

1. Perform the reverse transcription by using TaqMan Reverse Transcription Reagents (Applied Biosystems) according to the manufacturer's instructions. Mix 200 ng of the purified RNA with 5 μl of Random Hexamers (50 μM), 20 μl of dNTP Mixture (2.5 mM each dNTP), 25 μl of $MgCl_2$ Solution (25 mM), 10 μl of 10× RT Buffer, 2 μl of RNase Inhibitor (20 U/μl), and 2.5 μl of MultiScribe Reverse Transcriptase (50 U/μl), and add Nuclease-Free Water to be a total reaction volume of 100 μl.
2. Reaction with the condition as follows: 25 °C for 10 min, 37 °C for 60 min, 95 °C for 5 min, and then keep cDNA samples at 4 °C until the preparation of a real-time quantitative PCR (*see* **Note 7**).
3. Mix 3 μl of cDNA sample with primer mix for each gene (1 μM each) listed in Table 1, 10 μl of 2× SYBR Green PCR Master Mix (Applied Biosystems), and Nuclease-Free Water to be a total reaction volume of 20 μl.
4. Use StepOne Plus (Applied Biosystems) for real-time quantitative PCR with the cycling conditions as follows: 95 °C for 5 min followed by 40 cycles of 95 °C for 10 s and 60 °C for 15 s. After each run, verify amplification specificity and the absence of primer dimers using a dissociation curve acquired by heating the PCR products from 60 to 95 °C. The relative quantities of transcripts were determined using a standard curve and normalized against the geometric mean of two reference genes (*rrlA* of 23S rRNA and *rrsA* of 16S rRNA) (Fig. 2) (*see* **Notes 8–10**).

Table 1
Primers for quantitative PCR

Primer	Sequence (5′ to 3′)
acrA-F	GTCTATCACCCTACGCGCTATCTT
acrA-R	GCGCGCACGAACATACC
acrD-F	GTACCCTGGCGATTTTTTCATT
acrD-R	CGGTCACTCGCACATTCG
acrE-F	CGTGATTGCCGCAAAAGC
acrE-R	TTGGCGCAGTGACTTTGGTA
bcr-F	TGTTTTTCTGTTCGTGATGACCAT
bcr-R	GGAACATATTTAACGCGCCAAT
cusB-F	CGCTTACCGTGGGCGATA
cusB-R	TTCCACCCAGTCAGGAATGG
emrA-F	GCGAATATTGAGGTGCAGAAAA
emrA-R	GGCACACGGCGGTTGTA
emrD-F	GTGGATCCCCGACTGGTTT
emrD-R	CCCGGCACCGAAAAAGA
emrE-F	GGTATTGTCCTGATTAGCTTACTGTCAT
emrE-R	GCACAAATCAACATCATGCCTATAA
emrK-F	GCGCTTAAACGTACGGATATTAAGA
emrK-R	ACTGTTTCGCCGACCTGAAC
fsr-F	TGGTGTTGGCGCAAATCA
fsr-R	TCGTCGCTTTGGGTTTTCC
macA-F	CGGTGATTGCCGCACAA
macA-R	TTACCAGCATGGCGCTCAT
mdfA-F	CTTGCTGTTAGCGCGTCTGA
mdfA-R	GCCAGCCGCCCATAATAAT
mdtA-F	CGCCGTAGAACAGGCAGTTC
mdtA-R	TGCGCACCGTAACGGTATTA
mdtE-F	CCCCCGGTTCGGTCAA
mdtE-R	GGACGTATCTCGGCAACTTCAT
mdtF-F	TTACCGTCAGCGCTACCTATCC
mdtF-R	GCCATCAAGCCCATTCATATTT
mdtG-F	CGGTATTGTCTTCAGCATTACATTTT

(continued)

Table 1
(continued)

Primer	Sequence (5′ to 3′)
mdtG-R	GGCGAGTCCACCCCAAA
mdtH-F	TTTTCACCCTGATTTGTCTGTTTTAT
mdtH-R	CAGCGAAGCACTTAAGGTTTCA
mdtJ-F	TGATGAAAATTGCCGGGTTAA
mdtJ-R	CGCTTTACGGGTACCTGATTTTA
mdtK-F	CCGGTTATCGCGCAATTAAAT
mdtK-R	GAAACCTTGTCGCACCTGATG
mdtL-F	TATCCCGCCGGGATTGATAT
mdtL-R	CGCTTCGCTGGCATTGA
mdtM-F	CGTGATTTTAATGCCGATGTCA
mdtM-R	GCCATACCGCCAGCAAGAT
tolC-F	CCGGGATTTCTGACACCTCTT
tolC-R	TTTGTTCTGGCCCATATTGCT

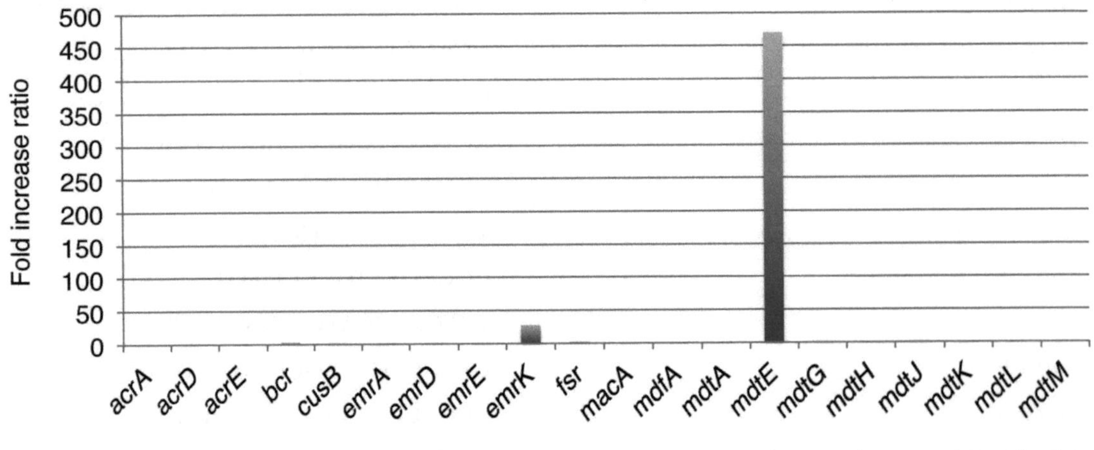

Fig. 2 EvgA activates the expression of *emrK* and *mdtE*. The amount of transcript was determined by quantitative real-time PCR. The fold increase ratio was calculated by dividing the expression level of the gene in the MG1655/pUC18-*evgA* strain by that in the MG1655/pUC18 strain

3.5 Drug Efflux Assay

1. Under sterile conditions, streak out a 10 μl loop of the *E. coli* MG1655Δ*acrB*, MG1655Δ*acrB*/pUC18, MG1655Δ*acrB*/pUC18-*evgA* strains onto fresh LB agar plates containing 100 μg/ml ampicillin.

2. On the following day, inoculate 5 ml of LB medium with single colony of the strains above. Grow the culture overnight at 37 °C with shaking at 200 rpm.
3. Inoculate bacterial cultures into the LB medium and grow them until an optical density at 600 nm of 0.6.
4. Harvest cells by centrifugation for 10 min at 8000 rpm and wash twice with 100 mM potassium phosphate buffer (pH 7.5) containing 5 mM $MgSO_4$.
5. For maximal accumulation of the fluorophore, incubate the cells (optical density at 600 nm of 1.0 for rhodamine 6G or 20 for doxorubicin) with 1 μM rhodamine 6G or 11.5 μM doxorubicin and 40 μM carbonyl cyanide *m*-chlorophenyl hydrazone at 37 °C for 1 h.
6. Collect cells by centrifugation for 10 min at 8000 rpm and resuspend in the potassium phosphate buffer containing 25 mM glucose to energize the cells, and subject them to fluorescence measurement.
7. To measure the active efflux of rhodamine 6G or doxorubicin from the cells, monitor the fluorescence of the compounds (with excitation at 529 nm and emission at 553 nm for rhodamine 6G, and with excitation at 478 nm and emission at 591 nm for doxorubicin) with a Hitachi model F-2000 fluorescence spectrophotometer.

4 Notes

1. Because the *acrB* deleted strain is more sensitive to chloramphenicol than the wild-type strain, use lower concentration of chloramphenicol such as 10 μg/ml for the selection when it is difficult to isolate transformants of the *acrB* deleted strain harboring the pHSG plasmids.
2. Because the pUC18 plasmids carry *bla* (Ap^r), use the pHSG plasmids to measure the effects of the *evgA*, *baeR*, and *cpxR* on the *E. coli* susceptibilities to β-lactams.
3. When purifying the genomic DNA or total RNA, do not put much number of bacterial cells into the columns in order to prepare good quality samples.
4. RNA Protect Bacteria Reagent can be stored at room temperature and the reagent is stable for at least 12 months under this condition.
5. When using RNA Protect Bacterial Reagent, RNA is stabilized before bacterial cells are lysed. After RNA stabilization by RNA Protect Bacteria Regent, bacterial cells can be pelleted by

centrifugation and the pellets can be frozen and stored at −20 °C for up to 2 weeks, or at −70 °C for up to 4 weeks.

6. Because RNases are very stable and they are difficult to inactivate, do not use any plasticware or glassware without first eliminating possible RNase contamination.
7. After reverse transcription, cDNA samples can be stored at −20 °C before quantitative PCR.
8. Overexpression of *evgA* activates the expression of the multidrug efflux genes, *mdtEF* (*yhiUV*) and *emrKY*, resulting in increasing multidrug resistance level of *E. coli* cells.
9. Overexpression of *baeR* activates the expression of the multidrug efflux genes, *acrD* and *mdtABC*, resulting in increasing multidrug resistance level of *E. coli.*
10. Overexpression of *cpxR* activates the expression of the multidrug efflux genes, *acrD* and *mdtABC*, resulting in increasing multidrug resistance level of *E. coli.*

Acknowledgments

I gratefully acknowledge Japan Agency for Medical Research and Development (AMED), the Center of Innovation Program from Japan Science and Technology Agency (JST), Japan Society for the Promotion of Science, Japan Science and Technology Agency (JSPS), the International Joint Research Program from Osaka University, the Cooperative Research Program of "Network Joint Research Center for Materials and Devices", the Nano-Macro Materials, Devices and System Research Alliance, the Ministry of Education, Culture, Sports, Science and Technology (MEXT) for funding.

References

1. Nishino K, Yamaguchi A (2001) Analysis of a complete library of putative drug transporter genes in *Escherichia coli.* J Bacteriol 183:5803–5812
2. Nishino K, Nikaido E, Yamaguchi A (2009) Regulation and physiological function of multidrug efflux pumps in *Escherichia coli* and *Salmonella.* Biochim Biophys Acta 1794:834–843
3. Li XZ, Nikaido H (2009) Efflux-mediated drug resistance in bacteria: an update. Drugs 69:1555–1623
4. Nikaido H (2009) Multidrug resistance in bacteria. Annu Rev Biochem 78:119–146
5. Routh MD, Su CC, Zhang Q, Yu EW (2009) Structures of AcrR and CmeR: insight into the mechanisms of transcriptional repression and multi-drug recognition in the TetR family of regulators. Biochim Biophys Acta 1794:844–851
6. Nishino K, Yamaguchi A (2008) Role of xenobiotic transporters in bacterial drug resistance and virulence. IUBMB Life 60:569–574
7. Piddock LJ (2006) Multidrug-resistance efflux pumps – not just for resistance. Nat Rev Microbiol 4:629–636
8. Hoch JA, Varughese KI (2001) Keeping signals straight in phosphorelay signal transduction. J Bacteriol 183:4941–4949
9. Parkinson JS (1993) Signal transduction schemes of bacteria. Cell 73:857–871

10. Mizuno T (1997) Compilation of all genes encoding two-component phosphotransfer signal transducers in the genome of *Escherichia coli*. DNA Res 4:161–168
11. Nishino K, Yamaguchi A (2001) Overexpression of the response regulator *evgA* of the two-component signal transduction system modulates multidrug resistance conferred by multidrug resistance transporters. J Bacteriol 183:1455–1458
12. Nishino K, Yamaguchi A (2002) EvgA of the two-component signal transduction system modulates production of the *yhiUV* multidrug transporter in *Escherichia coli*. J Bacteriol 184:2319–2323
13. Nagakubo S, Nishino K, Hirata T, Yamaguchi A (2002) The putative response regulator BaeR stimulates multidrug resistance of *Escherichia coli* via a novel multidrug exporter system, MdtABC. J Bacteriol 184:4161–4167
14. Hirakawa H, Nishino K, Hirata T, Yamaguchi A (2003) Comprehensive studies of drug resistance mediated by overexpression of response regulators of two-component signal transduction systems in *Escherichia coli*. J Bacteriol 185:1851–1856
15. Nishino K, Inazumi Y, Yamaguchi A (2003) Global analysis of genes regulated by EvgA of the two-component regulatory system in *Escherichia coli*. J Bacteriol 185:2667–2672
16. Hirakawa H, Nishino K, Yamada J, Hirata T, Yamaguchi A (2003) ß-Lactam resistance modulated by the overexpression of response regulators of two-component signal transduction systems in *Escherichia coli*. J Antimicrob Chemother 52:576–582
17. Nishino K, Honda T, Yamaguchi A (2005) Genome-wide analyses of *Escherichia coli* gene expression responsive to the BaeSR two-component regulatory system. J Bacteriol 187:1763–1772

Chapter 14

Study of the Expression of Bacterial Multidrug Efflux Pumps in Anaerobic Conditions

Jingjing Sun, Ziqing Deng, Danny Ka Chun Fung, and Aixin Yan

Abstract

Bacterial multidrug efflux pumps belong to a class of membrane transporter proteins that dedicate to the extrusion of a diverse range of substances out of cells including all classes of currently available antibiotics. They constitute an important mechanism of bacterial antibiotic and multidrug resistance. Since many ecological niches of bacteria and the infection foci in animal host display low oxygen tension under which condition bacterial pathogens undergo fundamental changes on their metabolic modes, it is necessary to study the expression profiles of drug efflux pumps under these physiologically and clinically relevant conditions. In this chapter, we first introduce procedures to culture bacteria under anaerobic conditions, which is achieved using screw-capped Pyrex culture tubes without agitation. We then introduce β-galactosidase activity assay using promoter-*lacZ* (encoding the β-galactosidase enzyme) fusion to measure the expression of efflux pumps at transcriptional level, and Western blot using chromosomal FLAG-tagged construct to examine the expression of these proteins at translational level. Applications of these gene expression studies to reveal the regulatory mechanisms of efflux genes expression as well as their physiological functions are also discussed.

Key words Multidrug efflux pump, Multidrug resistance, Anaerobiosis, β-galactosidase activity assay, SDS-PAGE, Western blot

1 Introduction

Bacterial multidrug efflux pumps belong to a class of membrane transporter proteins which dedicate to the extrusion of a diverse range of structurally unrelated substances out of bacterial cells. The substrate profiles of efflux pumps include virtually all classes of currently available antibiotics [1]. Consequently, they constitute an important mechanism of antibiotic resistance in bacteria. Unlike other resistance mechanisms, such as target mutation or drug modification which confer resistance to a certain class of antibiotics, overexpression of drug efflux pumps almost always leads to

Jingjing Sun, Ziqing Deng and Danny Ka Chun Fung contributed equally to this work.

Akihito Yamaguchi and Kunihiko Nishino (eds.), *Bacterial Multidrug Exporters: Methods and Protocols*, Methods in Molecular Biology, vol. 1700, https://doi.org/10.1007/978-1-4939-7454-2_14, © Springer Science+Business Media, LLC 2018

simultaneous resistance to multiple drugs, i.e. multidrug resistance (MDR). Thus, it is important to study the environmental or physiological factors that lead to efflux pump overexpression and its underlying regulatory mechanism.

A frequently encountered environmental signal in many natural ecological niches and animal host of bacteria is low oxygen tension [2]. In animal host, these include inflamed tissues, infected skins, animal intestine tract, and biofilm interior [3, 4]. In addition, antibiotics, host-derived antimicrobial peptides [5], reactive oxygen species [6], toxic metabolic products, etc. [7] are often present in these niches. Study of the expression profiles of efflux pump genes under these conditions will advance not only our understanding of the emergence of multidrug resistance but also the process of bacterial stress adaptation.

Indeed, in recent years, many clinical multidrug-resistant strains isolated from anaerobic niches of infections were found to display elevated efflux pump expression. For example, Schaible et al. reported that several clinical *Pseudomonas aeruginosa* strains isolated from chronically infected cystic fibrosis patients were found to display resistance to penicillins and cephalosporins due to MexE overexpression [8]. *Listeria monocytogenes* strains isolated from the anaerobic intestinal tract of patients in a hospital at Los Angeles, California, were found to display resistance to bile acid due to MdrT overexpression [9]. Bradford et al. reported that *Klebsiella pneumonia* strains isolated from complicated skin and intra-abdominal infections display resistance to tetracycline due to AcrB overexpression [10]. It was believed that efflux pump overexpression under these conditions contributes to both drug resistance and bacterial stress adaption. This phenomenon is not unexpected as many efflux pump genes are broadly distributed in bacterial genomes prior to the development of antibiotics and they have been recognized to have physiological roles in bacteria as a general means of detoxification. Indeed, increasing studies have demonstrated that efflux pumps are involved in a variety of physiological and pathogenic processes including invasion, proliferation in host, biofilm formation, and expelling nonmetabolizable waste byproducts. [7, 11–13]. Given the prevalence of low oxygen tensions in infection sites, it is necessary to systematically study the expression and physiological roles of multidrug efflux pumps in these environments.

Unlike under aerobic conditions where microorganisms exclusively respire oxygen, during anaerobiosis, metabolism is shifted toward either fermentation [14] or anaerobic respiration when alternative electron acceptors such as nitrate, fumarate, and sulfate are present [15]. To adapt to this diverse mode of anaerobiosis, bacteria must undergo global gene expression changes to coordinate the energy metabolism with various physiological activities including nutrients uptake and waste product extrusion which is largely carried out by efflux pumps. This often requires the involvement of various global and local regulators. Identification of these transcription regulators is essential to understand the induction of efflux pump genes during physiological adaptation.

In this chapter, we introduce methods to study the expression of efflux genes under anaerobic conditions using *Escherichia coli* as an example. We first introduce procedures to culture bacteria under anaerobic conditions, followed by the methods to study the expression of efflux pump genes both at transcriptional and translational levels using the well-established promoter-*lacZ* transcription fusion and chromosomally FLAG-tagged construct, respectively. Since the components of efflux systems include both periplasmic soluble proteins and integral membrane proteins, we use MdtE which is the periplasmic fusion protein of the MdtEF-TolC efflux system and CusC which is the outer membrane channel component of the CusCBA system as examples to illustrate the study of expression of the soluble and integral membrane protein component, respectively. Applications of these methodologies to reveal the regulators of efflux systems and their physiological roles will also be briefly described.

2 Materials

Prepare all solutions using milli-Q water and analytical-grade reagents. Prepare and store all reagents at room temperature unless indicated otherwise.

2.1 Materials for Construction of Promoter-lacZ Transcription Reporters

- iProof High-fidelity DNA polymerase (Bio-Rad, California, USA).
- Plasmid pNN387 (a gift from Prof. Kunihiko Nishino, Osaka University): single copy vector with NotI and HindIII cloning sites upstream of the promoterless *lacZ*.
- Restriction enzyme NotI and HindIII (New England Biolabs, Hitchin, UK).
- Quick Ligation Kit (New England Biolabs, Hitchin, UK).

2.2 Materials for Chromosomal Epitope Tagging (Using CusC as an Example)

- Plasmid pKD46: red recombinase expression plasmid.
- Plasmid pKD4: plasmid containing the desirable selection marker.
- IProof High-fidelity DNA polymerase (Bio-Rad, California, USA).
- PCR primers. Primers for construction of C-terminal CusC-FLAG are listed below as an example (Table 1) (*see* **Note 1**).
- L-Arabinose: 1 M solution in water.

Table 1
Primers used for chromosomal epitope tagging

Primer	Primer sequence (5–3′)[a]
CusC-FLAG-up	GGTTAACGAAATTTCTTTGTATACCGCACTTGGTGGCGGTGACTACA AGGACGACGATGACAAG**TGA**GAACCAACTCGCGAGGGGAT
CusC-FLAG-down	CGCAGCACATGCAACTTGAAGTATGACGAGTATATTGCTGGCCAGCC TTTCATGATATAT

[a]Underlined letters indicate the FLAG coding sequence. Bolded letters indicate the stop codon

- ECM 399E Electroporation system and cuvettes (BTX Harvard apparatus, MA, USA).
- NanoVue Plus Spectrophotometer (GE Healthcare, Little Chalfont, UK).
- Ice-cold sterile ddH_2O.

2.3 Anaerobic Bacteria Culture

- 9 mL Pyrex culture tubes with screw caps (Pyrex, NY, USA).
- M9 salt medium: M9 minimal medium (6 g/L Na_2HPO_4, 3 g/L KH_2PO_4, 0.5 g/L NaCl and 1 g/L NH_4Cl, pH 7.0) supplemented with 0.2% glucose, 1 mM $MgSO_4$, 0.1 mM $CaCl_2$, 0.01 mg/mL ferric ammonium citrate, 4 μg/mL thiamine, 0.2 μM ammonium molybdate, and 0.2% casamino acids (CAA). In the case of investigating the Cus efflux pump expression during anaerobic amino acid limitation, CAA is not supplemented.
- Spectrophotometer: Thermo Scientific SPECTRONIC 20D+ (Thermo Fisher Scientific, Madison, USA).

2.4 β-Galactosidase Activity Assay

- Z buffer: 60 mM $Na_2HPO_4{\cdot}7H_2O$, 40 mM $NaH_2PO_4{\cdot}H_2O$, 10 mM KCl, 1 mM $MgSO_4{\cdot}7H_2O$, 50 mM β-mercaptoethanol. To prepare the stock solution, 16.084 g $Na_2HPO_4{\cdot}7H_2O$, 5.52 g $NaH_2PO_4{\cdot}H_2O$, 0.75 g KCl, and 0.247 g $MgSO_4{\cdot}7H_2O$ are dissolved in 900 mL water. Adjust pH to 7.0 then bring the volume to 1 L. Store at 4 °C and add appropriate β-mercaptoethanol before use.
- 2-Nitrophenyl β-D-galactopyranoside (ONPG): 4 mg/mL solution in Z buffer.
- Sodium dodecyl sulfate (SDS): 0.1% solution in water.
- Na_2CO_3: 1 M solution in water.
- Chloroform (Merck KGaA, Darmstadt, Germany).
- U-2800 Spectrophotometer (Hitachi High-Technologies, Tokyo, Japan).

2.5 Protein Sample Preparation

- BugBuster protein extraction reagent (Novagen, MA, USA).
- 50 mg/mL lysozyme solution. Weigh 500 mg lysozyme powder (Sigma-Aldrich, St. Louis, USA) and dissolve it into 10 mL water. Store at −20 °C.
- DNase I (2 U/μL; Ambion, California, USA). Store at −20 °C.
- SDS sample buffer (4×): 250 mM Tris (pH 6.8), 40% glycerol, 8% SDS, 0.04% bromophenol blue. To prepare the stock solution, weigh 3.03 g Tris and dissolve it in 50 mL water. Adjust pH to 6.8 with HCl. Add 40 mL glycerol, 8 g SDS, 0.04 g bromophenol blue and mix to dissolve completely. Bring up to 100 mL with water. Store at room temperature. Add 10% (e.g. 100 μL for 1 mL) β-mercaptoethanol before use. Store at −20°C afterward.

2.6 SDS-PAGE Gel Components

- Resolving gel buffer: 0.5 M Tris–HCl, 0.4% SDS, pH 6.8. Weigh 6.05 g Tris and dissolve it in 85 mL water. Adjust pH to 6.8 with HCl. Add 0.4 g SDS and mix to homogeneity. Bring up to 100 mL with water.
- Stacking gel buffer: 1.5 M Tris–HCl, 0.4% SDS, pH 8.8. Weigh 18.2 g Tris and dissolve in 85 mL water. Adjust pH to 8.8 with HCl. Add 0.4 g SDS and mix. Bring up to 100 mL with water.
- 40% bis-acrylamide solution (Bio-Rad, California, USA). Store at 4 °C.
- 10% ammonium persulfate (APS). Weigh 1 g ammonium persulfate (Sigma-Aldrich, St. Louis, USA) and dissolve in 10 mL water. Store at 4 °C.
- N, N, N, N′-tetramethyl-ethylenediamine (TEMED) (Bio-Rad, California, USA).
- Isopropanol solution (Sigma-Aldrich, St. Louis, USA).

2.7 SDS-PAGE and Transfer

- SDS-PAGE running buffer (10×): 25 mM Tris, 192 mM glycine, 0.1% SDS. To prepare the stock solution, weigh 288 g glycine, 60.4 g Tris and dissolve them in 1.8 L water. Add 20 g SDS and mix to dissolve it completely. Bring up to 2 L with water.
- Transfer buffer (10×): 25 mM Tris, 192 mM glycine. To prepare the stock solution, weigh 288 g glycine, 60.4 g Tris and dissolve them in 1.8 L water. Bring up to 2 L with water and store at 4 °C.
- Nitrocellulose membrane (Bio-Rad, California, USA).
- Filter paper (3 M, Minnesota, USA).
- Methanol solution (Sigma-Aldrich, St. Louis, USA).
- Mini-PROTEAN Tetra Cell (Bio-Rad, California, USA).
- Mini-PROTEAN spacer plate with 1.5 mm integrated spacer and short plate (Bio-Rad, California, USA).

2.8 Western Blotting

- TBS buffer (10×): 0.5 M Tris–HCl (pH 7.6), 1.5 M NaCl. To prepare the stock solution, weigh 60.5 g Tris and 87.6 g NaCl and dissolve them in 800 mL water. Adjust pH to 7.6 with HCl. Autoclave and store at room temperature.
- TBST buffer: 50 mM Tris–HCl (pH 7.6), 150 mM NaCl, 0.05% Tween 20. Add 0.5 mL Tween 20 into 100 mL 10× TBS buffer, mix and bring to 1 L with water.
- Nonfat powdered milk: Blotting-Grade Blocker (Bio-Rad, California, USA).
- Primary antibody: Monoclonal ANTI-FLAG M2 antibody produced in mouse (Sigma-Aldrich, St. Louis, USA) is used for MdtE-FLAG. The ANTI-FLAG polyclonal affinity antibody produced in rabbit (Sigma-Aldrich, St. Louis, USA) is used for CusC-FLAG. Store at −20 °C.
- Secondary antibody: Goat Anti-Mouse IgG (H + L)-HRP Conjugate or Goat Anti-Rabbit IgG (H + L)-HRP Conjugate (Bio-Rad, California, USA). Store at −20 °C.
- Detection reagent (ECL Western Blotting Detection Reagents; GE Healthcare, Little Chalfont, UK). Store at 4 °C.
- X-Ray films (Super RX; Fujifilm, Tokyo, Japan).

3 Methods

3.1 Construction of Promoter-lacZ Fusions

1. Perform PCR with the genomic DNA of *E. coli* MG1655 as template to generate DNA fragments corresponding to the desired promoter regions. The pair of primers used in PCR includes protective bases (lowercase), restriction enzyme site (NotI or HidIII), followed by ~20 bp homologous to the 5′-and 3′-end of the desirable promoter regions respectively. The primers used to construct P*gadE-lacZ* to examine the transcription of the operon *gadE-mdtEF* are: P*gadE*-F: 5′-aaggaaaaaa GCGGCCGCTTACCCCGGTTGTCACCCGGAT-3′; P*gadE*-R: 5′-cccAAGCTTAACTTGCTCCTTAGCCGTTATC-3′.
2. Purify the PCR product using illustra GFX PCR DNA and Gel Band Purification Kit (GE Healthcare, Little Chalfont, UK) or ethanol precipitation and resuspend the purified DNA in ddH_2O. Confirm the quality and quantity of the purified DNA fragment using NanoVue Plus Spectrophotometer.
3. Both the DNA fragment and plasmid vector pNN387 are digested by NotI/HindIII (New England Biolabs, Hitchin, UK) and then purified using illustra GFX PCR DNA and Gel Band Purification Kit.

4. Ligate digested DNA fragments into pNN387 using Quick Ligase kit (New England Biolabs, Hitchin, UK). The ratio of the DNA insert vs. the vector is ~3:1.
5. Transform the above ligated plasmid into DH5α using standard $CaCl_2$ method, plate transformants on LB plates containing 25 μg/mL Chloramphenicol. Incubate plates at 37 °C overnight.
6. Perform colony PCR and DNA sequencing to confirm successful construction.
7. Transform the constructed plasmid into host strains of interest using standard $CaCl_2$ method, plate transformants on LB plates containing 25 μg/mL Chloramphenicol. Incubate plates at 37 °C overnight (*see* **Note 2**).

3.2 Chromosomal Epitope Tagging Using CusC-FLAG as an Example

The CusCFBA efflux system is a conserved tripartite efflux system in many gram negative bacteria. It is composed of the outer membrane protein CusC, the inner membrane protein CusA, and the periplasmic protein CusB (Fig. 3a), which forms a transporter complex spanning the inner and outer membrane of gram negative bacteria and is capable of removing excess Cu/Ag ions from the cytoplasmic or periplasmic space [16]. CusF is a periplasmic chaperone protein which binds Cu ions and transfers them to the CusCBA efflux transporter.

1. Perform PCR with the template plasmid pKD4 to generate DNA fragment for homologous recombination. Purify the fragment and resuspend the DNA in ddH_2O. Confirm fragment quality and quantity using NanoVue Plus Spectrophotometer.
2. Transform the Red recombinase expression plasmid (pKD46) into the host strain of interest using standard $CaCl_2$ method, plate transformants on LB plates containing 100 μg/mL Ampicillin. Incubate plates at 30 °C overnight.
3. Inoculate a single colony of the pKD46 transformant into LB medium with 100 μg/mL Ampicillin and grow overnight at 30 °C with shaking (220 rpm).
4. Subculture the overnight culture with 1/100 dilution ratio into LB containing 100 μg/mL Ampicillin and 10 mM L-arabinose, grow at 30 °C with shaking (220 rpm) until OD_{600} ~0.3.
5. Chill the cells on ice and pellet cells by centrifugation at 3000 × *g* for 10 min at 4 °C.
6. Resuspend the cells 1/100 original volume of ice-cold sterile ddH_2O.

7. Spin down the cells at 4 °C, wash the pellet three times with same volume of ice-cold sterile ddH_2O. Keep cells on ice constantly.
8. Mix 40 μL of cells with 50 μg of recombination fragment into electroporation cuvette.
9. Electroporate cells at 1.8 kV with a time constant of ~5 ms.
10. Immediately add 1 mL of LB (no antibiotics) into the cuvette and transfer cell suspension into a 15 mL culture tube.
11. Shake cells (220 rpm) at 37 °C for 4 h to allow outgrowth of the electroporated cell.
12. Plate 100 μL of cell suspension on LB-antibiotic plate depending on the selection marker of choice (e.g. 20 μg/mL Km for kanamycin resistance marker). Incubate plates at 37 °C overnight.
13. Streak-purify transformants at 42 °C to remove pKD46. Patch selective single colonies on LB-ampicillin plate to test the loss of resistance (*see* **Note 3**).
14. Perform colony PCR and DNA sequencing to confirm successful recombination.

3.3 Culture of Bacteria Under Anaerobic Conditions

Carry out all procedures in aseptic chamber or around the Benson Burner to avoid bacterial contamination.

1. Prepare bacteria overnight cultures one day before the experiment.
2. Fill the screw-capped Pyrex culture tubes with desirable growth medium (*see* **Note 4**).
3. Measure the OD of the overnight cultures and dilute it to a cell density of 10^6 cells/mL using sterilized PBS buffer. Inoculate bacterial culture into the 9 mL Pyrex culture tube filled with medium such that the final cell density is ~10^3 cells/mL (Fig. 1) (*see* **Note 5**).
4. Screw the cap tightly and incubate the tubes at desirable temperature without agitation (*see* **Note 5**).
5. Monitor cell density of the sealed culture using Thermo Scientific SPECTRONIC 20D+ (Thermo Fisher Scientific, Madison, USA).

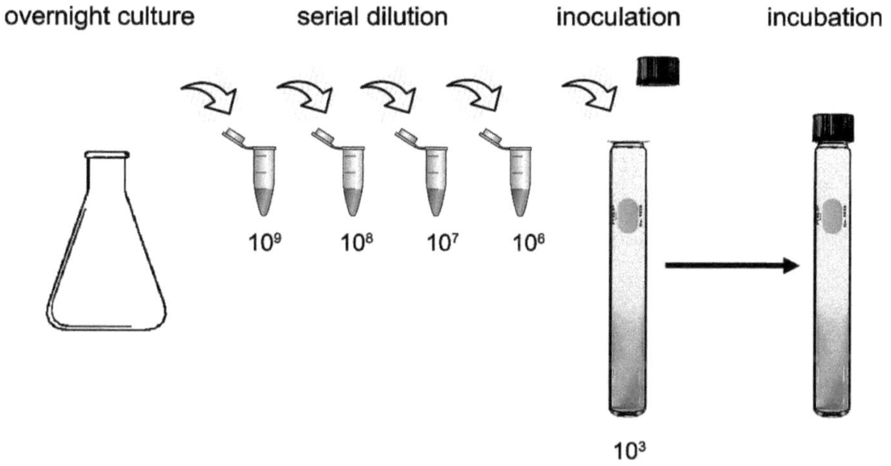

Fig. 1 Diagram of the anaerobic bacterial culture system. Overnight culture is prepared one day before the experiment. Following serial dilution, appropriate volume of the diluted inoculum is added into screw-capped Pyrex culture tube filled with growth medium with a final cell density of ~10^3 cells/mL. The inoculum is subject to incubation without agitation. The residual oxygen dissolved in the growth medium is rapidly consumed by the small amount of initial inocula, resulting in a complete anaerobic environment for cell growth

3.4 Analyzing the Expression of Efflux Component at Transcriptional Level Using β-Galactosidase Activity Assay (PgadE-lacZ as an Example to Measure the Transcription of mdtE in the gadE-mdtEF Operon)

1. Grow bacterial cells containing a plasmid-encoded promoter-*lacZ* fusion anaerobically as described in Subheading 3.3 to OD_{600} ~ 0.3.
2. Terminate cell growth and protein synthesis by adding tetracycline to a final concentration of 10 μg/mL.
3. Mix the cell culture and place it on ice till use.
4. 1 h prior to the reaction, dissolve appropriate amount of ONPG in Z buffer (final concentration 4 mg/mL) and turn on 28 °C water bath.
5. Add 800 μL of Z buffer and 200 μL well mixed cell culture to each tube and vortex until cells are resuspended homogeneously (*see* **Note 6**). Set up a blank by adding just Z buffer and the culture medium.
6. Add two drops of chloroform and one drop of 0.1% SDS to each tube followed by vortex for 10 s and incubate in 28 °C water bath for 5 min to lyse the cells.
7. Start the reactions by adding 200 μL ONPG, vortex thoroughly, and incubate in 28 °C water bath. Start timer or record the start time. Monitor the yellow color development in the reaction.
8. After yellow color developed or after 2 h of reaction if no visible color development occurs, add 500 μL freshly prepared 1 M Na_2CO_3 stop solution and vortex to stop the reaction. Record reaction time in minutes.

9. Measure the absorbance of reaction mixture at A_{420} and A_{550}, and the OD_{600} of the cell culture using U-2800 Spectrophotometer (Hitachi High-Technologies, Tokyo, Japan).
10. Calculate promoter activity and express in Miller Units (Fig. 2b) as follows:

$$\text{Miller Units} = 1000 \times [(A_{420} - 1.75 \times A_{550})]/(T \times V \times OD_{600})$$

where: T = elapsed time (in min) of incubation
V = volume of cells used

3.5 Protein Sample Preparation for Translational-Level Analysis: Soluble Component (MdtE-FLAG as an Example)

1. Culture bacteria under anaerobic condition as Subheading 3.3.
2. Harvest the cells when the OD_{600} reaches 0.3 (*see* **Note 7**). Chill cells on ice for 15 min then transfer 4 mL culture into a 15 mL falcon tube and centrifuge at 5000 × *g* for 10 min.
3. Remove supernatant. Proceed to cell lysis and protein extraction as described below. Or store cell pellets at −80 °C for future use (*see* **Note 8**).
4. Take cell pellets from −80 °C and thaw on ice (if applicable).
5. Carefully resuspend cell pellet in 40 μL 1× SDS sample buffer (10% β-mercaptoethanol) such that the final concentration is approximately 3×10^7 cells/μL (*see* **Note 9**). Mix by pipetting.
6. Incubate samples at 90 °C for 10 min and centrifuge at 16,000 × *g* for 5 min.
7. Samples are ready for SDS-PAGE application. Or store at −20 °C for future use (thaw at room temperature before loading).

3.6 Protein Sample Preparation for Translational-Level Analysis: Membrane Component (CusC-FLAG as an Example)

1. Take cell pellets (prepared as stated in Subheading 3.5) from −80 °C and thaw on ice (if applicable).
2. Prepare fresh lysis buffer by adding following components proportionally: 40 μL 50 mg/mL lysozyme and 4 μL DNase into 360 μL BugBuster reagents (*see* **Note 10**). Mix and avoid making bubbles.
3. Carefully resuspend cell pellet in 40 μL lysis mix and avoid making bubbles. The ratio of cells vs. lysis buffer is approximately 3×10^7 cells/μL.
4. Incubate at room temperature for 30 min, mix by tapping tubes occasionally.
5. Centrifuge lysate mixture at 16,000 × *g* for 20 min at 4 °C.
6. Transfer 30 μL supernatant to a clean eppendorf tube.
7. Add 10 μL 4× SDS sample buffer (10% β-mercaptoethanol). Mix by pipetting.

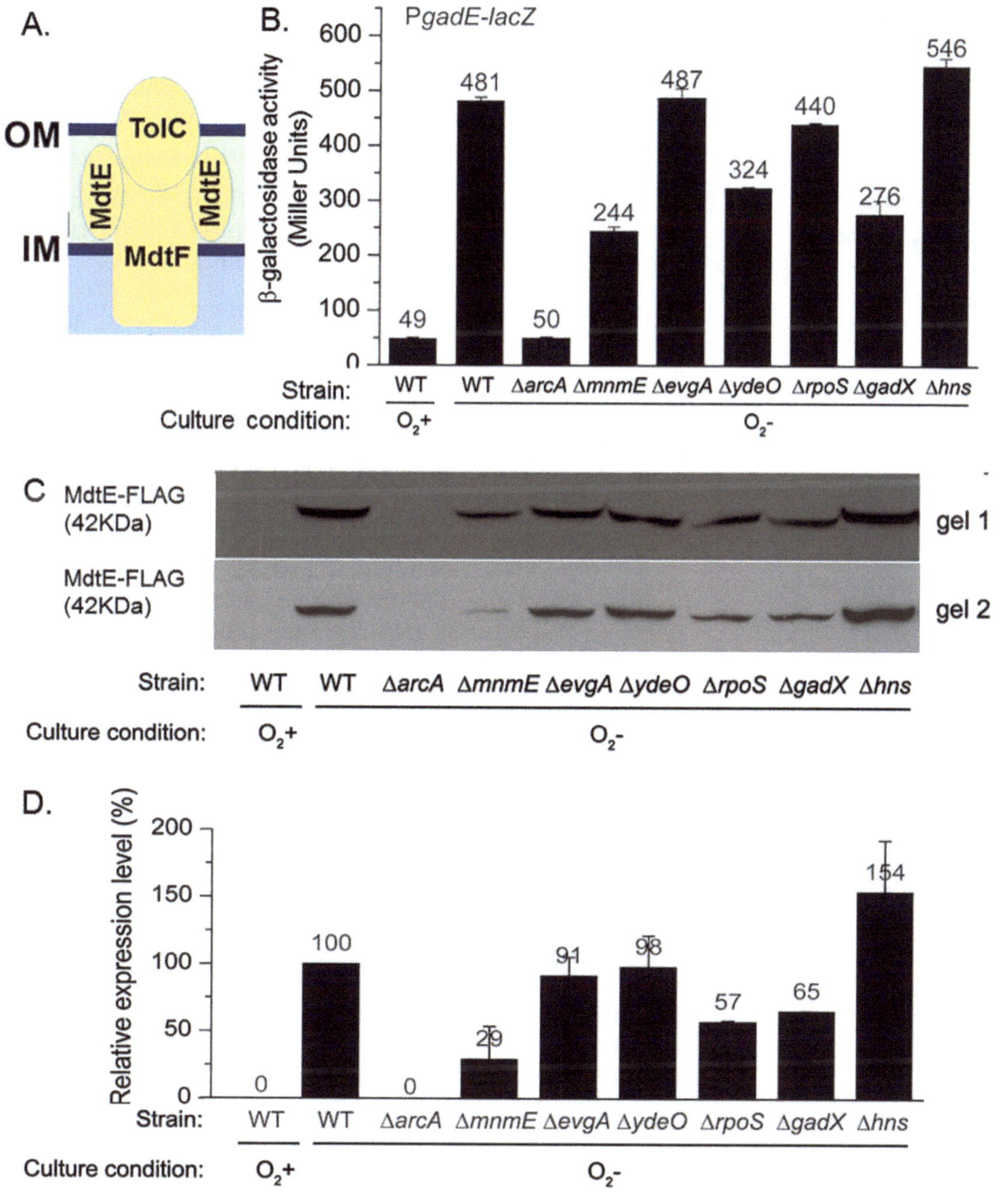

Fig. 2 Examination of efflux pump expression under anaerobic conditions and identification of the regulatory transcription factors using MdtEF as an example. (**a**) Schematic diagram of the MdtEF-TolC efflux system. (**b**) Transcription of the *gadE-mdtEF* operon from PgadE was determined by β-galactosidase activity of P*gadE-lacZ*. Δ*arcA* abolished the anaerobic induction of the *gadE-mdtEF* operon indicating its role in the transcription activation of *mdtEF* under this condition. O_2+, sample from aerobic culture; O_2−, sample from anaerobic culture. Error bars represent the standard errors of triplicate experiments. (**c**), Production of MdtE-FLAG under anaerobic condition from two separate Western blot analyses. (**d**) Relative production level of MdtE-FLAG from various samples quantified using ImageJ. Mean and errors are deviated from results of two biological replicates (Gel 1 and Gel 2)

8. Incubate samples at 55 °C for 25 min and centrifuge at 16,000 × *g* for 5 min.
9. Samples are ready for SDS-PAGE application. Or store at −20 °C for future use (thaw at room temperature before loading).

3.7 SDS-PAGE and Western Blot

SDS-PAGE and Western blot is conducted following the standard procedure described by the manufacture (Bio-Rad). Only the procedures require special attention is presented here.

1. If the gel is not used immediately, wrap the whole setup with a piece of plastic wrap and store at 4 °C for a few hours.
2. The volumes of protein samples loaded onto the gel can be in a range of 5-30 μL depending on the expression level of the proteins. In the cases described here, 10 μL was loaded to detect MdtE-FLAG and 30 μL to detect CusC-FLAG. Samples which protein contents are subject to comparison should be loaded onto the same gel (Fig. 2c).
3. The buffer for electrophoretic transfer is prepared in advance (during the gel electrophoresis) by mixing the following components: 100 mL 10× transfer buffer, 700 mL milli-Q water, 200 mL methanol. Store the buffer at 4 °C for about 60 min to allow it cool down.
4. 30 mL primary antibody solution is prepared in 1% nonfat powdered milk in TBST. To detect MdtE-FLAG, dilute monoclonal ANTI-FLAG M2 antibody solution to a working concentration of 10 μg/mL. To detect CusC-FLAG, dilute polyclonal ANTI-FLAG antibody solution to a working concentration of 10 μg/mL (*see* **Note 11**).
5. 30 mL secondary antibody solution is prepared in 1% nonfat powdered milk in TBST. To detect MdtE-FLAG, add 10 μL Goat anti-Mouse IgG (H + L) antibody into 30 mL solution (1:3000 dilution). To detect CusC-FLAG, add 10 μL Goat anti-Rabbit IgG (H + L) antibody into 30 mL solution (1:3000 dilution) (*see* **Note 11**).
6. The identity of the protein band on the developed film is determined by its migration in comparison of the protein marker, and the absence of the band in the non-FLAG control strain. The relative expression level of efflux pump proteins from anaerobic culture and aerobic culture is compared (Figs. 2c and 3b).
7. The relative expression level of samples is quantified using software ImageJ (Fig. 2d) and expressed as the mean of at least two replicates.

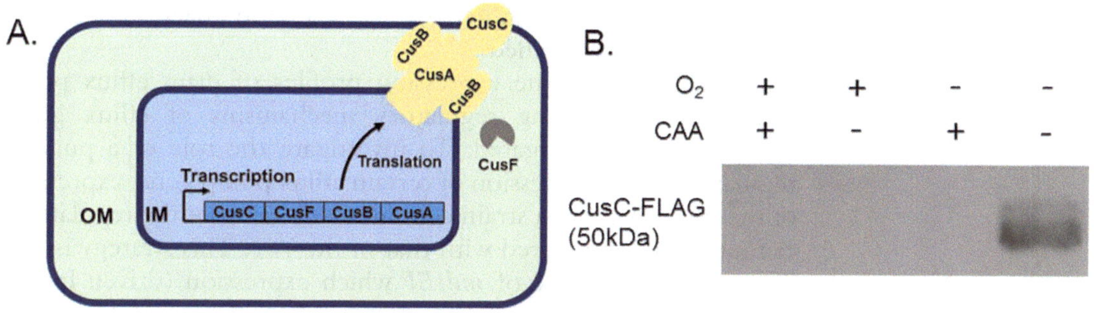

Fig. 3 Examination of efflux pump expression at translational level using the membrane integral protein CusC-FLAG as an example. (**a**) Schematic diagram of the CusCBA system. The Cus efflux system is composed of the CusCBA tripartite efflux transporter and the periplasmic Cu chaperone CusF. (**b**) Production of CusC-FLAG under oxygen and/or amino acids limitation (presence or absence of CAA) using the FLAG-tagged strain

4 Application

In the past decades, increasing evidence indicated that expression of efflux pumps are induced under a broad range of environmental and physiological conditions [17]. However, the physiological functions of these induced efflux pumps are only started to be revealed recently. This is because most studies were conducted under laboratory conditions, i.e. rich medium at 37 °C with agitation. However, it is known that microorganisms often exist in an environment of nutrient scarce, low oxygen tension, and are often challenged with surrounding competitors and host defense substances. Hence, investigating the expression profile of efflux pump genes under these conditions is necessary to reveal their physiological roles.

In this chapter, we described the methods to culture bacteria anaerobically which enables the investigation of efflux pump expression under anaerobic conditions. By using different culture medium and supplying different compounds such as the various electron acceptors to sustain anaerobic respiration, or depletion of certain nutrient components, such as CAA, many physiologically relevant anaerobic conditions could be mimicked for investigation in laboratory. For example, Pyrex anaerobic culture tubes containing artificial sputum medium was designed to mimic cystic fibrosis lung conditions for *P. aeruginosa* [18]. Owing to the O_2 dependency of the fluorescence emission of GFP and its variants, the common gene expression analysis using GFP reporter in aerobic cultures is not applicable under anaerobic conditions. Recently, Pu et al. reported a method of inserting a tetracysteine tag into the linker structure of the efflux gene *tolC* (FIAsH labeling) and measuring its expression using time-lapse fluorescent microscopy [19]. However, whether this method can be expanded to analyze all efflux genes and whether it is applicable in anaerobic bacterial cultures remain examined.

In addition to the expression profiles of drug efflux pump genes, the underlying regulatory mechanisms of efflux genes could also be investigated. To investigate the role of a putative regulator on the expression of certain efflux pump gene, expression of the efflux gene in a strain containing deletion of the regulator is examined and compared with that of the WT. This strategy is well illustrated in the case of *mdtEF* which expression (driven by the *gadE* promoter) is found to be activated by the anaerobic global transcription factor ArcA through antagonizing the H-NS-mediated repression [20].

Furthermore, results obtained from the expression analysis can provide information on the physiological functions of efflux pumps. In the case of MdtEF, the fact that its expression is subject to regulation by the global regulator ArcA which primarily represses genes involved in aerobic respiration and promotes anaerobic switch suggested that MdtEF may contribute to the energy metabolism of *E. coli* under this condition. This led to the identification of the physiological substrate of MdtEF, i.e. indole nitrosative derivatives generated during anaerobic respiration of nitrate [7]. In the case of CusCBA, the fact that depletion of CAA caused significant induction of the Cus system had led to the identification of its role in protecting Fe-S cluster biogenesis and its containing enzymes in the metabolic processes of amino acid biosynthesis and fumarate respiration [12]. Together, these studies illustrated the various applications of studying efflux pumps expression in revealing their regulatory mechanisms and physiological functions.

5 Notes

1. Forward primer includes (from 5′ to 3′) a 40–50 bp homology region immediately upstream of the stop codon of the target protein, followed by the FLAG coding sequence, stop codon, and the sequence specific to the selection marker in the template plasmid. Reverse primer includes (from 5′ to 3′) a 40–50 bp homology region immediately downstream of the stop codon of the target protein followed by sequence specific to the selection marker in the template plasmid.
2. For *E. coli* and other bacteria that contain additional β-galactosidase genes, chromosomal copy of *lacZ* should be deleted.
3. Occasionally this step needs to be repeated to completely remove the Red recombinase plasmid.
4. The volume of added growth medium and inocula depends on the volume of the tubes. For example, the volume of the

cultural tube we used is 9 mL, so we add about 9 mL medium into tube, and add 9 μL diluted overnight culture into the medium, which makes the final concentration of the inocula to be 10^3 cells/mL.

5. The concentration of initial inocula is set at a low level to ensure that these inocula consume the residual oxygen dissolved in the growth medium at a very early stage of cell growth, resulting in a complete anaerobic environment for the subsequent growth of the entire population. This anaerobic bacterial culture system illustrated here is suitable to facultative bacterial species but may not be applicable to obligate anaerobes due to the presence of residual oxygen in the medium. Culture medium for obligate anaerobes can be prepared by boiling to remove oxygen and be stored under oxygen-free condition. After flushing the culture tubes with oxygen-free gas, such as N_2 or Ar, culture medium is then added to fill up the tube followed by inoculation. Anaerobic manipulations should be conducted in anaerobic cabinet or glove box [21].
6. If calculated Miller Units >500, try to use less cells, such as 900 μL Z buffer and 100 μL cell culture.
7. For *E. coli* cells grown in M9 minimal medium under anaerobic condition, the saturation OD is about 0.8 and OD 0.3 corresponds to the mid-log phase. The exact OD of the culture prior to harvest should be recorded for normalization of protein contents in different samples.
8. We usually store cell pellets at −80 °C overnight and perform cell lysis and protein extraction on the second day.
9. We use 40 μL lysis mix to lyse 4 mL culture (OD as 0.3). According to the OD of harvested cells, adjust the volume of lysis mix such that the final concentration is approximately 3×10^7 cells/μL.
10. This volume of lysis mix can be used for 10 cell samples. The volume can be proportionally increased or decreased depending on the number of samples.
11. Can be stored at −20 °C and reusable twice. We found that reuse of primary and secondary antibody solution often led to decreased detection signal. So it is recommended to use freshly prepared antibody solution for those proteins expressed at a low level in bacteria or with unknown expression level.

Acknowledgment

We are grateful to Prof. Kunihiko Nishino (Institute of Scientific and Industrial Research, Osaka University) for the pNN387 vector and the P*gadE-lacZ* strain. The studies are supported by the Hong

Kong University Grants Council General Research Fund (HKU 17142316) and the Health and Medical Research Fund (HMRF) Hong Kong (No. 13120662) to A.Y.

References

1. Nikaido H (1998) Multiple antibiotic resistance and efflux. Curr Opin Microbiol 1:516–523
2. Marteyn B, West NP, Browning DF et al (2010) Modulation of Shigella virulence in response to available oxygen in vivo. Nature 465:355–358
3. Høiby N, Bjarnsholt T, Givskov M et al (2010) Antibiotic resistance of bacterial biofilms. Int J Antimicrob Agents 35:322–332
4. Sannasiddappa TH, Hood GA, Hanson KJ et al (2015) *Staphylococcus aureus* MnhF mediates cholate efflux and facilitates survival under human colonic conditions. Infect Immun 83:2350–2357
5. Urbán E, Nagy E, Pál T et al (2007) Activities of four frog skin-derived antimicrobial peptides (temporin-1DRa, temporin-1Va and the melittin-related peptides AR-23 and RV-23) against anaerobic bacteria. Int J Antimicrob Agents 29:317–321
6. Uysal B, Yasar M, Ersoz N et al (2010) Efficacy of hyperbaric oxygen therapy and medical ozone therapy in experimental acute necrotizing pancreatitis. Pancreas 39:9–15
7. Zhang Y, Xiao M, Horiyama T et al (2011) The multidrug efflux pump MdtEF protects against nitrosative damage during the anaerobic respiration in *Escherichia coli*. J Biol Chem 286:26576–26584
8. Schaible B, Taylor CT, Schaffer K (2012) Hypoxia increases antibiotic resistance in *Pseudomonas aeruginosa* through altering the composition of multidrug efflux pumps. Antimicrob Agents Chemother 56:2114–2118
9. Schwartz KT, Carleton JD, Quillin SJ et al (2012) Hyperinduction of host beta interferon by a Listeria monocytogenes strain naturally overexpressing the multidrug efflux pump MdrT. Infect Immun 80:1537–1545
10. Bradford PA, Sands DTW, Petersen PJ (2005) In vitro activity of tigecycline against isolates from patients enrolled in phase 3 clinical trials of treatment for complicated skin and skin-structure infections and complicated intra-abdominal infections. Clin Infect Dis 41(Supplement 5):S315–S332
11. Kalia NP, Mahajan P, Mehra R et al (2012) Capsaicin, a novel inhibitor of the NorA efflux pump, reduces the intracellular invasion of *Staphylococcus aureus*. J Antimicrob Chemother 67:2401–2408
12. Fung DKC, Lau WY, Chan WT et al (2013) Copper efflux is induced during anaerobic amino acid limitation in *Escherichia coli* to protect iron-sulfur cluster enzymes and biogenesis. J Bacteriol 195:4556–4568
13. Horiyama T, Nishino K (2014) AcrB, AcrD, and MdtABC multidrug efflux systems are involved in enterobactin export in *Escherichia coli*. PLoS One 9:e108642
14. de Sarrau B, Clavel T, Clerté C et al (2012) Influence of anaerobiosis and low temperature on Bacillus cereus growth, metabolism, and membrane properties. Appl Environ Microbiol 78:1715–1723
15. Yang Y, Xiang Y, Xia C et al (2014) Physiological and electrochemical effects of different electron acceptors on bacterial anode respiration in bioelectrochemical systems. Bioresour Technol 164:270–275
16. Osman D, Cavet JS (2008) Copper homeostasis in bacteria. Adv Appl Microbiol 65:217–247
17. Sun J, Deng Z, Yan A (2014) Bacterial multidrug efflux pumps: Mechanisms, physiology and pharmacological exploitations. Biochem Biophys Res Commun 453:254–267
18. Kirchner S, Fothergill JL, Wright EA et al (2012) Use of artificial sputum medium to test antibiotic efficacy against *Pseudomonas aeruginosa* in conditions more relevant to the cystic fibrosis lung. J Vis Exp 64:3857
19. Pu Y, Zhao Z, Li Y et al (2016) Enhanced efflux activity facilitates drug tolerance in dormant bacterial cells. Mol Cell 62:284–294
20. Deng Z, Shan Y, Pan Q et al (2013) Anaerobic expression of the gadE-mdtEF multidrug efflux operon is primarily regulated by the two-component system ArcBA through antagonizing the H-NS mediated repression. Front Microbiol 4:194
21. Yan A, Kiley P (2009) Techniques to isolate O_2-sensitive proteins: [4Fe-4S]-FNR as an example. Methods Enzymol 463:787–805

Chapter 15

Identification of a *Staphylococcus aureus* Efflux Pump Regulator Using a DNA–Protein Affinity Technique

Que Chi Truong-Bolduc and David C. Hooper

Abstract

In this chapter, we describe the step-by-step identification of a putative regulator protein and demonstrate the function of this protein as a repressor of the expression of a specific efflux pump, causing resistance to quinolones in *Staphylococcus aureus*. We show that the knockout gene mutant has an increase in transcript levels of the target efflux pump when compared to that of the *S. aureus* parental strain RN6390. We provide a detailed protocol that includes the identification of the DNA-binding transcriptional regulatory protein from *S. aureus* cell extracts using DNA sequences linked to magnetic beads. In addition, we describe the real-time qRT-PCR assays and MIC testing to evaluate the effects of the regulator on *S. aureus* drug resistance phenotype.

Key words *Staphylococcus aureus*, Regulator, Efflux pump, Gel-shift assay, Affinity binding assay, In-frame gene deletion, Protein purification, qRT-PCRs

1 Introduction

Efflux pumps are important for *S. aureus* survival and adaptation to diverse environments, and they can produce resistance or reduced susceptibility to antimicrobial compounds when overexpressed [1, 2]. The expression of efflux pumps is often modulated by transcriptional regulatory protein(s) that bind to the promoter DNA of the efflux pump structural gene. We use this specific property to isolate and identify the direct regulator(s) from crude cell extracts [3]. Because regulatory cascades and hierarchical networks are often involved in controlling pump expression, this approach can help to characterize such networks by determining those regulators that act directly on pump gene expression [4]. Proteins binding specifically to efflux pump promoter DNA can be either activators or repressors of pump structural gene expression, thereby affecting antimicrobial susceptibility [5]. To isolate the regulator from a whole cell extract, we use the promoter DNA of the pump gene as an affinity reagent for specific binding of

Akihito Yamaguchi and Kunihiko Nishino (eds.), *Bacterial Multidrug Exporters: Methods and Protocols*, Methods in Molecular Biology, vol. 1700, https://doi.org/10.1007/978-1-4939-7454-2_15, © Springer Science+Business Media, LLC 2018

regulator proteins found in cell extracts. For this purpose, we design a pair of primers that flanks the documented or candidate promoter DNA immediately upstream of the pump gene. The forward primer is labeled with biotin at the 5′ end for subsequent affinity binding and detection [6]. The two primers are used in a PCR reaction to amplify a ~150–200-bp biotinylated DNA fragment, which contains the entire promoter region of the efflux pump. The biotinylated DNA is mixed with streptavidin covalently coupled to magnetic beads to use in a DNA affinity column or slurry for protein isolation. The DNA is mixed with a crude cell extract in binding buffer, and then the mixture is washed prior to elution with saline. The isolated protein is blotted onto a PVDF membrane and a small amino acid sequence of ~14 amino acids is identified by N-terminal sequencing using the Edman degradation method. The resulting N-terminal amino acid sequence (or the mass spectrometry analysis data) is used to search a data bank of protein sequences from the *S. aureus* genome, NCTC 8325 or another relevant strain, and leads to the identification of a hypothetical open reading frame (ORF) on the chromosome. The structural gene of the ORF is then cloned in an expression vector, and the protein is expressed and purified from *Escherichia coli*. The purified protein is then used in a series of DNA–protein gel mobility shift assays designed to confirm the protein's DNA binding properties and to establish the specificity of DNA binding to the promoter sequence. This technique provides a rapid and unbiased method to identify direct regulators of efflux pump genes [4, 6].

We then create a deletion mutant of the regulatory gene to assess its effect on the transcript levels of the efflux pump gene using quantitative qRT-PCR assays and to assess effects on antimicrobial susceptibility, including known and other potential pump substrates. For this purpose, we use an in-frame deletion technique with the pIMAY plasmid and then perform allelic exchange to replace the wild-type gene copy by the mutated copy on the chromosome [7]. To verify that the drug resistance phenotype of the deletion mutant is linked to the increase in transcript level of the efflux pump, we introduce a plasmid harboring a wild-type regulator gene in order to demonstrate complementation of the mutant's effect on pump transcript levels and resistance phenotype [8, 9].

2 Materials

2.1 *Staphylococcus aureus* Strain

We use a derivative of the *S. aureus* strain RN6390, a derivative of NCTC8325 but other strains should be suitable. This strain is susceptible to the majority of antimicrobial agents used to treat staphylococcal infections, including methicillin. *S. aureus* RN4220 is a restriction minus strain used in transformation of *S. aureus* [10].

2.2 *Escherichia coli* Strains

We use *E. coli* Top10 (F– *mcr*A Δ(*mrr-hsd*RMS-*mcr*BC) Φ80*lac*ZΔM15 Δ*lac*X74 *rec*A1 *ara*D139 Δ(*ara leu*) 7697 *gal*U *gal*K *rps*L (StrR) *end*A1 *nup*G) for cloning and *E. coli* BL21 (*fhuA2 [lon] ompT gal [dcm] ΔhsdS*) for protein expression. *E. coli* DH10B (*F– mcrA Δ(mrr-hsdRMS-mcrBC) Φ80lacZΔM15 ΔlacX74 recA1 endA1 araD139 Δ(ara leu) 7697 galU galK rpsL nupG λ–*) is used in the construction of an in-frame deletion copy of the regulator gene [7].

2.3 *Plasmids*

Plasmid pGEM3zf^{+}, ampicillin resistant, is a standard cloning vector with a multicloning sites (MCS) (Promega). Plasmids pTrcHis A, B, C, ampicillin resistant, are vectors for expression of recombinant proteins in *E. coli*. These plasmids contain N-terminal 6×-histidine tags that are added to the expressed proteins (Life Technologies). The histidine-tagged proteins are purified by nickel affinity chromatography with elution of bound proteins using imidazole buffers. Plasmid pIMAY is an *E. coli–S. aureus* shuttle plasmid, constructed by Monk et al. [7]. This plasmid is temperature-sensitive and contains a chloramphenicol selection marker. Plasmid pLZ113 is a high copy-number plasmid and is used in the construction of a regulator complementing plasmid [11].

2.4 *Primers*

Primers P-1 and P-2 are designed to amplify the promoter region of a target efflux pump published in NCBI (Fig. 1a). Primers P-3 and P-4 are designed to amplify the structural gene of the regulator, flanking the start and the stop codons of the gene. Restriction sites are incorporated into the primer sequences to facilitate cloning (Fig. 3). Four primers P-5, P-6, P-7, and P-8 are used in the construction of an in-frame deletion mutant and are designed as follows: P-5 is a forward primer with restriction site 1 (RS1), P-6 is a reverse internal primer, P-7 is a forward internal primer that has a complementary portion of the primer P-6, and P-8 is a reverse primer with restriction site 2 (RS2) (*see* **Note 10** and Fig. 5). Primers P-9 and P-10 are designed from the region upstream and downstream of the regulator gene (Fig. 6). This pair of primers is used to verify the deletion of the regulator gene on the chromosome. Primers P-11 and P-12 are designed from the internal region of the regulator gene to generate a small DNA replicon of ~100-bp for quantitative qRT-PCR assays (Fig. 6). Primers *gmk*1and *gmk*2 are designed to amplify a 100-bp internal region of the housekeeping gene *gmk*, which is used as an internal control for the real-time qRT-PCR assays.

2.5 *Template DNAs*

Genomic DNAs are isolated from *S. aureus* RN6390 and the isogenic mutant(s) with the PureLink Genomic DNA Extraction kit (Life Technologies), following the manufacturer's recommendation.

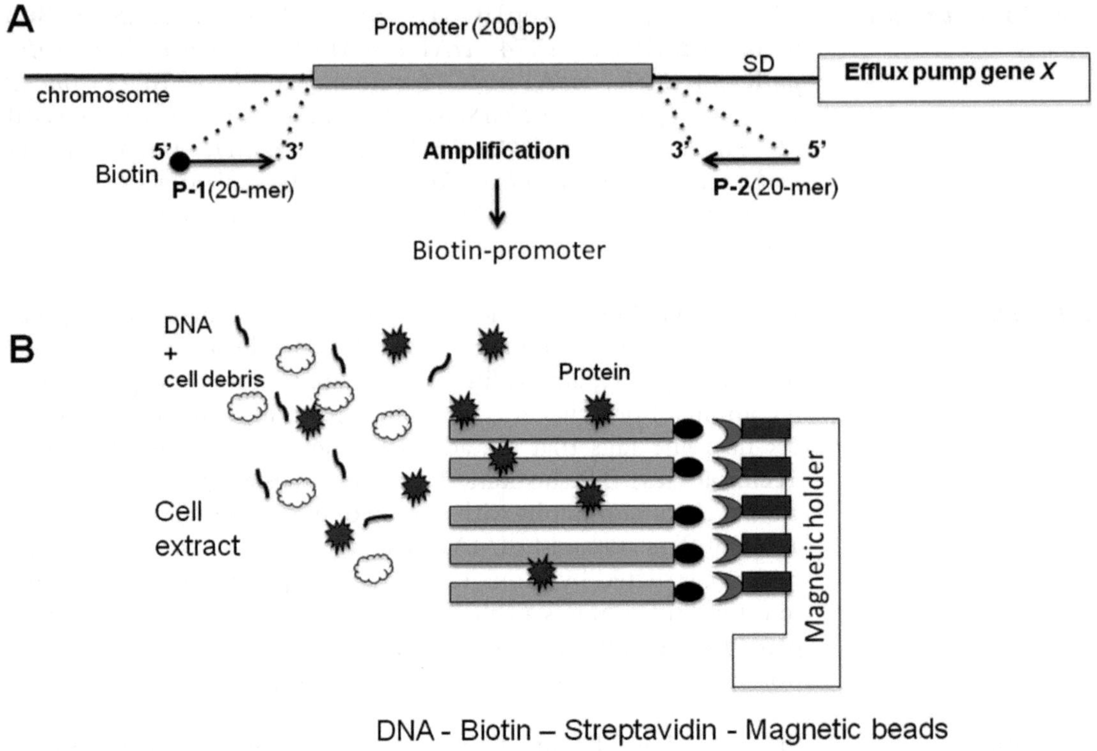

Fig. 1 Amplification of a biotin-labeled promoter. (**a**) Forward primer P-1 is labeled with biotin at its 5′-end. The PCR product encompasses the full length of the promoter region upstream of the Shine-Dalgarno region (SD). The length of the promoter is between 150–400 bp. In this example, the promoter is 200-bp in length. (**b**) DNA affinity column is immobilized via a magnetic holder (DynaMag magnet, Thermofisher Scientific) to facilitate successive washing steps. The DNA binding proteins are eluted using a solution of 0.5 M NaCl

2.6 Template RNAs

Total RNAs are isolated from *S. aureus* RN6390 and mutant(s) using the Qiagen RNA Extraction Kit (RNeasy Mini Kit, Qiagen).

2.7 Media

Luria-Bertani (LB), Brain Heart Infusion (BHI), or Mueller-Hinton (MH) in broth and agar.

2.8 Antibiotics

Ampicillin, chloramphenicol, anhydrotetracycline, kanamycin, norfloxacin, ciprofloxacin.

2.9 Chemical Compounds

Streptavidin-coupled magnetic beads (Dynabeads-M280 Streptavidin, Thermofisher Scientific), isopropyl β-D-1-thiogalactopyranoside (IPTG), imidazole, nickel resin for histidine-tag protein purification (His60-Ni resin, Takara Clontech), buffer Tris–HCl pH 7.6, sodium chloride (NaCl), lysostaphin, and 6× gel loading dye bromophenol blue.

2.10 Materials for Real-Time qRT-PCR Assays

Verso cDNA synthesis Kit (Thermofisher Scientific) and Eva Green-master mix for qPCR assays (Bio-Rad).

2.11 Restriction Enzymes

All restriction enzymes and Vent polymerase are purchased from New England Biolabs.

2.12 Materials for Protein Electrophoresis

Western blotting apparatus iBlot system for protein transfer (Life Technologies), XCell Surelock Mini Cell electrophoresis system for SDS-PAGE electrophoresis, NuPAGE MOPS-SDS running buffer (20×), NuPAGE 10% Bis-Tris gels, Coomassie blue G-250 Simply-Blue SafeStain (Thermofisher Scientific).

2.13 Gel Mobility Shift Assays

LightShift Chemiluminescent EMSA Kit for the detection of DNA binding proteins (Thermofisher Scientific).

2.14 Buffers

Buffer A (1×) for histidine-tag protein purification: 20 mM Tris–HCl pH 7.6, 150 mM NaCl, 10% glycerol).

Buffer B (1×) for cell extract: 20 mM Tris–HCl, 50 mM $MgCl_2$, 1 mM dithiothreitol, 0.1 mM EDTA, 5% glycerol.

Buffer C (1×) for cell extract DNA binding assay: 10 mM HEPES pH 8, 60 mM KCl, 4 mM $MgCl_2$, 0.1 mM EDTA, 0.1 mg/ml of bovine serum albumin, 0.25 mM dithiothreitol, 1 μg of poly(dI-dC), and 2 ng/μl herring sperm DNA.

Running buffer for agarose gel electrophoresis (TAE solution 1×): 40 mM Tris acetate and 1 mM EDTA, pH 8.

Sample buffer for SDS-PAGE: 0.1 M Tris–HCl pH 6.8, 10% SDS, 0.1% bromophenol blue, 5% glycerol, 4 M β-mercaptoethanol.

3 Methods

Outline of the Methods section: (1) isolation of putative regulatory protein(s) from crude cell extracts, (2) identification of the DNA binding protein, (3) cloning and purification of the protein, (4) protein–DNA affinity binding, (5) creation of an in-frame deletion mutant, (6) influence of the regulatory protein on efflux pump expression, (7) complementation study.

3.1 Isolation of Putative DNA-Binding Protein(s) from Crude Cell Extracts

*3.1.1 Primer Design (See **Note 1**)*

We retrieve the nucleotide sequence of the target efflux pump gene and its promoter from the genome of *S. aureus* NCTC8325 published in the National Center for Biotechnology Information (NCBI). We design two 20-nucleotide primers that are designated as the forward primer P-1 and the reverse primer P-2, located upstream and downstream from the promoter region of the target efflux pump gene, respectively (Fig. 1a). The forward primer P-1 is biotinylated at its 5′ end for two purposes: (1) attachment of the promoter DNA after amplification to the streptavidin-Dynabeads, and (2) detection of the binding ability of isolated proteins using a chemiluminescence technique [12].

3.1.2 Amplification and Purification of the PCR Fragment

We extract the genomic DNA from 3 ml of an overnight culture of *S. aureus* RN6390 (NCTC8325 background strain) in LB broth media, using the PureLink Genomic DNA Extraction kit. The genomic DNA is dissolved in sterile water at a concentration of 200 ng/ml. We design the PCR reaction with the following parameters to amplify a product of ~200-pb: one cycle of 94 °C for 3 min; 30 cycles of 94 °C for 30 s, T °C for 30 s (annealing temperature T is dependent on the Tm of the primer pair, *see* **Note 15**), 72 °C for 9 s; and one cycle of 72 °C for 10 min. The PCR product is ~150–200 bp in length and is biotinylated at its 5′-end. We analyze the PCR product by electrophoresis through a 1% agarose gel in TAE (1×) buffer, with the running conditions set at 100 V for 45 min. The PCR product is excised from the agarose gel, and ethanol precipitated to remove any contaminating materials. Details of the procedures are described below.

3.1.3 Extraction of PCR Product from Agarose Gel

The PCR product is excised from the agarose gel using a sterile razor, sliced into small fragments, and purified using the Freeze'N Squeeze DNA gel extraction spin column from Bio-Rad (Cat#732-6165, Bio-Rad Laboratories, Hercules, CA). The agarose-embedded DNA fragments are transferred sites to a Freeze'N Squeeze spin column, frozen at −20 °C for 5 min, and centrifuged at 15,000 × g in a microcentrifuge for 3 min. The collected DNA sample is then ethanol precipitated.

3.1.4 Ethanol Precipitation

For a volume of DNA sample of 50 μl, we add 250 μl of sodium chloride 5 M (NaCl) (5× of the sample), and 1000 μl of cold 100% ethanol (20× of the sample). We incubate the DNA/NaCl/ethanol mixture at −80 °C for 10 min, then centrifuge the frozen sample at 15,000 × g in a microcentrifuge for 20 min to collect the precipitated DNA. The DNA is quantified using the Nanodrop 2000 spectrophotometer (Thermofisher Scientific), as suggested by the manufacturer. The DNA concentration is adjusted to 200 ng/μl.

3.1.5 Preparation of a DNA Column (See ***Note 2****)*

Streptavidin-coupled Dynabeads are commercially available and are ready for immediate use following the manufacturer's recommendations (Fisher Scientific). We mix the biotinylated DNA with Streptavidin-Dynabeads in the proportion 1:1 (vol DNA/vol beads) for 20 min at room temperature under gentle rotation using a wheel or shaker. The mixture is washed three times with 300 μl of binding buffer C (*see* Subheading 2) to equilibrate the DNA column. After the final washing and buffer C removal, The DNA column is now ready for loading of the cell extract [3, 6].

3.1.6 Preparation of S. aureus *Cell Extract (See* ***Note 3****)*

We start a fresh *S. aureus* culture by adding 25 ml of an overnight culture to 1 L of fresh LB broth to have an initial optical density OD_{600} of 0.05. The new culture reaches an OD_{600} of 0.6 after ~4 h growth at 37 °C under shaking (220 rpm); we then harvest the bacteria by centrifugation at 2000 × *g*. The pellet is washed once in buffer B (*see* Subheading 2), and is resuspended in 10 ml of the same buffer supplemented with 0.1 mg/ml of lysostaphin. The pellet is incubated for 2 h on ice with occasional vortexing to lyse the bacteria. We prepare 6 ml of 2 M KCl in Buffer B from a 10 M stock solution, add this solution to the pellet, and then incubate the mixture on ice for an additional 30 min. The KCl will precipitate the DNA in the crude cell extract. The bacterial lysate is centrifuged at 40,000 × *g* for 30 min to remove debris and precipitated DNA. The supernatant (~15 ml) is dialyzed for 3 h against water, and dialyzed against buffer C for an additional 30 min prior to loading the column.

3.1.7 Isolation of DNA Binding Protein(s) (See ***Note 4****)*

We incubate the DNA (20 μg) prepared in Subheading 3.1.3 with 15 ml of protein extract in binding buffer C supplemented with herring sperm DNA (2 ng/μl) for 20 min at room temperature under gentle shaking. The magnetic beads are attached to a magnetic holder (DynaMag magnet, Thermofisher Scientific), which allows the proteins to remain attached to the DNA/beads complex during the washing steps (Fig. 1b). We wash the mixture proteins-DNA-Dynabeads three times with 2 ml of buffer C with herring sperm DNA (2 ng/μl), and three times with 2 ml of buffer C without herring DNA. We elute the DNA binding proteins with 500 μl of buffer C supplemented with 0.5 M NaCl. We then dialyze the eluted protein(s) against water, and separate them by SDS-PAGE.

3.2 Identification of the Binding Proteins

To identify a protein, we first visualize it on an acrylamide gel and then identify the protein by N-terminal sequencing and mass spectrometry.

3.2.1 SDS-PAGE of the Eluted Protein

We quantify the dialyzed proteins using the Bradford colorimetric assay from Bio-Rad (Bio-Rad Protein Assay), following the manufacturer's recommendations. We prepare a series of eppendorf tubes containing protein standard solutions ranging from 0.2 to 0.8 μg/μ l in a volume 100 μl. We prepare in parallel an eppendorf tube with 100 μl of the sample protein after dialysis. We add 1 ml of the dye reagent (Bio-Rad) to the proteins, vortex to mix them, and incubate the mixtures at room temperature for 15 min. We transfer the dye and protein mixture into plastic cuvettes and measure the absorbance at 595 nm. Since the quantification only gives the total amount of proteins eluted, we submit the sample to SDS-PAGE to check the number of proteins isolated by the DNA binding technique. We prepare protein samples containing: 20 μl of proteins (~ 150 ng), 20 μl of water, 10 μl of sample buffer (*see* Subheading 2), and boil them for 5 min. We load 35 μl of sample per lane onto a 10% acrylamide/bis-acrylamide minigel (Life Technologies), fill the electrophoresis apparatus with the MOPS-SDS

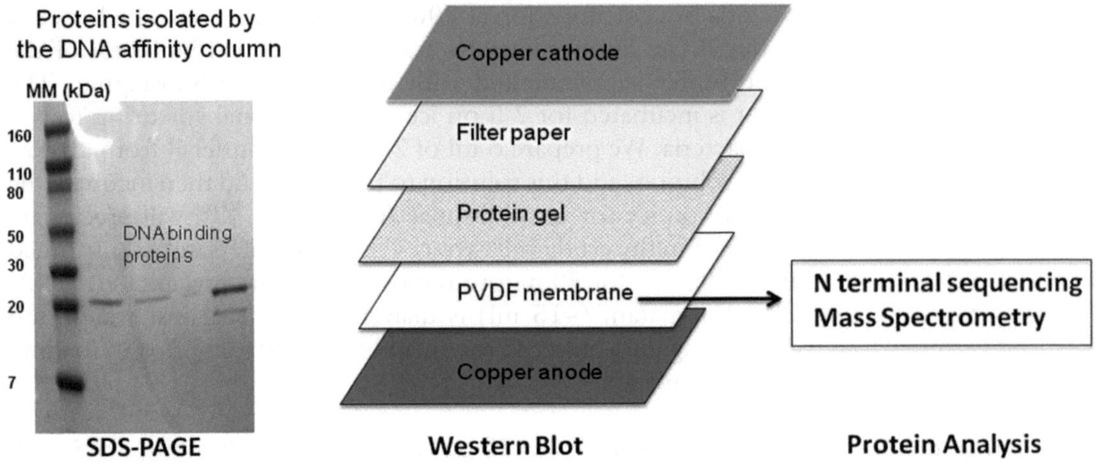

Fig. 2 Separation of eluted proteins via electrophoresis through SDS-PAGE gel. MM, molecular mass. Western blot: The proteins are transferred onto PVDF membrane and submitted to N-terminal sequencing for identification

running buffer (0.5×), and apply the current at 40 mA/110 V for 1 h. Proteins are visualized by staining with SimplyBlue Safe Stain (*see* Subheading 2) (Fig. 2).

3.2.2 Western Blot (See ***Note 5****)*

We prepare two sets of protein samples: one set for the SDS-PAGE and one set for Western blot. We carry out the sample preparation and electrophoresis through a 10% acrylamide/bis-acrylamide gel as described above. At the end of the electrophoresis, one gel is stained with the SimplyBlue stain and the other is submitted to Western blotting. We use the iBlot Gel Transfer device from Life Technologies, with copper anodes and cathodes from Fisher Scientific. We prepare the following stacks on the platform of the iBlot device: the copper anode and PVDF membrane (included in the copper anode), the protein gel without staining, a single layer of filter paper (Whatman, standard grade) cut to the size of the protein gel and pre-wet with the MOPS-SDS running buffer (0.5×), and place the copper cathode on the top of the stack. We run the iBlot device to transfer protein(s) from the gel into the PVDF membrane for 7 min (Fig. 2). Following the protein transfer step, we stain the membrane with a staining solution (0.1% Coomassie Blue R-250, 50% methanol, 7% acetic acid, and water for 50 ml) for 5 min, destain (50% methanol, 10% acetic acid) for 10 min or until the desired color, and air dry the membrane before submitting the transferred proteins to N-terminal sequencing by the Edman degradation method, and/or to mass spectrometry analysis at a core facility. The amount of protein necessary for these two analyses is ~0.5–1 μg on PVDF membrane. The Western blot needs to be carried out with appropriate precautions (gloves, sterile water, and sterile buffers) to avoid contamination.

3.2.3 N-Terminal Sequencing Analysis (See Note 6)

We analyze the 14 amino acids of the N-terminal sequence of the isolated protein by using the search tools of the NCBI web site (http://www.ncbi.nlm.nih.gov). We perform a BLAST search using the amino acid sequence against the data bank of the published genome of *S. aureus* NCTC8325 (http://www.ncbi.nlm.nih.gov/Blast.cgi). From the matching open reading frame (ORFs) generated by the web-based program, we select the ORF that shows the highest homology with the amino acid sequence for further analysis. Since we are looking for a putative regulator of a target efflux pump, we focus on ORFs that are ~500-bp in length with a hypothetical role in gene regulation.

3.3 Gene Cloning, E. coli Transformation, and Protein Purification

We amplify the target gene from the *S. aureus* chromosome and insert the amplified gene into a cloning plasmid for further study (Fig. 3).

3.3.1 Gene Cloning

3.3.2 Amplification of the Regulatory Gene

The genomic DNA is prepared as described in Subheading 3.1.2. We design the forward P3 and the reverse P4 primers that flank the start (ATG) and stop (TAA, TAG, TGA) codons of the chosen ORF on the genome of *S. aureus* NCTC8325. The primers are 20-mer oligonucleotides, and each primer harbors the restriction site chosen from the multicloning site of the vector plasmid pGEM3zf$^+$ for subsequent cloning (*see* Subheading 2). We choose two different restriction sites to ensure a directional insertion-ligation of the

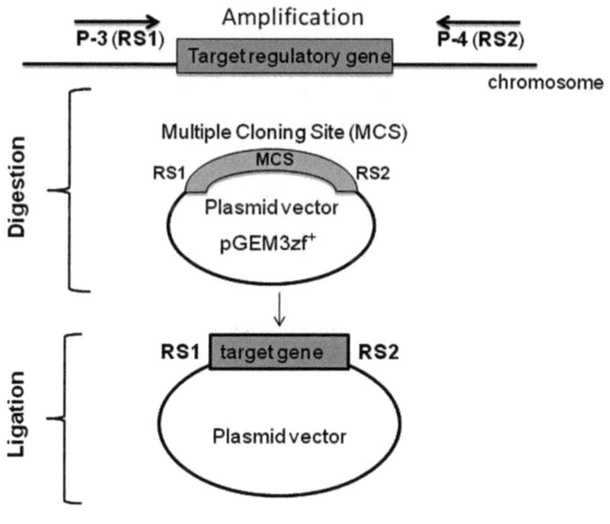

Fig. 3 Cloning of the regulatory gene into the cloning plasmid pGEM3zf$^+$. Both PCR product and the plasmid vector are digested with the same restriction enzymes and ligated together using T4 ligase

insert into the vector. We perform a PCR reaction to amplify a 500-bp gene using the following parameters with *T* as the annealing temperature, that is dependent on the T_M of the primer pairs: one cycle at 94 °C for 3 min; 30 cycles of one cycle at 94 °C for 50 s, one cycle at T°C for 50 s, one cycle at 72° for 30 s; and one cycle at 72 °C for 10 min. The PCR product is analyzed by electrophoresis through a 0.8% agarose gel in Tris-Acetate-EDTA (1× TAE, pH 8) buffer at 8 V/cm for 1 h.

3.3.3 Digestion of the PCR Product and the Cloning Plasmid

Plasmid pGEM3zf$^+$ is purified using the plasmid DNA extraction kit from Qiagen (Qiaprep Spin Miniprep Kit). The PCR product and the vector plasmid are digested with the same restriction enzymes (New England Biolabs). The digestion mixture is prepared as follows: 1 μg of DNA, 10 units of each restriction enzyme, 1× digestion buffer provided by the manufacturer, and sterile water to a final volume of 50 μl, followed by incubation at 37 °C for 1 h. The PCR product and the plasmid are submitted to electrophoresis through 0.8% agarose gel in TAE 1× buffer after digestion, excised from agarose gel as described in Subheading 3.1.3, and purified using the Gel Extraction kit (Qiagen).

3.3.4 Dephosphorylation of the Plasmid

The digested plasmid is treated with the alkaline phosphatase to dephosphorylate the 5′-end of the vector. The protocol for the dephosphorylation reaction is as follows: 500 ng of plasmid DNA in 10 μl of sterile water is mixed with 2 μl of dephosphorylation buffer (1x, included in the kit), 10 units of calf intestinal phosphatase (CIP, New England Biolabs), and sterile water to have a total volume of 20 μl. The mixture is incubated for 1 h at 37 °C. The 5′-end dephosphorylated vector pGEM3zf$^+$ is purified using the PCR purification kit (Qiagen).

3.3.5 Ligation of Digested Products (See ***Note 7****)*

The vector and PCR product are ligated covalently with DNA ligase enzyme. The procedure is as follows: the digested plasmid and the PCR product are mixed in the proportion of 1:4 (vol:vol). We add 2 μl of ligase buffer (T4 ligase, New England Biolabs), 10 units of enzyme T4 ligase, and sterile water to have a total volume of 20 μl. The ligation mixture is incubated at 16 °C overnight or alternatively the reaction can be done at room temperature for 20 min. The plasmid construct (pGEM3zf$^+$—insert) is ready to be electroporated into *E. coli*.

3.3.6 Transformation of E. coli Top10 with the Plasmid Construct, E. coli Competent Cell Preparation

We start a 500 ml fresh culture of *E. coli* Top10 in LB broth by adding 20 ml of an overnight culture to reach an OD_{600} of 0.05. After incubation at 37 °C at $OD_{600} = 0.5$, bacteria are harvested by centrifugation at 2000 × *g* for 15 min at 4 °C. The bacteria are kept on ice at all times. The supernatant is discarded, and the cell pellet is resuspended in 100 ml of ice-cold sterile water. The cells are harvested by centrifugation at 2000 × *g* for 15 min. We repeat this step four times using ice-cold water in decreasing volumes 50, 25, 10, and 5 ml. The pellets are resuspended in 500 μl of ice-cold water with 10% glycerol, and divided into 50 μl aliquots for storage at −80 °C. Cold-competent *E. coli* (50 μl) is mixed with DNA (500 ng in 5 μl) and transferred into an ice-cold 0.2-cm electroporation cuvette [13].

3.3.7 Transformation of E. coli Top10 by Electroporation

Electroporation is performed at 2.5 kV with 25 μF and 200 Ω using the Bio-Rad Gene Pulser. We add 1 ml of LB immediately to the electroporated *E. coli*, and incubate the bacteria at 37 °C for 30 min with shaking. We spread 100 μl of the culture on LB agar plates supplemented with the appropriate antibiotic selection (ampicillin 100 μg/ml for plasmid pGEM3zf+ selection), and incubate the plates at 37 °C overnight to select transformants. Genomic DNAs of five transformants that grow on LB plates plus 100 μg/ml ampicillin, are extracted as described in Subheading 3.1.2 and used as templates for PCR reactions with primers P3 and P4. PCR products are purified with the PCR purification kit (Qiagen) and submitted for DNA sequencing for verification of the cloned gene.

3.3.8 Protein Purification, Selection of an Expression Vector

The cloned gene needs to be transferred into the appropriate expression vector pTrcHisA, B, or C, induced by isopropyl β-D-1-thiogalactopyranoside (IPTG), and purified with a nickel (Ni^{2+}) affinity column [12, 14]. We construct the protein expression clone by inserting the DNA sequence of the target gene to the multiple cloning sites (MCS) of the expression vector pTrcHisA,B, C. Each vector pTrcHisA, pTrcHisB, and pTrcHisC has an extra short sequence of nucleotides to allow the insert to be in frame with the vector, leading to the expression of the complete protein with a sequence of six histidines at its 5′-end (Fig. 4a). We use the Open Reading Frame Finder tool from NCBI (http://www.ncbi.nlm.nih.gov/gorf/) to verify the translation of the construct. We verify the presence and the orientation of the two restriction sites that were previously inserted into the target gene for its cloning into the pGEM3zf+ vector. For this study, we use the expression vector pTrcHisA.

3.3.9 Protein Purification, Construction of the Expression Clone

We purify the plasmid pTrcHisA using the Qiaprep Spin Miniprep Kit form Qiagen. We digest the pTrcHisA and the clone pGEM3zf+-target gene with the same two restriction enzymes. The digestion products are analyzed by electrophoresis through 0.8% agarose gel. The pTrcHisA plasmid and the insert are purified with the Qiagen Gel Extraction kit, and are ligated together with T4 ligase. The new construct pTrcHisA-target gene is introduced into *E. coli* BL21 by electroporation for protein expression. The

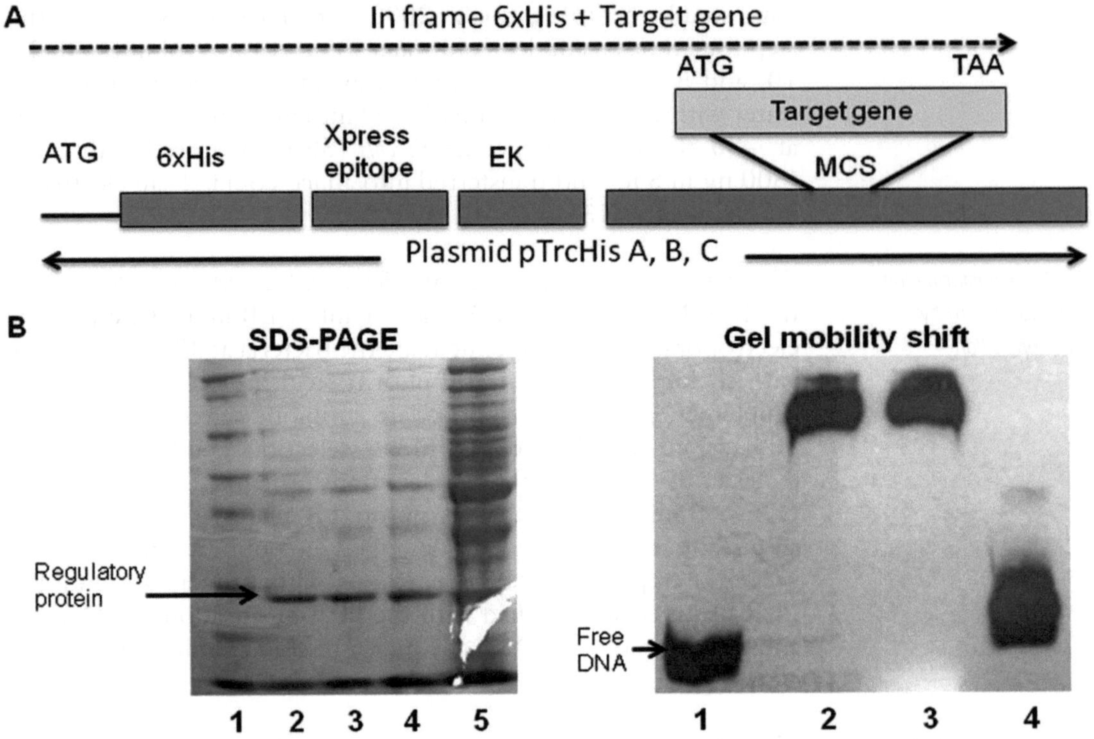

Fig. 4 Cloning and purification of the regulatory protein. The regulatory gene is cloned into the protein expression plasmid pTrcHis that generates an inframe hybrid protein which harbors the 6×-histidine tag at the N-terminal end of the regulatory protein. The hybrid protein contains the Xpress epitope which can be used as an antibody target in Western hybridization to verify the expression of the cloned gene. Enterokinase can be used to cleave the His tag from the protein. *EK* enterokinase cleavage site, *MCS* mutiple cloning site. SDS-PAGE: *Lane 1*, molecular weight marker; *lane 2*, proteins eluted with 200 mM imidazole; *lane 3*, proteins eluted with 150 mM imidazole, *lane 4*, proteins eluted with 100 mM imidazole, *lane 5*, cell extract. Gel mobility shift assay: *Lane 1*, free biotinylated promoter DNA; *lane 2*, biotinylated promoter DNA and protein; *lane 3*, biotinylated promoter DNA and protein in the presence of 100-fold in excess of herring DNA; *lane 4*, biotinylated promoter DNA and protein in the presence of 100-fold in excess of unlabeled promoter DNA

procedures are the same as those described in Subheadings 3.3.1 and 3.3.7.

3.3.10 Purification of the Target Protein Using a Nickel Affinity Column (See ***Note 8****)*

We start a 1 L culture of transformant *E. coli* BL21(pTrcHisA-target gene) from 25 ml of overnight culture, and grow them until $OD_{600} = 0.5$. We add 1 ml of IPTG stock solution 1 M to the growing culture to have a final concentration of 1 mM IPTG. The induction continues for an additional 3 h, and bacteria are harvested by centrifugation at 2000 × *g*. Note that optimal induction conditions can vary with the protein being expressed and may require empiric testing of varying conditions of growth and inducing IPTG concentrations. The cell pellet is resuspended in 20 ml of buffer A (*see* Subheading 2) and kept on ice. We add 2 ml of lysozyme (stock solution 10 mg/ml) and one tablet of cocktail protease inhibitors (Life Technologies) to the cell suspension and continue to incubate the mixture on ice with occasional vortexing for 1 h. The cells are centrifuged at 10,000 × *g* for 30 min to separate cell debris from the supernatant. The supernatant is filtered through a 0.45 μm filter cup and is ready for the purification step.

A 1 ml nickel column (Fisher Scientific) is connected to a peristaltic pump, and is adjusted to a flow of 1 ml per min. The column is washed twice with 5 ml of sterile water to eliminate traces of ethanol, and is equilibrated with 5 ml of buffer A supplemented with 10 mM imidazole. We apply the supernatant to the column and collect the flow-through for later analysis by SDS-PAGE. The 6× histidine tag of the protein binds to the nickel column matrix and can be eluted using a series of imidazole solutions. Once the supernatant is loaded on the column, we wash the column successively with 5 ml of buffer A + 50 mM imidazole, 5 ml of buffer A + 100 mM imidazole, 5 ml of buffer A + 150 mM imidazole, and 5 ml of buffer A + 200 mM imidazole. We collect a series of 1 ml fractions for each buffer, and analyze the purified proteins by SDS-PAGE to evaluate their amount and homogeneity (Fig. 4b) [6].

3.4 Protein–DNA Affinity Binding

We use the promoter DNA described in Subheading 3.1 without the streptavidin coupled magnetic Dynabeads. Purified protein is dialyzed against water overnight and dialyzed against buffer A for an additional 30 min. Protein is quantified using the Bradford colorimetric assay from Bio-Rad as described in Subheading 3.2.1.

3.4.1 Mobility-Shift Assays (See Note 9)

The mobility-shift assays are carried out using the LightShift Chemiluminescent EMSA Kit from ThermoFisher Scientific. Biotin 5′-end labeled promoter DNA and protein are incubated together in the proportion 2 ng of DNA, 200 ng of purified protein, binding buffer C 1× (*see* Subheading 2), combined in a total volume of 20 μl. The reaction mixture is incubated at room temperature without shaking for 30 min. We add 5 μl of loading dye (Bromophenol Blue, EMSA kit) to the mixture prior to electrophoresis through a 5% acrylamide nondenaturing gel.

3.4.2 Polyacrylamide Nondenaturing Gel

We prepare the running gel as follows: for a 25 ml gel solution, we combine 4.2 ml of acrylamide/bis-acrylamide (29/1%), 1.25 ml TBE 10× (Tris-base pH 8.3, boric acid, EDTA), 275 μl ammonium persulfate (APS) 10%, 25 μl tetramethylethylenediamine (TEMED), and 19 ml of water. The gel solution is transferred immediately into the gel casting (minigel casting apparatus, Bio Rad) and allowed to polymerize at room temperature. The samples are loaded onto the wells, and we run the electrophoresis in TBE 0.5× at 10 V/cm for 45 min. The gel is transferred onto a nitrocellulose membrane using the Mini Trans-Blot cell from Bio-Rad and TBE 0.5× as transfer buffer, at 100 V for 2 h. The membrane is washed and developed following the manufacturer's recommendations (EMSA Kit, ThermoFisher Scientific).

3.4.3 Competition Assays

We perform competition assays to verify the specificity of the DNA binding ability of the protein. We prepare four reaction mixtures as described above with the following modifications: reaction 1, only biotin-DNA is added; reaction 2, protein and biotin-DNA are added; reaction 3, protein and biotin-DNA are added plus herring DNA in excess (100×); reaction 4, protein and biotin-DNA are added plus unlabeled specific promoter DNA in excess (100×). The four reactions are incubated at room temperature and analyzed by electrophoresis through a 5% nondenaturing polyacrylamide gel (Fig. 4b).

3.5 Construction of an In-Frame Deletion Mutant

We describe the technique published by Monk et al. [7], that generates a marker-less mutant lacking the major part of the gene but remains in-frame with the adjacent genes on the bacterial chromosome (Fig. 5).

3.5.1 PCR Reactions (See Note 10)

The designs of the four primers P-5, P-6, P-7, and P-8 are described in Subheading 2, and illustrated in Fig. 5. The PCR reaction is carried out as follows: one cycle at 94 °C for 3 min;

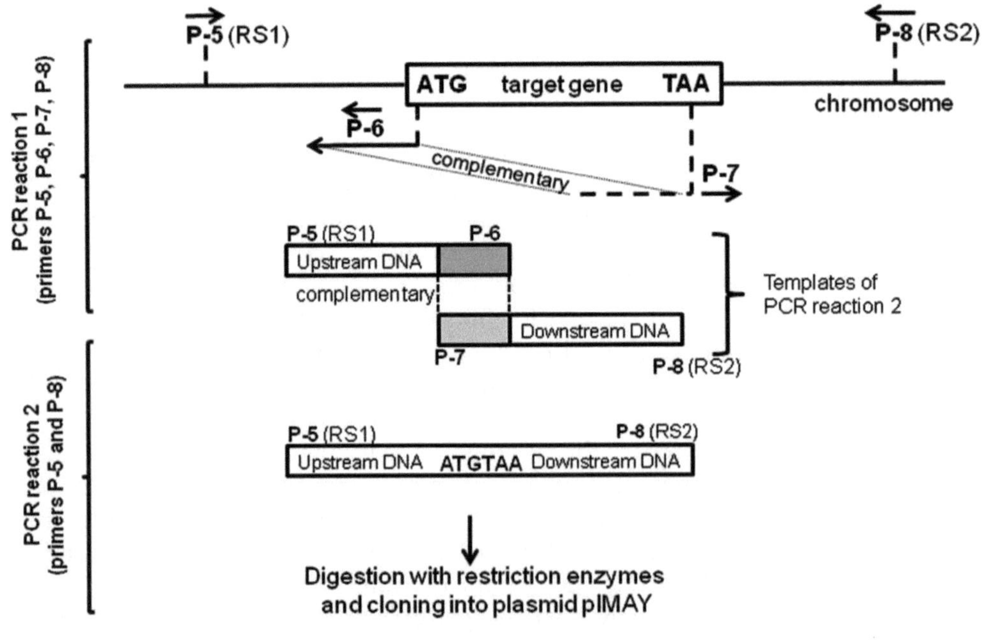

Fig. 5 Construction of an inframe deletion mutant of a regulatory gene. The products of the first PCR reaction harbor a short complementary fragment and are used as templates for the second PCR reaction. The PCR product of the second reaction contains a region upstream and a region downstream of the target gene. The regulatory gene is deleted except for the ATG and TAA codons. The PCR product is digested by restriction enzymes that match the two inserted restriction sites and cloned into the shuttle *E. coli–S. aureus* plasmid pIMAY for subsequent integration and allelic exchange

30 cycles of one cycle at 94 °C for 50 s, one cycle at 50 °C for 50 s, one cycle at 72 °C for 1 min; one cycle at 72 °C for 10 min. Primers P-5 and P-6 are used to amplify a 500-bp upstream sequence of the regulator gene. Primers P-7 and P-8 are used to amplify a 500-bp downstream sequence of the gene. Each PCR product is diluted 20-fold, and 1 μl of each is used as the template (2 μl total) for a second PCR reaction with primers P-5 and P-8, using the same parameters as above for the second PCR reaction.

3.5.2 Cloning of the PCR Product

The 1-kb PCR product carries a deleted regulator gene with RS1 and RS2 restriction sites in the flanking end regions. The PCR product is digested by restriction enzymes RE1 and RE2 as described in Subheading 3.3.3, then is ligated into the *E. coli–S. aureus* shuttle plasmid pIMAY as described in Subheading 3.3.5, and the final construct pIMAY-*Δregulatory gene* is transformed into *E. coli* DH10B as described in Subheading 3.3.7. The transformants are grown at 37 °C on LB plates containing chloramphenicol 10 μg/ml.

3.5.3 Transformation of S. aureus

The construct pIMAY-*Δregulatory gene* in *E. coli* DH10B is extracted and electroporated into *S. aureus* RN4220 and then into *S. aureus* RN6390 for subsequent allelic exchange.

3.5.4 Preparation of S. aureus Competent Cells for Electroporation

We start a 500 ml culture of *S. aureus* in LB broth by adding 10 ml of an overnight growth to have an OD_{600} of 0.05. As the OD_{600} reaches 0.25–0.3, bacterial cells are harvested by centrifugation at 2000 × *g* for 15 min, and the cell pellet is kept on ice at all times. We prepare a solution of sucrose 0.5 M in water by dissolving 171.15 g of sucrose in 1 L of sterile water and then filter the solution and keep it on ice. The *S. aureus* pellet is resuspended in 100 ml of ice-cold sucrose solution, centrifuged at 2000 × *g* for 15 min. We discard the supernatant and resuspend the pellet in another 100 ml of ice-cold sucrose solution. This step is repeated four more times with decreasing volumes of ice-cold sucrose water of 50, 25, 10, and 5 ml. The final pellet is resuspended in 1 ml of sucrose water and divided into 200 μl aliquots for electroporation.

3.5.5 Electroporation of S. aureus

Staphylococcus aureus RN4220 and 1 μg of DNA are mixed together and incubated on ice for 15 min prior to electroporation, using 0.2-cm cuvettes. The parameters are 2.5 kV, 25 μF, and 200 Ω. The electroporated *S. aureus* are resuspended in 1 ml of LB and incubated under shaking at 37 °C for 2 h. The bacteria are plated on LB agar supplemented with chloramphenicol 10 μg/ml, and incubated at 37 °C overnight to select for transformants. Plasmid construct is extracted from the RN4220 transformants using the Qiaprep Spin Miniprep kit, verified by DNA sequencing, and re-electroporated into *S. aureus* RN6390.

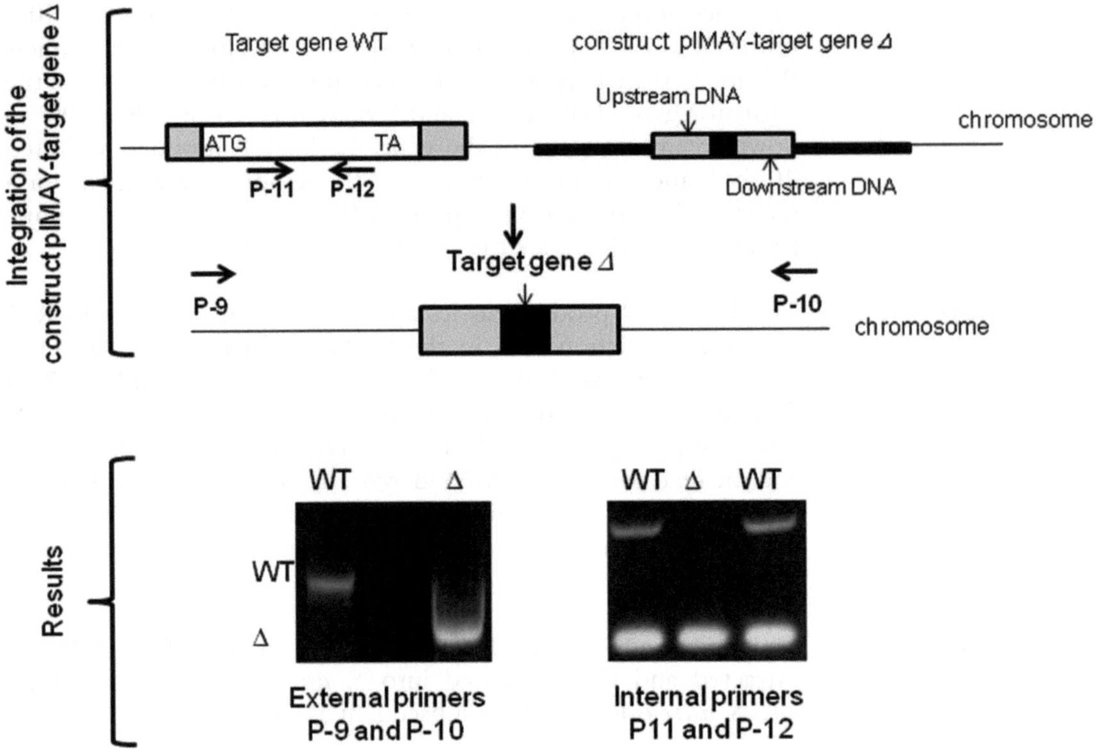

Fig. 6 Integration of the plasmid construct and the allelic exchange. PCR reaction using the external primers P-9 and P-10 generates a smaller product for the mutant due to the lack of the target gene. PCR reaction using the internal primers P-11 and P-12 generates a negative result for the mutant

3.5.6 Integration of the Construct and Allelic Exchange (See Note 11)

Staphylococcus aureus transformants are grown at 28 °C on LB plates supplemented with chloramphenicol 10 μg/ml. To integrate the construct pIMAY-*Δregulatory gene* into the chromosome, the transformants are diluted 100-fold and plated on LB plates plus chloramphenicol 10 μg/ml, then incubated at 37 °C for 24 h. Chloramphenicol-resistant colonies (~100 colonies) are selected and plated on LB agar plates supplemented with anhydrotetracycline 1 μg/ml and are grown at 28 °C for 48 h. We patch 100–200 colonies on LB plates and LB plus chloramphenicol 10 μg/ml plates to identify chloramphenicol-sensitive colonies. We verify the absence of the regulatory gene of the mutant by DNA sequencing using PCR with external primers P-9 and P-10 (*see* **Note 10**) (Fig. 6).

3.6 Influence of the Regulatory Protein on Efflux Pump Expression

We evaluate the result of an absence of the regulator on the transcription of the efflux pump and the subsequent drug resistance phenotype.

3.6.1 Real-Time qRT-PCRs, RNA Extraction (See ***Note 12****)*

We perform quantitative real-time RT-PCR to assess the change in target efflux pump transcript level. Primers P-11, P-12, *gmk*-1, and *gmk*-2 designed for the real-time assays are described in Subheading 2 [8]. We isolate the total *S. aureus* RNA using the RNeasy mini kit from Qiagen following the manufacturer's protocol. We start a 50 ml culture of *S. aureus* in LB broth by adding 1 ml of an overnight culture to have an OD_{600} of 0.05. When the OD_{600} of the culture reaches 0.6, *S. aureus* cells are collected by centrifugation at 2000 × *g* for 15 min. We resuspend the cell pellet in 1 ml of Tris buffered saline pH 7.6 (TBS: 50 mM Tris–HCl and 150 mM NaCl), and we lyse the cells with lysostaphin at a final concentration of 0.1 mg/ml. We add 700 μl RLT buffer plus beta-mercaptoethanol (10 μl of beta-mercaptoethanol/1 ml of RLT buffer) to every 500 μl of cell lysate in a 1.5-ml microcentrifuge tube. We mix the suspension by vortexing and add 500 μl of ethanol 70%. We mix the cell lysate/RLT/ethanol by vortexing and apply the suspension to the spin column provided with the RNeasy kit. The supernatant is discarded after a 30 s centrifugation at 16,000 × *g* in a microcentrifuge. We add 500 μl of RW1 buffer (RNeasy kit), centrifuge the spin columns for 30 s at 16,000 × *g*, and discard the supernatant. We add 500 μl RPE buffer with ethanol (RNeasy kit) and repeat the same centrifugation step as above. We centrifuge the spin column for an additional 2 min to eliminate all traces of ethanol from the RPE buffer. We add 40 μl of RNAse-free water to the column and centrifuge for 1 min at 16,000 × *g* to elute the RNA. The eluted RNA is treated with DNAse I (5 units of DNAse I per 1 μg RNA) to eliminate all contaminating DNA in the RNA samples [15].

3.6.2 Real-Time RT-PCR (See ***Note 13****)*

We perform the cDNA synthesis using the Verso cDNA kit (Fisher Scientific), the extracted total RNA, and the primers P-11, P-12, *gmk*-1, and *gmk*-2, according to the manufacturer's instructions. The reaction mixture contains the following components: 1 ng of RNA template, cDNA synthesis buffer 1×, 500 μM of each primer, dNTP mix (RNeasy kit), PCR grade water, and Verso enzyme (RNeasy kit) in a total volume of 20 μl. The cDNA synthesis reaction is one cycle at 42 °C for 30 min. We use the newly synthesized cDNAs as template for the quantitative PCR amplification with the Bio-Rad CFX96 Real-Time PCR apparatus. The products are labeled with the dye EvaGreen Master Mix (Bio-Rad), as recommended by the manufacturer. The reaction parameters are as follows: one cycle at 94 °C for 15 min; 40 cycles of one cycle at 94 °C for 15 s, one cycle at 55 °C for 15 s, one cycle at 72 °C for 15 s. The relative expression of the target efflux pump gene is calculated using the $\Delta\Delta C_T$ method. In this method, the amount of target cDNA, which is normalized to the housekeeping gene *gmk* and relative to an in vitro reference, is set by the variable $2^{-\Delta\Delta CT}$, in this case C_T is the cycle number of the detection threshold [15].

3.7 MIC Testing

Most frequently, a bacterium that overexpresses the efflux pump shows an increase in its minimum inhibitory concentration (MIC) of a certain drug or family of drugs. This phenomenon is usually corroborated by demonstration of a decrease in MIC in a pump gene deletion mutant. The magnitude of differences in MIC in the plasmid overexpressor or deletion mutant are affected by the level of pump gene expression in the reference parental strain as well as the level of gene expression from the plasmid. The MIC is measured by two methods: the microbroth dilution technique and the agar dilution technique [16].

3.7.1 Microbroth Dilution Technique

This technique is convenient to measure the MICs of several antibiotics of a small number of bacterial strains. We prepare all antibiotic solutions in distilled sterile water. Starting with the highest concentration of drug, we make a series of twofold dilutions in a 96-well microtiter plate with a 100 μl of final volume of drug. We start a fresh 10 ml culture of bacteria in Mueller-Hilton (MH) broth from 1 ml overnight culture to obtain $OD_{600} = 0.01$. Grow the bacteria until the $OD_{600} = 0.5$ and dilute the culture 100-fold. Distribute 100 μl of the diluted bacterial suspension into each of the wells previously filled with 100 μl of antibiotic concentrations in MH broth. Each well has a final volume of 200 μl. We incubate the 96-well plate at 37 °C for 24 h before MIC reading. The first well (lowest concentration) that shows no visible growth indicates the MIC of the corresponding antibiotic. Growth controls using no antibiotic are included.

3.7.2 Agar Dilution Technique

This technique is suitable to measure the MIC of a particular antibiotic for a larger number of bacterial strains. The series of antibiotic solutions (usually a series of twofold dilutions) are prepared in 1 ml of sterile water and are mixed with 24 ml of MH agar previously autoclaved and cooled to 55 °C. Each agar plate contains 25 ml of MH agar and antibiotic. We prepare the bacteria as described above and distribute 100 μl of bacteria into a 96-well plate. Using a 48 solid pin multi-blot replicator (Thermofisher Scientific), which delivers 10 μl of bacteria per application, to deposit the sample on the agar plate plus antibiotics. We incubate the agar plates at 37 °C overnight or at least 18 h before reading. The lowest concentration of antibiotic at which a pin spot is considered the MIC for the strain.

3.8 Complementation Study (See Note 14)

To verify that the cause of an increase in the MIC of drugs is due to the absence of the regulator, we reintroduce the wild-type regulator gene into the deletion mutant and measure the MICs. We carry out the cloning of the regulatory gene into the *E. coli*–*S. aureus* shuttle plasmid pLZ113. The selection marker of this plasmid in *E. coli* is ampicillin 100 μg/ml and the selection marker in *S. aureus* is kanamycin 50 μg/ml. Another plasmid available for the complementation test is the *E. coli–S. aureus* thermosensitive shuttle plasmid pSK950. The selection marker for this plasmid in *E. coli* is spectinomycin 25 μg/ml, and the selection marker in *S. aureus* is tetracycline 5 μg/ml. pSK950 is thermosensitive, and the incubation should be done at 30 °C for *S. aureus* [8, 9].

We perform the gene cloning, bacterial electroporation, real-time qRT-PCRs (*See* **Note 15**), and MIC testings as described above to assess the expression of the efflux pump and the level of drug sensitivity of the complementing *S. aureus.*

4 Notes

1. Genomic DNA

   Primers are designed mainly from the reference genome of *S. aureus* NCTC 8325 published under the Microbial Genome page of the NCBI website (http://www.ncbi.nlm.nih.gov). The complete DNA sequence of 2.82 Mb in length can be found using the RefSeq NC_007795.1. To avoid the polar effects on the downstream gene expression, primers are designed in a manner that the subsequent deletion of the target gene would not result in a frame-shift in the adjacent downstream DNA region. Confirmation is carried out by submitting the DNA fragment (up to 2000 bp) that contains the gene deletion design, to the Open Reading Frame Finder (ORF Finder) tool of NCBI. The translation of the entire fragment including the in-frame deletion gene should be the same as that of the sequence without deletion.

2. DNA Column

   The amount of streptavidin coupled with magnetic Dynabeads is calculated based on the molarity of the biotin-DNA fragment. Depending on the size and the base content of the DNA, we calculate the amount of DNA in μmol/μl using the molecular weight (Da), the amount (μg), and the volume of DNA (μl). The ratio of biotin-DNA and streptavidin-beads is 1:1. We prepare a DNA column with at least 5 μg of DNA to maximize the amount of DNA binding proteins retained.

3. Crude Cell Extracts

   We add potassium chloride (2 M KCl) to the cells prior to centrifugation to separate proteins and DNA in the crude extract. After centrifugation, DNA fragments in the preparation are precipitated in the cell pellet with other cell debris. The supernatant is filtered with a filter cup (0.45 μm) and is ready for the purification step. The amount of cell extracts may need to be scaled up if the target proteins are not abundant, such as the case of several regulatory proteins. The bacterial culture volume generally varies between 1 L and 5 L per purification.

4. DNA Binding Proteins

   The number of DNA binding proteins that shows affinity for the DNA column depends on the promoter of the target efflux pump gene. One promoter can show affinity toward one or more regulator proteins. In this case, these proteins are either competing for the same promoter region or are binding to their own binding sequences within the promoter. The size of the biotin-labeled promoter will determine this factor, and it varies between 150 and 400 bp.

5. Western Blot

   The amount of total proteins for this step should be ~500 ng–2 μg if possible, since there may be a mix of several proteins that vary in abundance. The protein separation is more efficient with 15% SDS-PAGE gels or gradient gels. The SDS-PAGE gel for Western blot should not be stained after electrophoresis but used immediately for the protein transfer step. Each protein is transferred onto PVDF membrane and submitted for N-terminal sequencing. The staining with Coomassie Blue should not be more than 5 min, and destaining for 10–15 min. Silver staining is not recommended for N-terminal sequencing. Manipulation of the PVDF membrane should be done with care to prevent contamination with other DNA or proteins.

6. Protein/DNA Web-Based Analysis Tools

   We use online resources such as the National Center for Biotechnology Information (NCBI), which contains links to all databases, including microbial genomes, nucleotides, proteins, and molecular biology analysis tools. Promoter searching is done using the website Neural Network Promoter Prediction Server (http://promotor.biosino.org).

7. DNA Ligation

   Ligation can also be done with the Quick T4 DNA ligase from the Quick Ligation Reaction kit (New England BioLabs). The DNAs are mixed together in the presence of the ligation buffer 1× and the Quick T4 DNA ligase. The reaction is performed at 25 °C for 5 min. The reaction mixture is then ready for electroporation.

8. Nickel Affinity Chromatography

   The nickel affinity column can either be purchased ready-to-use or prepared in the lab. To make a nickel column, we purchase 1 ml resin columns and prepare a solution of nickel sulfate ($NiSO_4$) 1 M. The resin column is connected to a peristaltic pump and washed with 5 ml of water regulated at a speed of 1 ml per min to eliminate the ethanol in the column. The nickel solution is connected to the pump and 1 ml of the solution is applied to the column drop by drop. The resin column changes to a blue coloration once it is saturated with nickel. We wash the nickel-resin column with 5 ml of water, and equilibrate the column with 5 ml of buffer A supplemented with 10 mM of imidazole. The affinity column is ready for protein purification.

9. Gel Mobility Shift Assay

   The LightShift Chemiluminescent EMSA kit from Thermofisher Scientific provides a very sensitive assay that requires a minimal quantity of labeled DNA. The biotinylated DNA diluted to ~15–20 fmol per assay is sufficient to yield a detectable signal following 10 s exposure to Kodak X-OMAT autoradiography film.

10. Primers for In-Frame Deletion

   The forward primer P-5 is located at 500-bp upstream from the start codon (ATG) of the target gene. This primer is 20-mer in length and has the restriction site RS1 incorporated at nucleotide 6 from the 5′-end of the primer. The reverse primer P-6 (20-mer) begins at the start codon and continues upstream of the ATG. The forward primer P-7 (30-mer) has two parts: the first part (10-mer) contains the stop codon (TAA) of the target gene and seven nucleotides downstream from the stop codon, the second part contains a nucleotide sequence of 20-mer that is the complementary sequence of primer P-6. Primer P-8 is located at 500 bp downstream from the target gene and has a restriction site RS2 inserted at the sixth nucleotide of the primer.

11. In-Frame Deletion Mutant

   The most critical step is the integration of the pIMAY construct and the allelic exchange. The temperatures (28 and 37 °C) should be precise and remain constant with limited disturbances (avoid frequent opening/closing of the incubator) over the course of the experiment. Screening of colonies is done by serial patching using toothpicks on LB plates with or without chloramphenicol. From 200 colonies selected, we are able to get ~2–3 probable mutants. All candidates are submitted to DNA sequencing for verification.

12. RNA Extraction

   The most common problems with RNA extraction are the presence of genomic DNA in the final RNA sample and the degradation of RNA. To eliminate traces of the genomic DNA, we treat the RNA sample with DNase I for 10 min at 37 °C after elution from the RNeasy spin column. Alternatively, we can use the on-column DNase I digestion kit that is purchased with the RNeasy kit. In this method, DNase I is added directly to the column at room temperature for 15 min and the column is washed with buffers RW1 and RPE of the extraction kit. The degradation of RNA can lead to a low yield of RNA and the

RNA bands appear smeared on an agarose gel. In this case, the bacterial pellet needs to be lysed quickly and kept on ice as much as possible. The *S. aureus* pellet is lysed with lysostaphin at 100 μg/ml for 10 min at 37 °C. The lysis buffer RTL should contain β-mercaptoethanol (10 μl/ml). The RNA should be eluted in sterile and RNase-free water. If the RNA bands appear intact on an agarose gel but the RNA yield is low, the bacterial cells first need to be mechanically disrupted (beads beater) and lysed again with lysostaphin.

13. Real-Time RT-PCR Assay

    Primers for real-time PCR assays should be designed with an online primer design program (e.g. GenScript Real-time PCR Primer Design). A housekeeping gene whose expression level remains relatively constant for the strains or conditions being tested should be chosen as normalization internal control (e.g. the *gmk* gene).

14. Complementation Study

    The plasmid pLZ113 is a high copy number plasmid that could lead to potential problem if the cloned gene is toxic to the bacterial cells. Another plasmid that could be used in a complementation study is pSK950. This plasmid is a medium to low copy number plasmid, thermosensitive (30 °C), and is resistant to tetracycline (5 μg/ml). The multiple cloning site has a small number of restriction sites for cloning.

15. Calculation of the Annealing Temperature

    Based on the Tm of the primers, the annealing temperature is (Tm—5) °C. We choose the lower Tm of the two primers to calculate the annealing temperature and increase the value by 1 or 2 °C if necessary to get a specific product.

Acknowledgments

This work was supported by the U.S. Public Health Service Grants R37-AI23988 from the National Institutes of Health to D.C.H.

References

1. Andersen JL, He GX, Kakarla P, Kc R, Kumar S, Lakra WS, Mukherjee MM, Ranaweera I, Shrestha U, Tran T, Varela MF (2015) Multidrug efflux pumps from Enterobacteriaceae, *Vibrio cholerae* and *Staphylococcus aureus* bacterial food pathogens. Int J Environ Res Public Health 12:1487–1547
2. Floyd JL, Smith KP, Kumar SH, Floyd JT, Varela MF (2010) LmrS is a multidrug efflux pump of the major facilitator superfamily from *Staphylococcus aureus*. Antimicrob Agents Chemother 54:5406–5412
3. Fournier B, Aras R, Hooper DC (2000) Expression of the multidrug resistance transporter NorA from *Staphylococcus aureus* is modified by a two-component regulatory system. J Bacteriol 182:664–671
4. Villet RA, Truong-Bolduc QC, Wang Y, Estabrooks Z, Medeiros H, Hooper DC (2014) Regulation of expression of *abcA* and

its response to environmental conditions. J Bacteriol 196:1532–1539

5. Chien YT, Manna AC, Projan SJ, Cheung AL (1999) SarA, a global regulator of virulence determinants in *Staphylococcus aureus*, binds to a conserved motif essential for *sar*-dependent gene regulation. J Biol Chem 274:37169–37176
6. Truong-Bolduc QC, Zhang X, Hooper DC (2003) Characterization of NorR protein, a multifunctional regulator of *norA* expression in *Staphylococcus aureus*. J Bacteriol 185:3127–3138
7. Monk IR, Shah IM, Xu M, Tan MW, Foster TJ (2012) Transforming the untransformable: application of direct transformation to manipulate genetically *Staphylococcus aureus* and *Staphylococcus epidermidis*. MBio 3:e00277-11
8. Ding Y, Onodera Y, Lee JC, Hooper DC (2008) NorB, an efflux pump in *Staphylococcus aureus* MW2, contributes to bacterial fitness in abscesses. J Bacteriol 190:7123–7129
9. Ding Y, Fu Y, Lee JC, Hooper DC (2012) *Staphylococcus aureus* NorD, a putative efflux pump coregulated with the Opp1 oligopeptide permease, contributes selectively to fitness *in vivo*. J Bacteriol 194:6586–6593
10. Truong-Bolduc QC, Ding Y, Hooper DC (2008) Posttranslational modification influences the effects of MgrA on *norA* expression in *Staphylococcus aureus*. J Bacteriol 190:7375–7381
11. Zhang L, Fan F, Palmer LM, Lonetto MA, Petit C, Voelker LL, St John A, Bankosky B, Rosenberg M, McDevitt D (2000) Regulated gene expression in *Staphylococcus aureus* for identifying conditional lethal phenotypes and antibiotic mode of action. Gene 255:297–305
12. Truong-Bolduc QC, Hooper DC (2010) Phosphorylation of MgrA and its efect on expression of the NorA and NorB efflux pumps of *Staphylococcus aureus*. J Bacteriol 192:2525–2534
13. Chung CT, Niemela SL, Miller RH (1989) One-step prepration of competent *Escherichia coli*: transformation and storage of bacterial cells in the same solution. Proc Natl Acad Sci U S A 86:2172–2175
14. Truong-Bolduc QC, Villet RA, Estabrooks ZA, Hooper DC (2014) Native efflux pumps contribute resistance to antimicrobials of skin and the ability of *Staphylococcus aureus* to colonize skin. J Infect Dis 209:1485–1493
15. Truong-Bolduc QC, Hsing LC, Villet R, Bolduc GR, Estabrooks Z, Taguezem GF, Hooper DC (2012) Reduced aeration affects the expression of the NorB efflux pump of *Staphylococcus aureus* by posttranslational modification of MgrA. J Bacteriol 194:1823–1834
16. Truong-Bolduc QC, Dunman PM, Strahilevitz J, Projan SJ, Hooper DC (2005) MgrA is a multiple regulator of two new efflux pumps in *Staphylococcus aureus*. J Bacteriol 187:2395–2405

Chapter 16

High-Throughput Flow Cytometry Screening of Multidrug Efflux Systems

Mark K. Haynes, Matthew Garcia, Ryan Peters, Anna Waller, Pietro Tedesco, Oleg Ursu, Cristian G. Bologa, Radleigh G. Santos, Clemencia Pinilla, Terry H. Wu, Julie A. Lovchik, Tudor I. Oprea, Larry A. Sklar, and George P. Tegos

Abstract

The resistance nodulation cell division (RND) family of proteins are inner membrane transporters that associate with periplasmic adaptor proteins and outer membrane porins to affect substrate transport from the cytosol and periplasm in Gram-negative bacteria. Various structurally diverse compounds are substrates of RND transporters. Along with their notable role in antibiotic resistance, these transporters are essential for niche colonization, quorum sensing, and virulence as well as for the removal of fatty acids and bile salts. As such, RNDs are an attractive target for antimicrobial development. However, while enhancing the utility of antibiotics with an RND inhibitor is an appealing concept, only a small core of chemotypes has been identified as efflux pump inhibitors (EPIs). Thus, our key objective is the development and validation of an efflux profiling and discovery strategy for RND model systems. Here we describe a flow cytometric dye accumulation assay that uses fluorescein diacetate (FDA) to interrogate the model Gram-negative pathogens *Escherichia coli*, *Franscisella tularensis*, and *Burkholderia pseudomallei*. Fluorochrome retention is increased in the presence of known efflux inhibitors and in RND deletion strains. The assay can be used in a high-throughput format to evaluate efflux of dye-substrate candidates and to screen chemical libraries for novel EPIs. Triaged compounds that inhibit efflux in pathogenic strains are tested for growth inhibition and antibiotic potentiation using microdilution culture plates in a select agent Biosafety Level-3 (BSL3) environment. This combined approach demonstrates the utility of flow cytometric analysis for efflux activity and provides a useful platform in which to characterize efflux in pathogenic Gram-negative bacteria. Screening small molecule libraries for novel EPI candidates offers the potential for the discovery of new classes of antibacterial compounds.

Key words Flow cytometry, RND efflux transporters, Antibiotic enhancement, Biological Warfare bacterial agents, *Burkholderia* sp., *Francisella* sp.

1 Introduction

Several bacterial proteins have been identified that utilize either ATP or secondary electrochemical gradients as energy sources to move molecules from the cytosol to the periplasmic space and

Akihito Yamaguchi and Kunihiko Nishino (eds.), *Bacterial Multidrug Exporters: Methods and Protocols*, Methods in Molecular Biology, vol. 1700, https://doi.org/10.1007/978-1-4939-7454-2_16, © Springer Science+Business Media, LLC 2018

ultimately to the extracellular environment. Active efflux of a number of antimicrobials has been documented for both Gram-positive and Gram-negative bacteria [1, 2]. While Gram-negative bacteria express a number of efflux systems, *R*esistance-*N*odulation-*D*ivision (RND) transporter proteins are common and function within a tripartite complex that includes membrane fusion proteins and outer membrane porins [3, 4]. RND pumps are considered the most relevant transporters involved in multidrug efflux and antibiotic resistance [2]. Structural studies have determined that RND trimers form the inner membrane transporter that binds efflux substrates [5–8]. RNDs transport various unrelated compounds including most classes of antibiotics, biocides, cationic antimicrobial peptides, bile salts, detergents, and dyes [2, 5]. The major mechanism proposed for transit through the RND pump and into the efflux channel is by functional rotation where monomers of the asymmetric trimer cycle through conformational states that open and collapse a distal binding pocket simultaneous with proton exchange which shuttles substrates into the TolC exit channel [7, 8]. Orthologues of this proton-motive-force (PMF)-driven system are expressed in numerous pathogens including *Pseudomonas aeruginosa*, *Francisella tularensis*, *Salmonella typhimurium*, *Yersinia pestis*, *Mycobacterium tuberculosis*, and *Burkholderia* species [9–13].

The lack of new antibacterials and the increasing prevalence of antibiotic resistance underscore the need for new antimicrobial therapies. Because transporters are integral to bacterial growth, virulence, and pathogenicity, inhibition of transporter activity has the potential to both restore antibiotic activity of compounds to which efflux pumps confer resistance, and reduce the ability of bacteria to colonize their host. This approach is underscored by the increased antibiotic susceptibility of bacteria that express mutations in, or deletions of efflux pump genes [14]. A limited number of efflux pump inhibitors (EPIs) have been identified. Phenyl-arginine-beta-naphthylamide (PaβN) was initially described by Lomovskaya et al. as a pump inhibitor based on antibiotic susceptibility and fluorescent substrate retention in *P. aeruginosa* [15]. Most recent studies have utilized fluorescent compounds as efflux reporter molecules. For example, phenothiazines and 1-(1-naphthyl-methyl)-piperazine (NMP) inhibit efflux of ethidium bromide in *E. coli* [16, 17] and it is interesting to note that efflux of ethidium is inhibited by phenothiazines but not by PaβN [18], whereas efflux of other fluorescent substrates (e.g. nile red, fluorescein-di-ß-galactopyranoside) can be inhibited by PaβN [19–21]. Whatever the physiological explanation, these findings indicate that different substrate/inhibitor pairings can impact EPI identification suggesting the need for both novel dye-substrates as well as new candidate EPIs.

To address these needs, we developed a dye-retention assay for RND efflux based on previous efforts used to interrogate mammalian *A*TP-*B*inding-*C*assette (ABC) transporters [22, 23]. 3′,6′-diacetyl fluorescein (FDA) is used as a fluorescent reporter to measure efflux in model Gram-negative bacteria. Applying this assay to a High-Throughput Flow Cytometry (HTFC) platform offers the ability to compare families of fluorescent substrates with the validated FDA compound in a single plate or series of plates. Moreover, validated fluorescent reporters can be used to screen chemical libraries for candidate EPIs. Identified EPIs are then tested for antibacterial activity and antibiotic enhancement using standard microbiological assays. Microdilution plate assays can be modified to accommodate testing several candidate EPIs in dose response with a number of relevant active antibiotics. Both efflux screening and growth inhibition are amenable to studying Gram-negative pathogens in the more restrictive biosafety level 3 (BSL3) environment. Growth inhibition data are analyzed using the inclusion-exclusion model for chemical mixture to assess antibiotic/EPI synergy [24].

2 Materials

All solutions are made with ultra-pure 18 MΩ cm water or reagent-grade dimethyl sulfoxide (DMSO, Honeywell). Cation-adjusted Mueller-Hinton broth, LB broth and agar, and chocolate agar plates are standard bacteriologic growth formulations (Thermo-Fisher Scientific). Kanamycin is used as a selection additive for RND single-deletion *E. coli*. The panel of fluorophore substrates are available from Life Technologies (Grand Island, NY) and Sigma-Aldrich. Antibiotics used are available from Sigma-Aldrich and Santa Cruz Biotechnology. Chemical libraries are from Prestwick Chemical (Illkirch-Graffenstaden, France) and Spectrum Chemical (New Brunswick, NJ). Individually prepared libraries and assay plates are discussed below.

2.1 Bacterial Strains

1. *E. coli* BW25113 (K12-wild type) and the *E. coli* isogenic single-deletion mutant strains (JW0451-2, *ΔacrA*; JW0452-3, *ΔacrB*; JW5503-1, *ΔtolC*) were obtained from the Keio collection [25].
2. *B. pseudomallei* 1026b (Bp1026b), a clinical prototype strain; *B. pseudomallei* 2650a (Bp2650a), a clinical isolate from Thailand; *B. pseudomallei* 320 (Bp320), an RND deletion strain derived from Bp1026b; and *B. pseudomallei ΔpurM* (Bp82), an attenuated strain of Bp1026b that lacks purine synthetic capability [26], were generously supplied by Herbert P. Schweizer (University of Florida).

3. The parental and Δ*tolC* deletion mutant of the live vaccine strain (LVS) of *Francisella tularensis* were generously provided by David G. Thanassi, (Stony Brooke University).

2.2 Reagents, Buffers, Growth Medium

1. Cation-Adjusted Muellar Hinton Broth (CA-MHB) supplemented as described below.
2. LB agar with or without 50 μg/ml kanamycin and chocolate agar plates.
3. Dye retention assay buffer: 5 mM HEPES, 137 mM NaCl, 5.4 mM KCl, 0.25 mM Na_2HPO_4, 0.44 mM KH_2PO_4, 1.3 mM $CaCl_2$, 1 mM $MgSO_4$, pH 7.4.
4. 50 mM phenyl-arginine β naphthylamide (PaßN) and carbonyl cyanide 3-chlorophenylhydrazone (CCCP) in DMSO, aliquot and store at −80 °C.
5. 50 mM DMSO stock solutions of individual fluorophores. Certain fluorophores are provided as 10 mM stock solutions from suppliers.
6. Potential EPIs identified by ligand-based virtual screening are stored as 10 mM DMSO stocks at −30 °C. Source plates are made with single and multichannel pipets.
7. Commercial libraries, provided as 10 mM solutions in 96-well plates, are reformatted into 384-well 5 mM source plates using a liquid handling robotics work station.
8. Antibiotic solutions used for potentiation assays are made immediately prior to use. Potentiation and combination matrix plates that are used to evaluate bacterial growth inhibition are assembled in a biological safety cabinet using an enclosed liquid handling work station.
9. Calibration bead sets for flow cytometry (Bangs Laboratories, IntelliCyt).
10. 384-Well plates (Greiner Bio-One), nonsterile, polypropylene, 100 μl volume.
11. 96-Well culture plates (Corning), sterile, flat-bottomed, 250 μl volume.
12. Sealing covers for plates (Gene Mate). Plate seals are available in a number of different formats.

2.3 Equipment

1. Plate mixers and rotation devices suitable for standard microtiter plates.
2. Spectrophotometers for optical density measurements in cuvettes (Biomate 3, ThermoFisher Scientific) and microtiter plates (Powerwave HT, BioTek).
3. Flow cytometers; (CyAn ADP, Beckman-Coulter; Accuri C6, Becton-Dickenson). For dye profiling assays, both 488 and

635 nm lasers are required. The data acquisition software (e.g. Summit) must include a time parameter capable of binning data at 100-ms intervals continuously for 15 min or more.

4. Biomek FX^P (Beckman-Coulter) multitip dispensing instrument equipped with a pin tool device (V&P Scientific).
5. Liquid dispenser (MultiFlo, BioTek) capable of filling various microtiter plate formats.
6. The HyperCyt™ platform (IntelliCyt). This instrument consists of an autosampler (e.g. Gilson Gx-274), a peristaltic pump (Gilson Minipuls 3), tubing, and an inlet probe that connects to a compatible flow cytometer. The associated software includes HyperSip™ which controls the autosampler and is used to compose microtiter plate templates, and HyperView™ which is used to bin the time-resolved data files stored in flow cytometry standard 2.0 or 3.0 formats. The platform is set up as described [27].
7. Liquid handling work station (epMotion, Eppendorf) suitable for use in a biological safety cabinet.
8. Containment enclosure (BioBubble, Inc.) to house the HyperCyt™ platform during BSL3 operation.

3 Methods

3.1 Rational

Transport phenomena have traditionally been evaluated by measuring reporter loss from preloaded specimens either by assessing accumulation in the surrounding medium or reduction from preloaded cells [20, 28]. This real-time perspective is typically measured using spectrophotometry, either in cuvettes or with plate readers. Dye retention assays, which are amenable to flow cytometry, measure fluorescent pump substrates that accumulate within cells against a gradient maintained by efflux. These methods avoid dye preloading in the presence of a chemical inhibitor and the subsequent excessive wash steps that can be cumbersome when dealing with the unique challenges of high-throughput analysis. While traditional flow cytometry has an individual sample design where single data files are generated after analyzing thousands of cells from a single sample tube, HTFC generates single data files acquired from an entire microtiter plate of individual samples. Identified EPIs are then assayed for antibacterial effects in order to evaluate an EPI's clinical potential. Using determined minimal inhibitory concentrations (MIC) of antibiotics as a guide, combination plates can be used to assess antibiotic potentiation of identified EPIs. Statistical analysis of the growth inhibition data from combination plates can involve isobologram analysis [29] as well as a consideration of chemical mixtures using the inclusion/exclusion model [24].

Table 1
Lead structures used for virtual screening

Structure	Activity-antibiotics	Target efflux pump	Reference
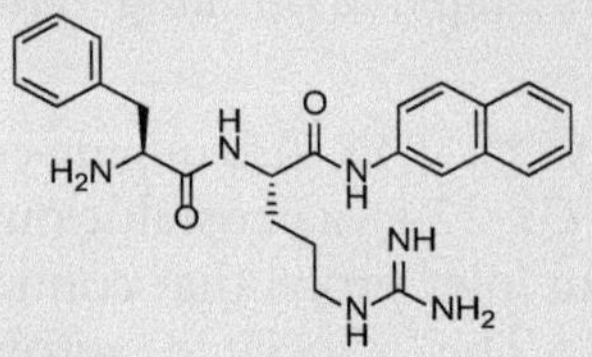 PaβN	10 μg/ml Levofloxacin	PAM1032, MexAB-OprM is overexpressed PAM1033, MexCD-OprJ is overexpressed PAM1034, MexEF-OprN is overexpressed	[30]
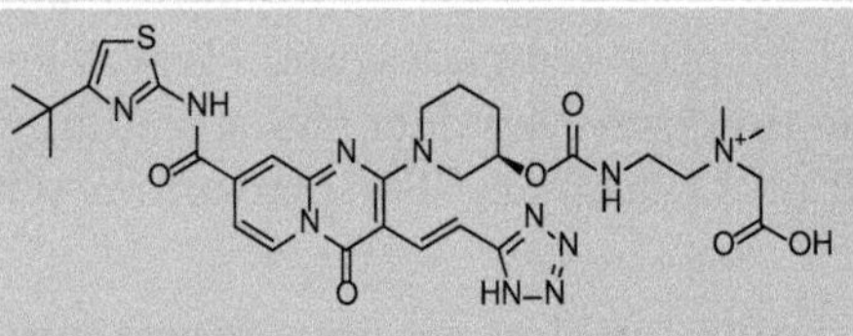 D13-9001	2 μg/ml Levofloxacin and aztreonam	*P. aeruginosa*, PAM1723 MexAB-OprM pump is overexpressed and the MexCD-OprJ and MexEF-OprN pumps are disrupted	[31]
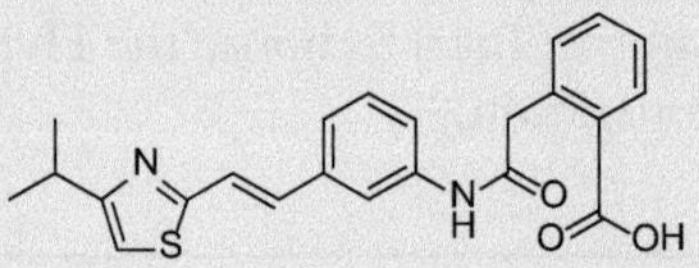 1	0.63 μg/ml Levofloxacin and aztreonam	PAM1032, a MexAB-OprM is overexpressed	[32]
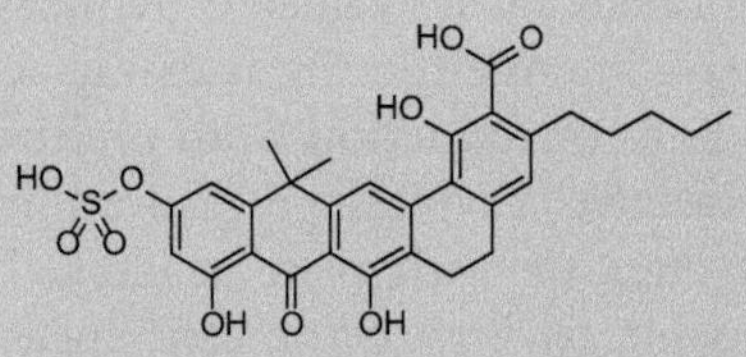 EA-371α	2.5 μg/ml Levofloxacin	PAM1032, MexAB-OprM is over-expressed	[33]

3.2 Efflux Inhibitor Library Selection

The ChemNavigator library, comprising 5.8 million commercially available compounds, was used in ligand-based virtual screening against six potent bacterial EPIs [30–33] shown in Table 1. The methodology consisted of: (1) 3D shape and pharmacophore screening using OpenEye ROCS; (2) 2D extended connectivity fingerprints (ECFP) screening [34]; and (3) OpenEye FRED docking. Similarity hits were assigned weighted scores and the top 100 hits for each query structure were used to assemble a library comprising 281 lead compounds. Emphasis was placed on identifying small organic molecules. Identified compounds were purchased from various suppliers. In-house and commercial chemical library source plates are assembled in 384-well microtiter plates and contain 5 mM stock solutions of test compounds (*see* **Note 1**).

3.3 Bacterial Culture

1. Single colonies used to propagate broth cultures for assays are selected from appropriate LB (*E. coli* and *B. pseudomallei*) or chocolate agar plates (LVS *F. tularensis*). Agar plates for the *E. coli* deletion strains contain 50 μg/ml kanamycin.
2. Bacteria are cultured overnight at 37 °C with aeration as follows: *E. coli* K-12, *B. pseudomallei* 1026b, *B. pseudomallei* Bp320, and *B. pseudomallei* 2650a are cultured with un-supplemented CA-MHB; *E. coli* RND deletion strain cultures are supplemented with 50 μg/ml kanamycin; *B. pseudomallei* *ΔpurM* (Bp82) cultures are supplemented with 80 μg/ml adenine; *F. tularensis* (LVS) cultures are supplemented with 250 mM glucose, 1.5 mM cysteine, and 150 μM ferric pyrophosphate.
3. Procedures using virulent *B. pseudomallei* strains are performed in an approved CDC registered BSL3 facility using select-agent-compliant procedures and protocols. Bp82 is excluded from select-agent regulations (www.selectagents.gov/SelectAgentsandToxinsExclusions.html). All other procedures are performed in BSL-2 facilities with Institutional Biosafety Committee approval.
4. Antibiotic susceptibility and MIC determination are performed using Clinical Laboratory Standards Institute guidelines [35]. Antibiotic MIC assays and potentiation testing of BSL3 *B. pseudomallei* strains are performed using the direct colony suspension method.

3.4 Flow Cytometric Measurement of Dye Retention

1. Collect overnight cultures by centrifugation (2 min/2000 × *g*) and wash twice (2 min/2000 × *g*). Use cultures at 0.6–0.8 OD_{600}; cultures with $OD_{600} > 1.0$ can be diluted into fresh medium for 2–4 h before use.
2. Resuspend the bacteria to a suitable OD_{600} in a sufficient quantity of assay buffer to accommodate the number of individual samples planned. A final concentration of 0.25 OD_{600} is used throughout (*see* **Note 2**).
3. Pre-incubate bacteria for 20–30 min with a standard efflux inhibitor (we use CCCP and PaβN) prior to addition of 10 μM FDA and incubate for an additional 60–90 min at room temperature (*see* **Note 2**).
4. Transfer cells to a compatible cytometry tube without dilution or washing.
5. Cytometer settings are initially determined using validation bead sets followed by evaluation of bacteria from an overnight culture (*see* **Note 3**).
6. Collect an appropriate number of events for each sample; 10–25,000 events are adequate for standard cytometry and are easily reached using an $OD_{600} = 0.25$.

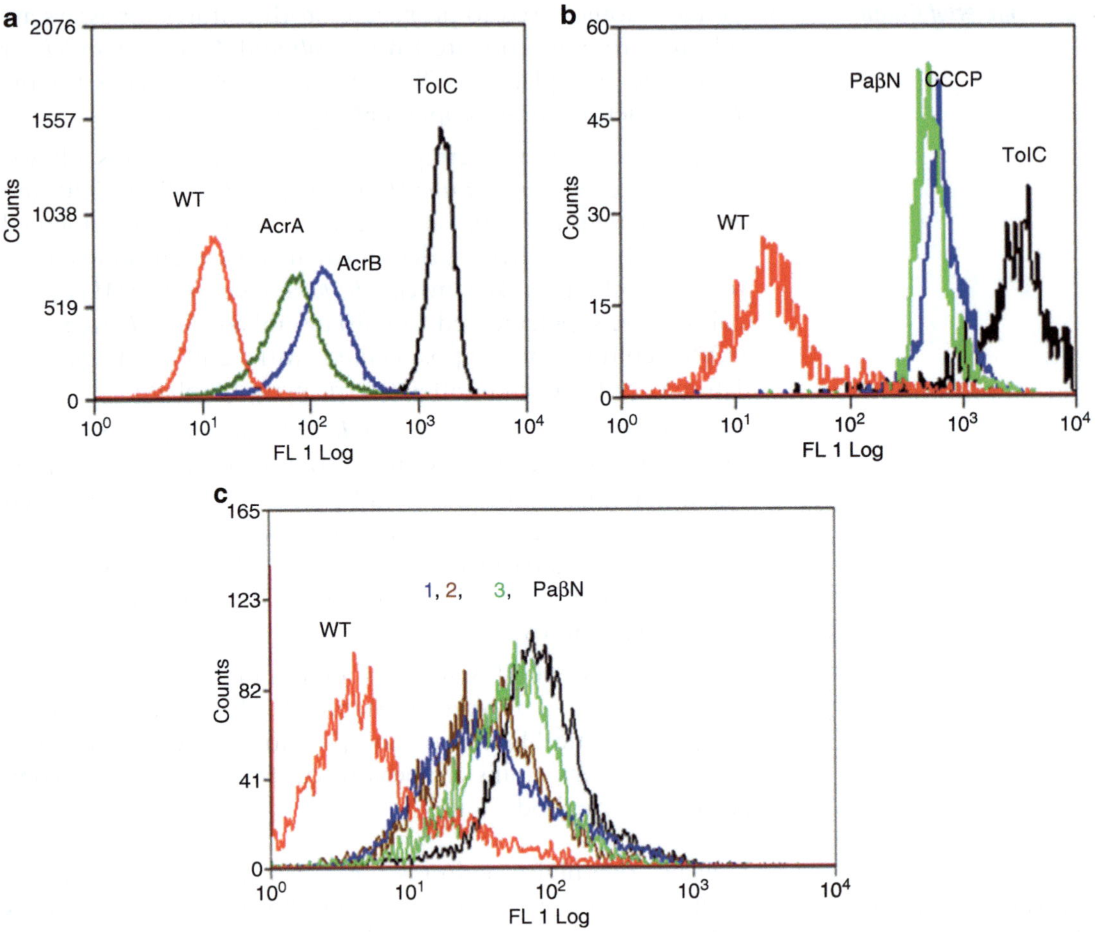

Fig. 1 (**a**) *E. coli* K-12 and RND single-deletion strains are incubated with 10 μM FDA for 60 min prior to analysis. Each histogram represents greater than 25,000 events. MFI values: WT, 39; AcrA, 108; AcrB, 334; TolC, 1915. (**b**) *E. coli* K-12 was incubated with 1% DMSO (WT), 500 μM PaβN or 50 μM CCCP for 30 min prior to the addition of 10 μM FDA. MFI values: WT, 69; PaβN, 647; CCCP, 779; TolC, 3300. (**c**) *E. coli* K-12 was incubated with 1% DMSO or 100 μM of an inhibitor compound for 30 min prior to addition of 10 μM FDA. MFI values: WT, 12; PaβN, 123; compound #1, 65; compound #2, 79; compound #3, 74

Data evaluating FDA efflux in the *E. coli* deletion strains is shown in Fig. 1. Compared to *E. coli* K-12, RND deletion strains exhibit increased dye retention; the mean fluorescence intensity (MFI) of *ΔacrB* and *ΔtolC* strains increased 10- to 100-fold, respectively (Fig. 1a) indicating that FDA is an RND substrate. Additional evaluation of RND-associated efflux entails examination of dye retention in the presence of known inhibitors. PaβN exhibits RND efflux inhibition [19–21] and CCCP disrupts the proton motive force rendering RND transporters nonfunctional. When *E. coli* K-12 is pre-incubated with an excess concentration of PaβN (500 μM) or CCCP (50 μM) prior to FDA incubation, MFI increased approximately tenfold compared to untreated

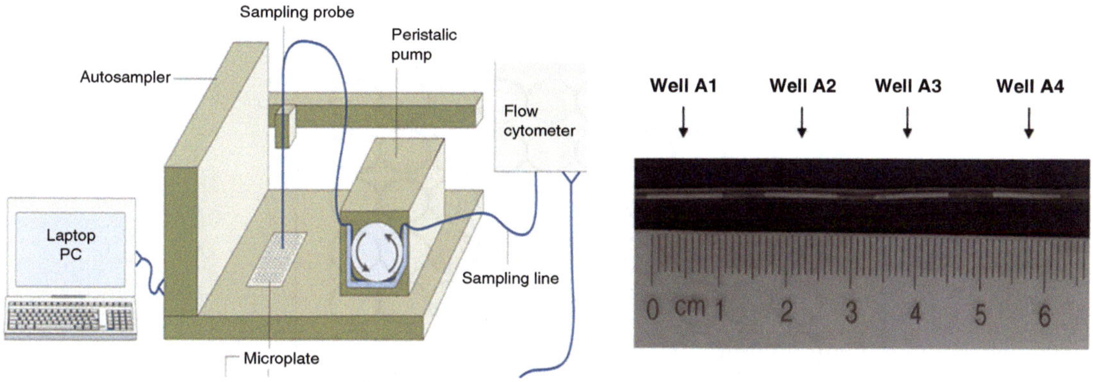

Fig. 2 A representation of the HyperCyt™ High-Throughput Flow Cytometry Platform is shown. HyperSip™ software controls the autosampler and pump that delivers samples to the cytometer. Constant aspiration generates an individually segmented flow of samples separated by air bubbles

bacteria (Fig. 1b). Finally, analysis of PaβN-like molecules identified with ligand-based virtual screening demonstrate the assay's ability to distinguish differing levels of compound inhibition (Fig. 1c).

3.5 High-Throughput Flow Cytometry (HTFC)

Careful consideration of sample size and cell concentration are necessary when assays are transferred to microtiter plates and analyzed in a no-wash condition for HTFC. Various plate formats (96, 384, and 1536) are compatible with HyperCyt™. During operation an inlet probe moves from well to well aspirating 1–2 μl samples. Individual samples are temporally separated by a bubble of air as they are delivered to the cytometer (Fig. 2, *see* **Note 4**). Single plate data, acquired as time-resolved files, are analyzed with HyperView™ software which parses the data into separated bins (red demarcation lines in Fig. 3b) and tabulates the acquisition parameters of interest. Each cluster is associated with an individual well. Data is compiled as comma separated files that can be further analyzed with standard analytical software (e.g. Microsoft Excel®, GraphPad Prism®).

3.5.1 Dye Profiling

Initial analysis of dye profiling uses the *E. coli* RND deletion strains to screen dyes for substrates that differentially accumulate due to a nonfunctional AcrAB-TolC efflux system. Sample wells are separated by wash wells to reduce carryover of fluorescent dyes.

1. 2 μM solutions of each potential RND substrate are made in assay buffer (*see* **Note 5**).
2. Place 25 μl of each dye into separate rows of a microtiter plate as depicted in Fig. 3a. Rows of identically colored wells represents a single dye.
3. Collect, wash, and resuspend overnight cultures of *E. coli* deletion strains in an appropriate volume of assay buffer at 0.5

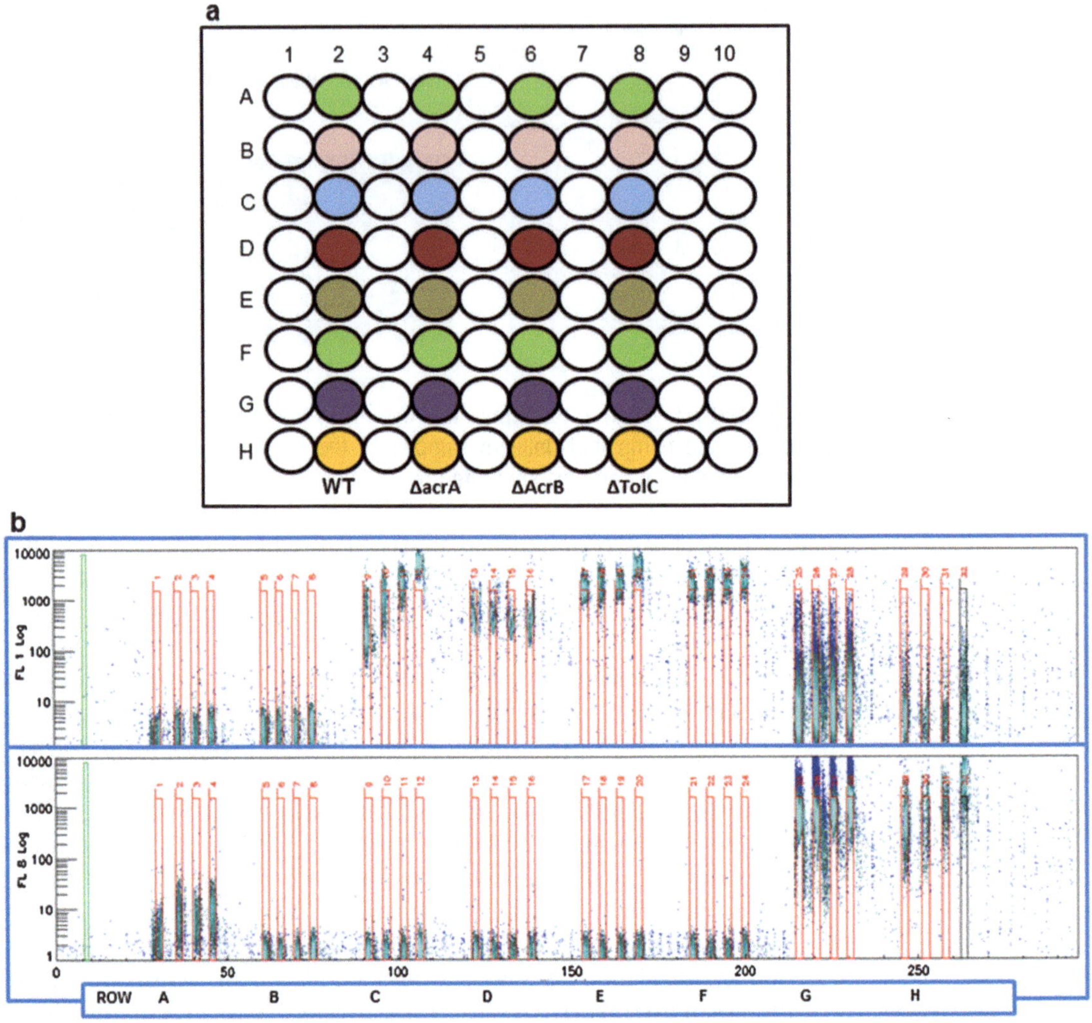

Fig. 3 Fluorescent dye plate (**a**) contains a series of different dyes arranged in rows A–H and the *E. coli* RND deletion strains arranged in separated columns of a 96-well microtiter plate. Final concentrations of cells and dye-substrates are 0.25 OD_{600} and 1 μM, respectively. (**b**) Representative plots of time vs. fluorescence are shown (*upper panel*, FL1; *lower panel*, FL8). Individual wells with associated fluorescence are defined by *red bins*. Ruthenium red, row A; eFluxx-ID-green, row B; $DiOC_2$, row C; JC-9, row D; $DiOC_5$, row E; $DiOC_6$, row F; $DiSC_3$, row G; and $DiIC_1$, row H

OD_{600}. Add 25 μl of bacteria into individual wells containing the separate dye substrates.

4. Add 50 μl of buffer into unused wells.
5. Mix and incubate microtiter plate for 60 min at room temperature. Use a rotation device during incubation to minimize settling.
6. Flow cytometer settings need to be set for the dye substrates being profiled. Data in Fig. 3b used 550–650 V voltage setting

for the FL1 channel (488-nm excitation, 530-nm emission) and a 650–700 voltage setting for the FL8 channel (635-nm excitation, 665-nm emission, as discussed in **Note 3**).

7. Analyze using HyperCyt™. Once bins are satisfactory, template files are used to assign well identification to individual bins. CSV files containing the data parameters of interest are generated for further analysis.

Figure 3b is representative of HTFC data analyzed with HyperView™ software where eight dyes having two separate excitation/emission characteristics are screened for RND efflux involvement. The sampling sequence used to collect the plate data shown in Fig. 3b is by row starting with well A1. Each 4-bin set, corresponding to wells 2, 4, 6, and 8 of a single row, depicts the dye retention signature associated with *E. coli* K-12, *ΔacrA*, *ΔacrB*, and *ΔtolC*, respectively (as depicted in Fig. 3a). In the plate sample shown, six of the eight dyes tested are carbocyanine-containing molecules (rows C-H), but only $DiOC_2$ (upper panel, row C) exhibits significantly increased dye retention in the *ΔtolC* and *ΔacrB* deletion strains (11- and 7-fold, respectively). Two other carbocyanine dyes exhibit mild increases in retention ($DiOC_5$, upper panel row E; $DiIC_1$, lower panel, row H). No other carbocyanine dye exhibits RND efflux activity, nor does ruthenium red (Row A), or eFluxx ID green (Row B). A 96-well plate (32 samples plus 64 wash wells) can be sampled in less than 5 min. This experimental design is applicable to comparative analyses in a 384-well platform that can include dose response analysis of individual dyes or analyses of different strains and/or genus.

Figure 4 shows profiling analysis from two 384-well dye profiling plates, where dye retention in *E. coli*, LVS *F. tularensis*, and their respective *ΔtolC* deletion strains are compared. Eight dyes are tested for differential activity. FDA, bodipy-modified (BD) histamine, and Syto 13 exhibit RND responsive activity when the parental strains are compared to the *ΔtolC* deletion strains. In contrast, the carbocyanine dyes $DiOC_2$ and $DiIC_1$ only report in the *E. coli* strains whereas BD-verapamil and Syto 16 had better reporter activity in *F. tularensis*-LVS. Using these data, fluorescent dyes can be chosen for further analysis in combination with inhibitors. Analysis of groups of potential substrates has the potential to identify chemical moieties that impact RND activity. For instance, retention of bodipy-modified histamine efflux was increased in *E. coliΔtolC* compared to *E. coli* K-12 whereas bodipy-modified verapamil retention appears unaffected by deletion of the TolC outer membrane protein.

3.5.2 Chemical Library Screening

In this instance, assay plates mimic chemical library source plate configurations and use a single dye and bacterial strain throughout the plate. Two standard configurations of plates are shown in

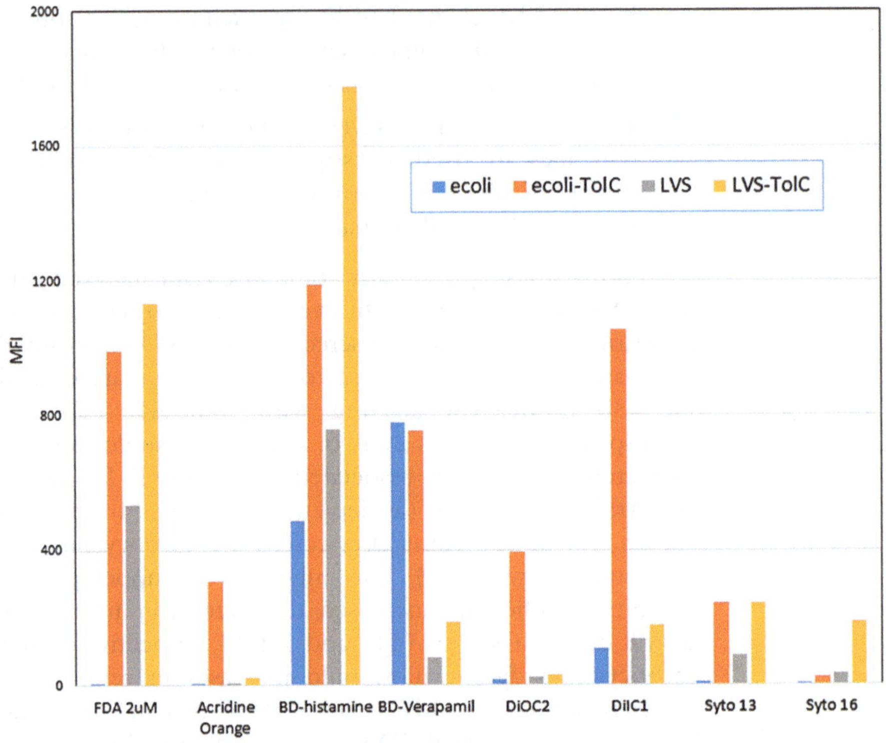

Fig. 4 Thirty-two dyes are tested for RND responsivity using *E. coli* and *LVS F. tularensis* and their respective *ΔtolC* strains. Data are collected from two 384-well plates where a total of 31 dyes in limited dose response are compared with FDA. Data shown compare 2 μM FDA with a selected set of profiled dyes at 500 nM. Total sample time/plate is less than 600 s

Fig. 5. Wells of a single-point source plate shown in Fig. 5a contain 5 mM concentrations of test compounds in columns 3–22 and DMSO in columns 1, 2, 23, and 24. Dose response plates (Fig. 5b) contain multiple 8-point dilution series' with a number of unoccupied wells that are used for controls. Assay plates using these designs are assembled using robotic work stations and microtiter plate liquid handlers. The procedure below is used to assemble both types of assay plates.

1. Dispense 10 μl of assay buffer into columns 2–24 of a 384-well microtiter plate using a BioTek microflo liquid dispenser. Column one is left empty for subsequent addition of the positive control (50 μM CCCP).
2. Test compounds from the source plate (100 nl) are added to the assay plate with the Biomek FX^P workstation and the pintool attachment. All assay plates to be screened should be made at one time (*see* **Note 6**).
3. Add 10 μl of 100 μM CCCP into column one of the assay plates.
4. Collect, wash, and resuspend overnight culture of bacteria to 0.5 OD_{600}.

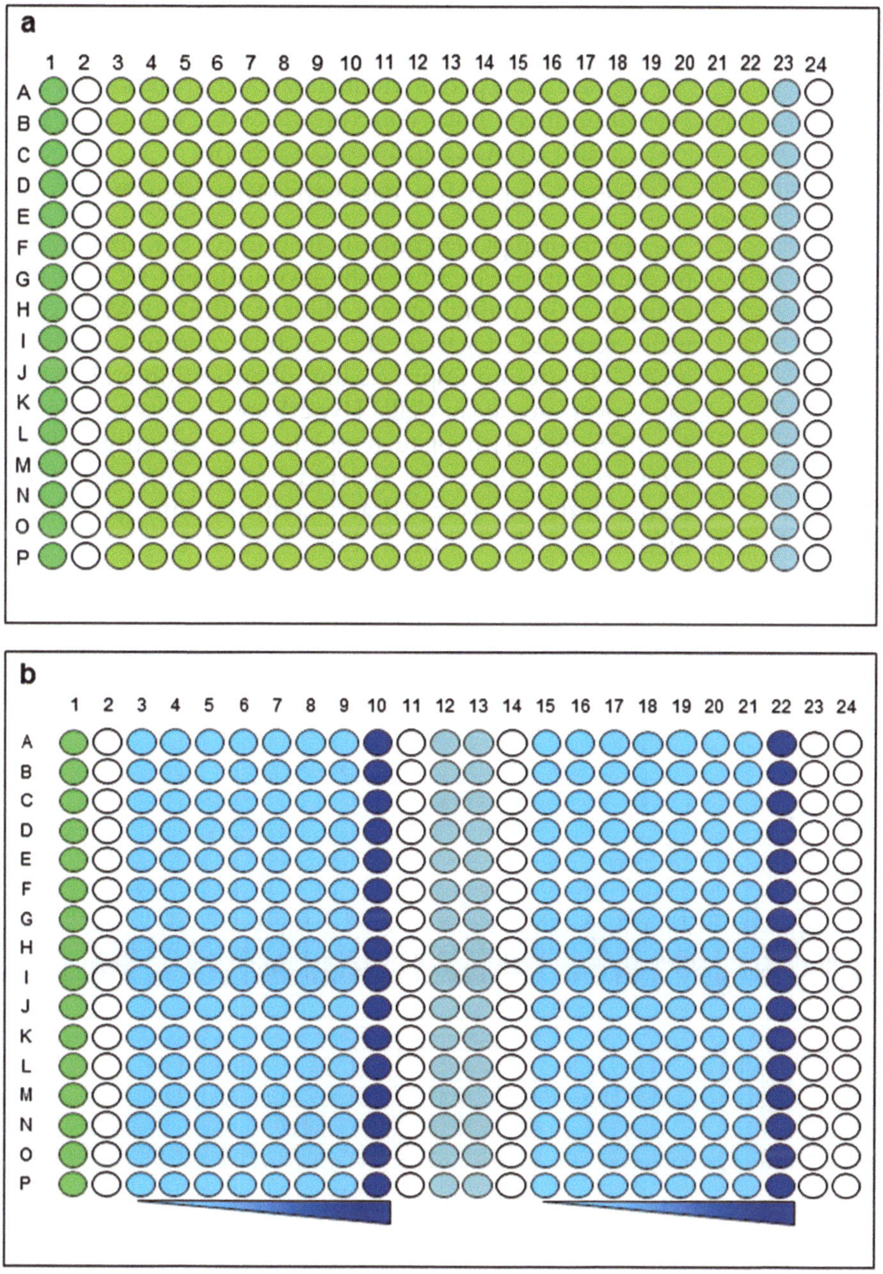

Fig. 5 Chemical screening plates for (**a**) single point assays have 16 positive controls (column 1), 16 negative controls (column 23), 320 test wells (columns 3–22), and two wash well columns (2, 24). (**b**) Dose response plates typically contain 32 8-point dilution series, numerous wash-well columns, positive controls (column 1) and negative controls (columns 12 and 13)

5. Add 5 μl of bacteria to columns 1 and 3–23 of the assay plate using the BioTek microflo liquid dispenser (*see* **Note 7**). Cover plates with plate seals and incubate for 30 min at RT. To ensure optimal mixing, plates can be rotated during incubations (*see* **Note 7**).

6. Add 5 μl of 10 μM FDA to all wells that contain bacteria using the BioTek microflo liquid dispenser (columns 1 and 3–23). Re-seal assay plates and incubate with rotation for an additional 60 minutes. Final concentrations are as follows: 0.25 OD_{600}, 2.5 μM FDA, 25 μM test compound.
7. Collect and analyze data with HyperView™.

Assembled plates contain one positive control column designed to mimic maximal efflux inhibition (50 μM CCCP) and negative controls containing library diluent (0.5% DMSO). Buffer-only wells separate negative and positive controls wells from assay wells. This design improves bin identification and reduces particle carry-over into assay wells. HyperView™ software merges cytometry FCS files with plate templates that identify individual wells. Parameter results from individual wells, corresponding to single compounds, are assigned and tabulated into spreadsheet files for further analysis. Figure 6 shows various HyperView™ annotated

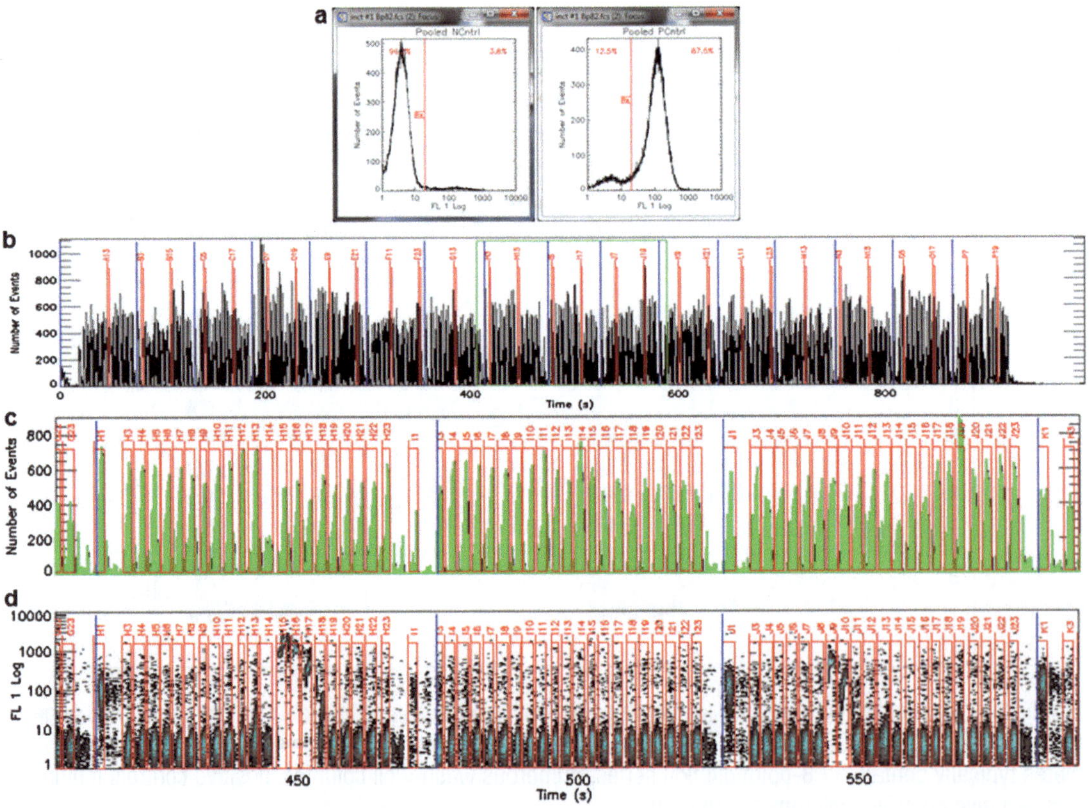

Fig. 6 Single-point plate analysis using HyperView™ is shown. (**a**) Histograms of combined untreated (NCntrl) and CCCP-treated (PCntrl) control wells. NCntrl are defined as a population with less than 5% of the cells exhibiting increased fluorescence. (**b**) A time versus events plot is displayed for the entire 384-well plate. Enlarged views of three rows of binned wells depicts event count (**c**) and MFI values (**d**). Note the space between bin H1 and H3 representing the empty H2 wash well. Fluorescence signals in wells H1, I1, J1, and K1 indicate dye retention due to CCCP. Signals from wells J9 and H15 indicate increased dye retention in test wells

data plots generated from a single-point screening plate of the in-house assembled EPI library. Pooled data from negative control wells are used to define regions in a histogram analysis plot such that <5% of the collected events from negative control wells exhibit increased fluorescence (Fig. 6a). Applying these defined regions to the pooled positive control wells identifies a shifted population resulting from RND inhibition. In the example shown in Fig. 6a, the percentage of cells in negative control wells exhibiting elevated FL1 fluorescence values is 3.8% whereas incubation with 50 μM CCCP results in 87.5% of the cells with increased FL1 fluorescence. Positive and negative controls are used to calculate Z and Z' assay quality statistics [36]. Plate data shown in Fig. 6 yielded a Z' value of 0.831 which indicates a robust assay. Assay plates that have Z' scores below 0.5 are repeated. Dose response plates are similarly processed. Once dose response data are acquired, they can be fitted by Prism® software (GraphPad Software, Inc.) using nonlinear least-squares regression analysis. Curve fit statistics can be used to determine a range of useful parameters such as the maximal effect (EC_{50}), confidence intervals of EC_{50} estimates, the Hill slope, and correlation coefficients.

Screening a library containing known actives such as the library assembled using virtual screening is a useful gauge of assay validity. In this instance PaβN, NMP, and two catechin gallates, previously identified EPIs in other systems, were all scored as actives (i.e. these wells demonstrated increased dye retention) [15, 16, 37]. Following the in-house library screen, *B. pseudomallei* Bp82 was screened against two commercial libraries containing over 2000 unique chemical entities. The total data set comprising ten plates is shown in Fig. 7. The average Z' value was 0.739 and over 160 compounds increased bacterial fluorescence after FDA incubation. Secondary analysis of the identified active compounds and additional compounds selected based on chemical similarity to the identified active compounds are retested in single point and confirmed hits are evaluated in dose response. The most prominent compound classes identified as actives were antifungal azoles and polyphenolic catechin gallate derivatives. This screening campaign identified 391 compounds that were subsequently screened for activity using select agent BSL3 pathogens.

Flow cytometry using select agents requires unique containment environments. In this instance, prior approval from CDC was necessary and biocontainment was validated according to CDC recommendations (*see* **Note 8**). Figure 8a shows the HyperCyt™ platform housed in a BioBubble HEPA filter containment unit within the BSL3 facility. BioBubble HEPA filter units must be certified prior to use and annually. To check for aerosols that might be produced from the HTFC platform, an entire 384-well microtiter plate containing *E. coliΔtolC* is sampled (Fig. 8b). Agar

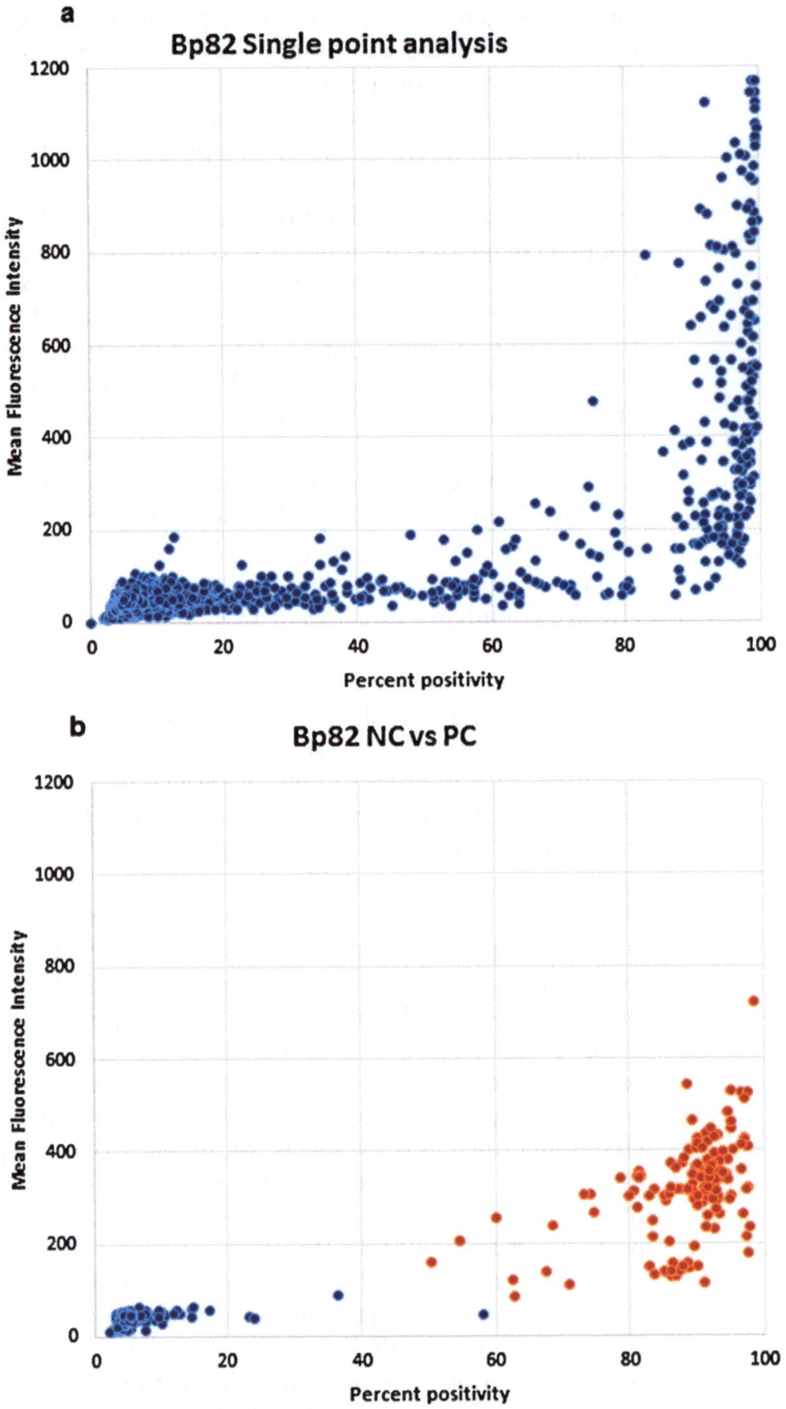

Fig. 7 Individual data points from ten plates screened against strain Bp82 are shown. (**a**) Single point values of MFI versus percent positivity for the assay wells from ten screened plates. (**b**) Negative (NC) (●) and positive (PC) control (●) values from the ten screened plates

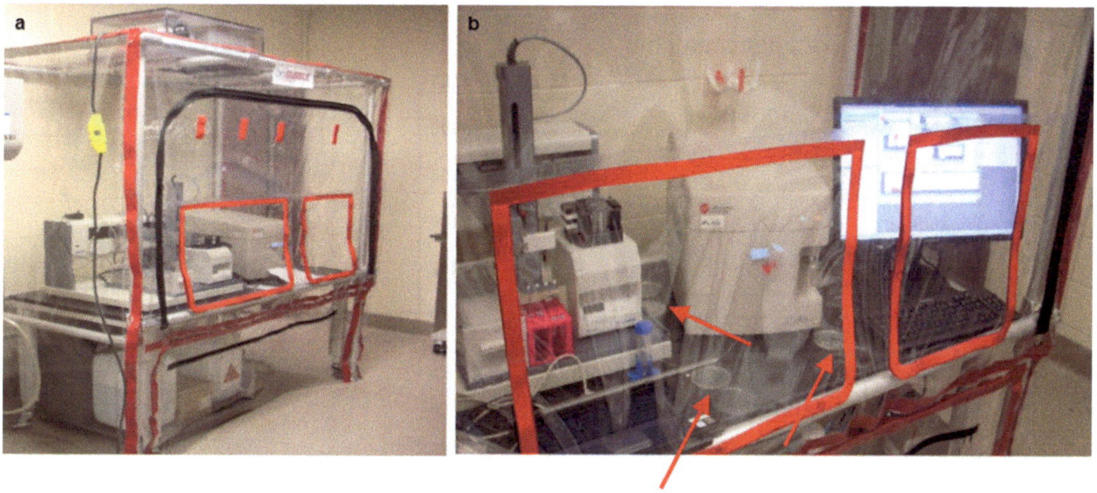

Fig. 8 (**a**) The BSL3 BioBubble containment of the HyperCyt™ platform connected to a CyAn cytometer and sheath management system. (**b**) A validation experiment is shown with *arrows* pointing to pairs of agar plates set to capture potential aerosolized bacteria

plates, within and outside the BioBubble, are positioned in various locations where aerosolized droplets might be expected to settle. Lids are removed prior to each operational step. Following each step, the lids are closed and the plates are incubated for up to 96 h to verify that no growth occurs.

To facilitate screening BSL3 level pathogens, assay plates with compounds are brought to the restricted facility where BSL3 select agent bacteria are cultured. Figure 9 shows examples of the comparative data that can be obtained using this approach. Dose response data of a catechin gallate and an azole are shown. Figure 9a shows data from the Thai clinical isolate of *B pseudomallei* (Bp 2650a) and data obtained from the Bp 1026b prototype strain is in Fig. 9b. Thirty-five compounds were identified that demonstrated dose effects from the total screening set, 25 of which had EC_{50} values of 10 μM or less representing a hit rate of 1.5%.

3.6 HTFC in a BSL3 Environment

3.6.1 Antimicrobial Susceptibility

Growth analysis in the presence of known or candidate antimicrobial compounds is a common microbiologic procedure. The assembled antibiotics panel was tested against the *B. pseudomallei* strains in MIC assays using CLSI guidelines [35]. Sub-inhibitory concentrations can then be paired with EPIs in preliminary potentiation studies that are used to triage combination pairs allowing for the selection of a subset of antibiotics and EPIs to test in matrix combination experiments. Limited potentiation and matrix combination studies are performed similarly.

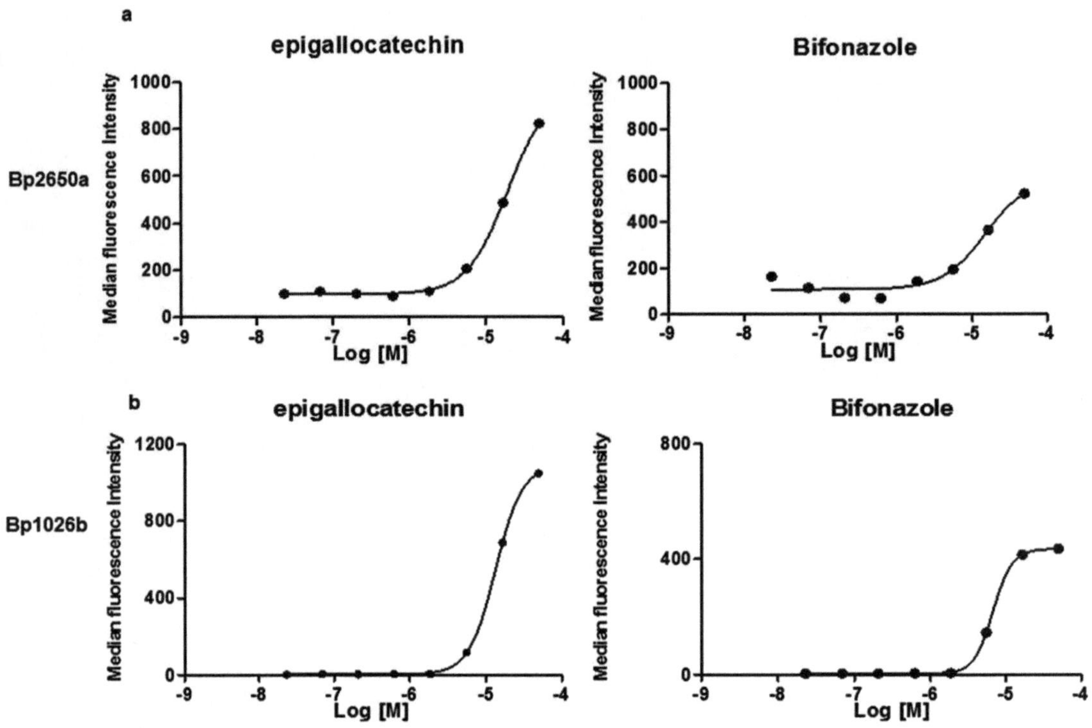

Fig. 9 Dose response analysis of selected members of the two chemical classes that exhibited potent EPI activity in *B. pseudomallei* strains, polyphenols and azoles. (**a**) shows compound effects on the Thai clinical isolate Bp2650a and (**b**) shows dose effects of EPIs on the prototype strain Bp1026b

1. Streak LB agar plates with *B. pseudomallei* strains. Culture overnight at 37 °C.
2. Resuspend isolated single colonies of each test strain to 0.1 OD_{600} in CA-MHB.
3. Dilute each strain to a final concentration of 5×10^4 CFU/well (*see* **Note 9**).
4. Assemble growth microdilution plates in a sterile biosafety cabinet. Use multichannel pipettes, a liquid handling device, and single-use tips. An epMotion liquid handling work station can be housed in a standard biological safety cabinet.
5. Add bacteria to preassembled potentiation plates or combination plates (*see* **Note 9**).
6. Incubate plates overnight at 37 °C and record OD_{600} with plate reader.

Potentiation analysis does not require combination plate assembly. These plates contain a limited number of combinations. For example, two sub-inhibitory concentrations of an antibiotic can be combined with two concentrations of the candidate EPI (*see* **Note 10**). Table 2 lists eight EPI compounds identified using

Table 2
Lead chemotype-potentiators

ID	Name	Structure	MW
1259	Desferrioxamine	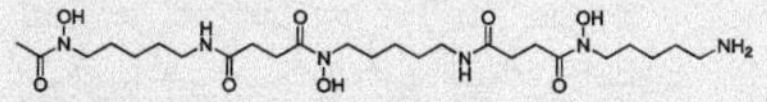	561
1277	Naphthylmethyl piperazine; NMP	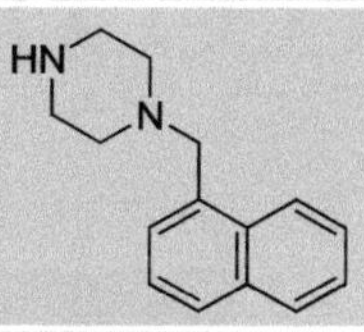	226
1342	Naphthyl piperazine NP	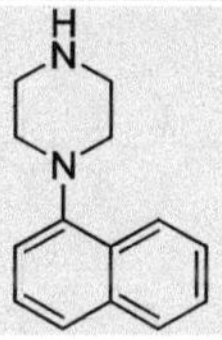	212
1453	Diaveridine	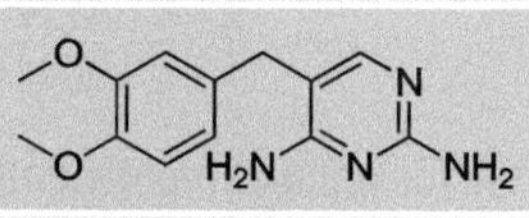	260
7514	Dichlorophene	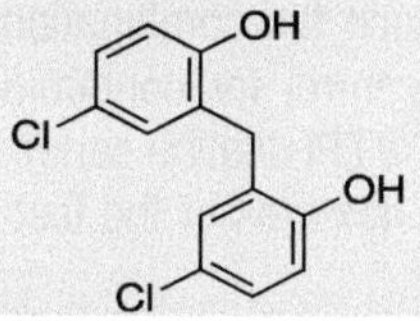	269
8227	Bithionate sodium	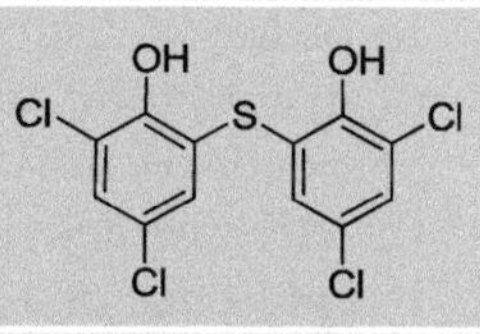	400
8518	Levamisole	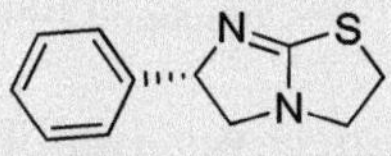	241
9316	Econazole	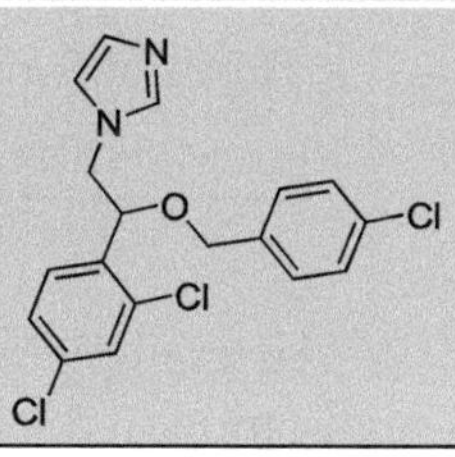	445

limited potentiation that decreased growth of *B. pseudomallei* clinical isolates when compared to growth in the presence of sub-inhibitory antibiotics alone. Compounds identified in this manner are further tested using a checkerboard matrix microdilution assay. Combination plates are assembled as depicted in Fig. 10.

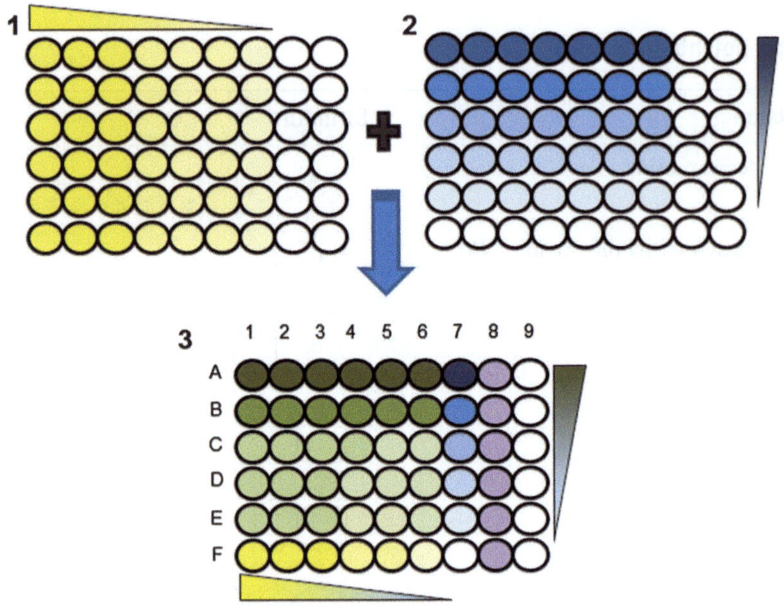

Fig. 10 Combination plates for antibiotic potentiation are assembled using dilution series plates **1** and **2**, as shown. Bacteria are added once the combination plate is assembled. In the combination plate (**3**), column 9 contains only medium, a control for contamination, column 8 is the growth control and only receives bacteria, column 7 contains the EPI dilution series alone and row F (1–6) contains the antibiotic dilution series alone. All other wells contain mixtures of the test antibiotic and EPI at varying ratios

3.6.2 Determination of EPI Potentiation Using Mixture Modeling

In order to distinguish compound activity from true potentiation of an antibiotic, we use the inclusion-exclusion model for a mixture of independently acting compounds [24]. In particular, we can find the effective activity of the antibiotic at a given dose $\text{Dose}_{\text{Antibiotic}}$ in the presence of a potentiating compound at a given concentration $\text{Dose}_{\text{Compound}}$ using the equation:

$$\text{Effective}\%\text{Inhibition}_{\text{Antibiotic Alone}}(\text{Dose}_{\text{Antibiotic}}) = \frac{\%\text{Inhibition}_{\text{Antibiotic \& Compound}}(\text{Dose}_{\text{Antibiotic}}, \text{Dose}_{\text{Compound}}) - \%\text{Inhibition}_{\text{Compound Alone}}(\text{Dose}_{\text{Compound}})}{1 - \%\text{Inhibition}_{\text{Compound Alone}}(\text{Dose}_{\text{Compound}})}$$

An example of two checkerboard assay results before and after application of the mixture model is shown in Fig. 11 where antibiotic dose increases in twofold increments along the horizontal axis and compound doses increase in twofold increments along the vertical axis. Normalization of the OD_{600} values using control replicates allows for the calculation of percent inhibition where MIC-90% represents an antibiotic concentration that corresponds to 90% growth inhibition. In Fig. 11a, after the model accounts for the compound's activity, the MIC-90% is essentially static, indicating

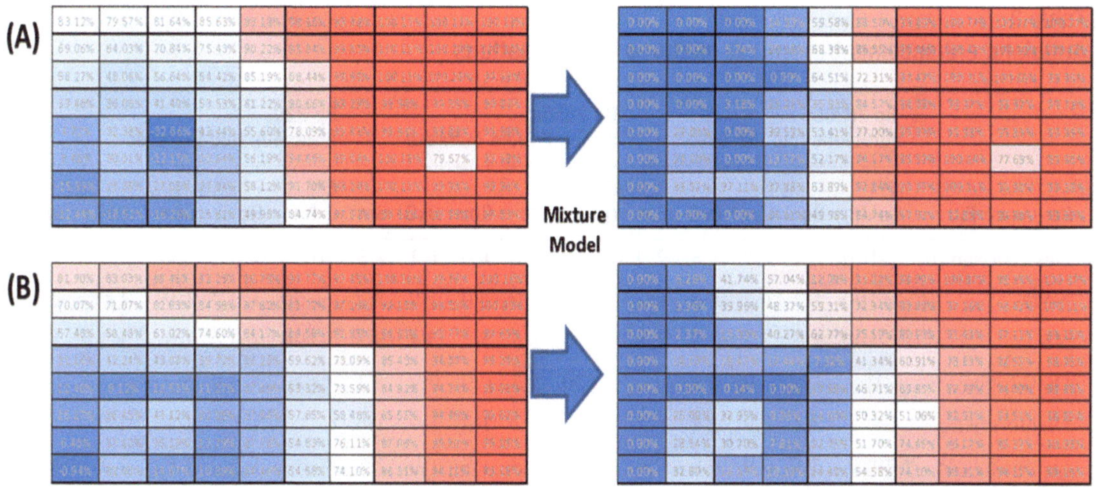

Fig. 11 A representative mixture model analysis of combination plate data is shown. Raw data from a twofold dilution series of antibiotics (*horizontal*) and EPIs (*vertical*), evaluated for growth inhibition, are normalized to percent inhibition using bacteria only wells. MIC 90% = antibiotic concentration producing 90% growth inhibition. Two examples are depicted. (**a**) A static example, the MIC-90% is unchanged after application of the model and (**b**) where the model predicts potentiation of antibiotic activity in the presence of the tested EPI

no potentiation of the compound for this antibiotic. In Fig. 11b, the MIC-90% for the antibiotic is achieved at a lower concentration indicative of EPI potentiation. Using this model in concert with checkerboard matrixes, we can "subtract off" the activity of a potentiating compound, leaving the antibiotic dose response curve at each dose of the compound. From this calculation we then take the ratio of the MIC-90% of the antibiotic alone to the MIC-90% of the antibiotic with the compound to determine overall fold potentiation. This approach led to a set of lead compounds with clear potentiation profiles of least two antibiotics from the panel (structures in Table 2).

Data for the eight compounds described in Table 2 are given in Table 3. Potentiation that reached threefold or more is considered significant; fold differences appear more dramatic for bactrim with desferrioxamine, doxycycline with NMP, and streptomycin with bithionate sodium. Desferrioxamine is an iron chelator and bacterial siderophore that was first identified in the actinobacteria *Streptomyces pilosus*. It has a demonstrated broad range of antimicrobial activity [37]. NMP is a well characterized pump inhibitor effective in *K. pneumoniae*, *E. coli*, *A. baumannii*, and several other *Enterobacteriaceae* sp. [16, 38, 39]. Naphthyl piperazine is an NMP analog; diaveridine is an antiprotozoal dihydrofolate reductase inhibitor [40]. Econazole is a topical antifungal with broad antimicrobial potential [41]. Bithionate sodium or bithionol is an anthel mintic [42] as is levamisole which also has immunomodulating properties [43, 44]. Dichlorophene is another broad spectrum antimicrobial agent.

Table 3
Antibiotic enhancement/chemical mixtures determinations

	Bp 1026b Antibiotic fold increase					Bp 2650a Antibiotic fold increase					Bp 320 Antibiotic fold increase				
Candidate EPI	EPI (μM)	Bc	Cpx	DxC	Sm	EPI (μM)	Bc	Cpx	DxC	Sm	EPI (μM)	Bc	Cpx	DxC	Sm
Desferrioxamine	>100	*6.9*	1.2	1.5	2.9	>100	*4.2*	1.0	1.2	2.5	96	2.0	1.1	1.1	2.3
NMP	>100	1.2	1.2	2.3	1.2	>100	1.0	1.0	*7.9*	*3.2*	>100	1.0	1.9	1.9	2.3
NP	>100	2.0	*3.2*	2.3	2.1	>100	1.5	*3.1*	*3.8*	*3.5*	>100	1.0	1.8	1.2	1.4
Diaveridine	71	1.1	0.9	0.9	*4.6*	>100	1.3	0.9	0.9	*3.1*	<1.6	N/A	N/A	N/A	N/A
Dichlorphene	54	*3.1*	1.9	1.4	*3.1*	92	1.2	1.3	1.9	1.6	6	1.1	1.1	1.0	1.2
Bithionate Na	29	*3.5*	0.9	0.8	*7.0*	45	2.8	1.5	1.0	2.8	8	1.2	1.9	1.1	1.1
Levamisole	>100	2.9	2.2	1.9	1.9	>100	2.2	*3.9*	2.7	1.8	>100	1.1	1.1	1.3	1.2
Econazole	>100	*3.3*	1.7	1.7	*3.9*	>100	*4.9*	2.2	2.0	2.4	16	1.0	0.9	2.2	*4.3*

Checkerboard data analysis for the eight lead potentiators. Included are the MIC of the EPI compound and the overall efficacy expressed as MIC-90% fold increase compared to the antibiotic alone. *Italic numbers* reflect fold efficacy greater than 3.0. Antibiotics are *Bc* bactrim, *Cpx* ciprofloxacin, *DxC* doxycycline, *Sm* streptomycin

4 Notes

1. Commercial libraries are supplied in 96-well plate formats as 10 mM stock solutions. The BioMek FX^{P} or a comparable liquid handling robotics device can be used to reformat these libraries into 384-well plates used as library source plates. Source plates are stored at −30 °C. Reformatted source plates are limited to four freeze-thaw cycles.
2. Log-phase cultures are used throughout. Preliminary experiments should be performed to determine suitable culture conditions and cell concentration for efflux evaluation. It is not uncommon for different bacterial cultures diluted to the same OD_{600} value to contain different cell concentrations. It is important to pay attention to the final concentration of DMSO when stock solutions are prepared for addition into assays. Concentrations above 2% DMSO can effect membrane integrity and influence efflux. In general, experiments described herein are performed at 0.5–1% DMSO. HTFC is performed as a no-wash procedure and it is best to mimic that requirement when performing preliminary, set-up studies.
3. Calibration beads are used to determine instrument settings. Beads allow the user to set voltage as well as light scatter (size) parameters. While new generation cytometers (e.g. BD Accuri) have pre-optimized detector settings, validation beads should still be used to check cytometer operations prior to running

samples. Adjust forward and side scatter parameters to bring particles onto scale. Run assay buffer to check instrument noise and to set the data acquisition threshold. This is an important consideration when dealing with bacterial analysis. This threshold is often determined using forward scatter. Calibration beads that are similar in size to bacteria are available and can aid in setting this threshold. Use both FITC and APC bead sets for cytometers with two or more lasers. Adjust FL1 (FITC) and FL8 (APC) voltage channels to accurately report bead intensity. Once the cytometer is set up, it is useful to analyze a sample from an overnight culture that is diluted in assay buffer. This will aid in identifying live cell region gates.

4. The peristaltic pump, when set to 15 rpm, results in a sample delivery of 1–2 μl/s. A faster or slower speed is typically suboptimal and may also result in increased particle carryover between samples. Autosampler times in and out of the wells is also adjustable; between 1–2 s is optimal. When adjusted properly, the pump clamping pressure should result in uniform air bubbles (*see* Fig. 2). Prime the tubing with assay buffer prior to sampling from a plate. If the bubbles are broken up on the flow cytometer side of the pump, the tension on the tubing is too great and should be appropriately adjusted. Excess events/sample can also lead to particle carryover. This can also be addressed by changing the final concentration of cells. During multiplate experiments, rinsing the inlet probe and tubing with assay buffer in between plate acquisition should reduce signal noise.

5. Certain dyes can leave residue resulting in unwanted carryover. Use multichannel pipettes and single use tips to deliver dyes and bacteria to individual wells. We find most dye substrates work well at a final concentration of 1 μM or lower. Dye profiling plates are also useful for dilution studies to arrive at optimal dye concentrations. Dye profiling plates can be assembled using either 96 or 384-well plates.

6. Pintool pins use a slit design of defined volume to transfer small volumes from the source plate to the assay plate. In order to maintain a reproducible slit volume across the plate, the pintool should only transfer test compounds into assay plates containing assay buffer without cells or beads. This eliminates the possibility of particulates becoming lodged in a pin. Once test chemicals are added to assay plates, clean the pintool by running a cleaning program. Cleaning programs move the pintool through a series of solutions that include detergent, water, and ethanol; each stage is separated by a wicking station. Dose response dilution series source plates are assembled using the span-8 device on the BioMek FX^{P}.

7. Rinse and autoclave the BioTek Microflo tubing and manifold to ensure sterility after use. End over end rotation of plates minimizes settling of particles (i.e. bacteria or cells). Make sure plate seals are properly affixed to isolate individual well contents. In general, gentle inversion of the plate does not result in spillage.
8. This procedure is offered as an example and should not be considered as a definitive validation. Individual institutions will have their own requirements. CDC requirements are subject to amendment.
9. Preliminary experiments are performed on each strain to determine colony forming units/OD_{600}. The CFU/well is crucial. Final plate well volumes vary depending upon a particular plate set-up. Growth inhibition plates and limited potentiation plates containing 2× concentrations of test compounds are inoculated with 50 μl of CA-MHB containing 5×10^4 CFU of bacteria. Combination plates are inoculated with 25 μl of CA-MHB containing 5×10^4 CFU of bacteria. This is due to the increased volume of CA-MHB/well that results when combination plates are assembled (*see* Fig. 10).
10. Choose two concentrations of an antibiotic that are lower than the determined MIC and in similar fashion choose two EPI concentrations based on the calculated EC_{50} determined from dose response efflux inhibition analyses.

Acknowledgments

This work was supported by The Department of Threat and Reduction Agency (HDTRA1-13-C-0005 to G.P.T. and L.A.S.) and The UNM Clinical and Translational Science Center (UL1TR001449 to L.A.S.). The authors thank the Italian Research Council for a Fellowship to Pietro Tedesco.

References

1. Piddock LJ (2006) Clinically relevant chromosomally encoded multidrug resistance efflux pumps in bacteria. Clin Microbiol Rev 19:382–402
2. Piddock LJ (2006) Multidrug-resistance efflux pumps? not just for resistance. Nat Rev Microbiol 4:629–636
3. Tikhova EB, Zgurskaya HI (2004) AcrA, AcrB and TolC of *Escherichia coli* form a stable intermembrane multi drug efflux complex. J Biol Chem 279:32116–32124
4. Symmons MF, Bokma E, Koronakis E et al (2009) The assembled structure of a complete tripartite bacterial multidrug efflux pump. Proc Natl Acad Sci U S A 106:7173–7178
5. Yu EW, Aires JR, McDermott G et al (2005) A periplasmic drug-binding site of the AcrB multidrug efflux pump: a crystallographic and site-directed mutagenesis study. J Bacteriol 87:6804–6815
6. Murakami S, Nakashima R, Yamashita E et al (2006) Crystal structures of a multidrug transporter reveal a functionally rotating mechanism. Nature 14:173–179
7. Nakashima R, Sakurai K, Yamasaki S et al (2011) Structures of the multidrug exporter

AcrB reveal a proximal multisite drug-binding pocket. Nature 27:565–569
8. Eicher T, Cha HJ, Seeger MA et al (2012) Transport of drugs by the multidrug transporter AcrB involves an access and a deep binding pocket that are separated by a switch-loop. Proc Natl Acad Sci U S A 10:5687–5692
9. Schell M, Zhao P, Wells L (2011) Outer membrane proteome of *Burkholderia pseudomallei* and *Burkholderia mallei* from diverse growth conditions. J Proteome Res 10:2417–2424
10. Middlemiss JK, Poole K (2004) Differential impact of MexB mutations on substrate selectivity of the MexAB-OprM multidrug efflux pump of *Pseudomonas aeruginosa*. J Bacteriol 186:1258–1269
11. Mima T, Schweizer HP (2010) The BpeAB-OprB efflux pump of *Burkholderia pseudomallei* 1026b does not play a role in quorum sensing, virulence factor production, or extrusion of aminoglycosides but is a broad-spectrum drug efflux system. Antimicrob Agents Chemother 54:3113–3120
12. Gil H, Platz GJ, Forestal CA et al (2006) Deletion of TolC orthologs in *Francisella tularensis* identifies roles in multidrug resistance and virulence. Proc Natl Acad Sci U S A 22:12897–12902
13. Sandhu P, Akhter Y (2015) The internal gene duplication and interrupted coding sequences in the MmpL genes of *Mycobacterium tuberculosis*: towards understanding the multidrug transport in an evolutionary perspective. Int J Med Microbiol 305:413–423
14. Sulavik MC, Houseweart C, Cramer C et al (2001) Antibiotic susceptibility profiles of *Escherichia coli* strains lacking multidrug efflux pump genes. Antimicrob Agents Chemother 45:1126–1136
15. Lomovskaya O, Warren MS, Lee A et al (2001) Identification and characterization of inhibitors of multidrug resistance efflux pumps in *Pseudomonas aeruginosa*: novel agents for combination therapy. Antimicrob Agents Chemother 45:105–116
16. Kern WV, Steinke P, Schumacher A et al (2006) Effect of 1-(1-naphthylmethyl)-piperazine, a novel putative efflux pump inhibitor, on antimicrobial drug susceptibility in clinical isolates of *Escherichia coli*. J Antimicrob Chemother 57:339–343
17. Rodrigues L, Wagner D, Viveiros M et al (2008) Thioridazine and chlorpromazine inhibition of ethidium bromide efflux in *Mycobacterium avium* and *Mycobacterium smegmatis*. J Antimicrob Chemother 61:1076–1082
18. Viveiros M, Martins A, Paixão L (2008) Demonstration of intrinsic efflux activity of *Escherichia coli* K-12 AG100 by an automated ethidium bromide method. Int J Antimicrob Agents 31:458–462
19. Coldham NG, Webber M, Woodward MJ et al (2010) A 96-well plate fluorescence assay for assessment of cellular permeability and active efflux in *Salmonella enterica* serovar Typhimurium and *Escherichia coli*. J Antimicrob Chemother 65:1655–1663
20. Bohnert JA, Karamian B, Nikaido H (2010) Optimized Nile Red efflux assay of AcrAB-TolC multidrug efflux system shows competition between substrates. Antimicrob Agents Chemother 54:3770–3775
21. Matsumoto Y, Hayama K, Sakakihara S et al (2011) Evaluation of multidrug efflux pump inhibitors by a new method using microfluidic channels. PLoS One 6:e18547
22. Strouse JJ, Ivnitski-Steele I, Khawaja HM et al (2013) A selective ATP-binding cassette subfamily G member 2 efflux inhibitor revealed via high-throughput flow cytometry. J Biomol Screen 18:26–38
23. Strouse JJ, Ivnitski-Steele I, Waller A et al (2013) Fluorescent substrates for flow cytometric evaluation of efflux inhibition in ABCB1, ABCC1, and ABCG2 transporters. Anal Biochem 1:77–87
24. Hoel DG (1987) Statistical aspects of chemical mixtures. In: Vouk VB, Butler GC, Upton AC, Parke DV, Asher SC (eds) Methods for assessing the effects of mixtures of chemicals. Wiley, New York, pp 369–377
25. Baba T, Ara T, Hasegawa M et al (2006) Construction of *Escherichia coli* K-12 in-frame, single-gene knockout mutants: the Keio collection. Mol Syst Biol 2:2006.0008
26. Propst KL, Mima T, Choi KH et al (2010) A *Burkholderia pseudomallei* Δ*purM* mutant is avirulent in immunocompetent and immunodeficient animals: candidate strain for exclusion from select-agent lists. Infect Immun 78:3136–3143
27. Edwards BS, Young SM, Oprea TI et al (2006) Biomolecular screening of formylpeptide receptor ligands with a sensitive, quantitative, high-throughput flow cytometry platform. Nat Protoc 1:59–66
28. Lamping E, Monk BC, Niimi K et al (2007) Characterization of three classes of membrane proteins involved in fungal azole resistance by functional hyperexpression in *Saccharomyces cerevisiae*. Eukaryot Cell 6:1150–1165
29. Tallarida RJ (2006) An overview of drug combination analysis with isobolograms. J Pharmacol Exp Ther 6:1–7

30. Renau T, Léger R, Flame EM et al (1999) Inhibitors of efflux pumps in *Pseudomonas aeruginosa* potentiate the activity of the fluoroquinolone antibacterial levofloxacin. J Med Chem 42:4928–4931
31. Yoshida K, Nakayama K, Ohtsuka M et al (2007) MexAB-OprM specific efflux pump inhibitors in *Pseudomonas aeruginosa*. Part 7: Highly soluble and in vivo active quaternary ammonium analogue D13-9001, a potential preclinical candidate. Bioorg Med Chem 15:7087–7097
32. Nakayama K, Ishida Y, Ohtsuka M et al (2003) MexAB-OprM-specific efflux pump inhibitors in *Pseudomonas aeruginosa*. Part 1: Discovery and early strategies for lead optimization. Bioorg Med Chem 13:4201–4204
33. Lee MD, Galazzo JL, Staley AL et al (2001) Microbial fermentation-derived inhibitors of efflux-pump-mediated drug resistance. Farmaco 56:81–85
34. Rogers D, Hahn M (2010) Extended-connectivity fingerprints. J Chem Inf Model 50:742–754
35. Clinical Laboratory Standards Institute (2007) M7-A7. Methods for dilution antimicrobial susceptibility tests for bacteria that grow aerobically; approved standard, 7th edn. CLSI, Wayne, PA
36. Zhang J-H, Chung TDY, Oldenburg KR (1999) A simple statistical parameter for use in evaluation and validation of high throughput assays. J Biomol Screen 4:67–73
37. Kurinčič M, Klančnik A, Smole-Možina S (2012) Epigallocatechin gallate as a modulator of *Campylobacter resistance* to macrolide antibiotics. Int J Antimicrob Agents 40:467–471
38. Eto D, Watanabe K, Saeki H et al (2013) Divergent effects of desferrioxamine on bacterial growth and characteristics. J Antibiot 66:199–203
39. Coban AY, Tanriverdi-Cayci Y, Erturan Z et al (2009) Effects of efflux pump inhibitors phenyl-arginine-beta-naphthylamide and 1-(1-naphthylmethyl)-piperazine on the antimicrobial susceptibility of *Pseudomonas aeruginosa* isolates from cystic fibrosis patients. J Chemother 21:592–594
40. Schumacher A, Steinke P, Bohnert JA et al (2006) Effect of 1-(1-naphthylmethyl)-piperazine, a novel putative efflux pump inhibitor, on antimicrobial drug susceptibility in clinical isolates of Enterobacteriaceae other than *Escherichia coli*. J Antimicrob Chemother 57:344–348
41. Cirioni O, Giacometti A, Scalise G (1997) In-vitro activity of atovaquone, sulphamethoxazole and dapsone alone and combined with inhibitors of dihydrofolate reductase and macrolides against *Pneumocystis carinii*. J Antimicrob Chemother 39:45–51
42. Alsterholm M, Karami N, Faergemann J (2010) Antimicrobial activity of topical skin pharmaceuticals – an in vitro study. Acta Derm Venereol 90:239–245
43. Sun W, Park YD, Sugui JA et al (2013) Rapid identification of antifungal compounds against Exserohilum rostratum using high throughput drug repurposing screens. PLoS One 8:e70506
44. Hansen T, Nejsum P, Friis C et al (2014) *Trichuris suis* and *Oesophagostomum dentatum* show different sensitivity and accumulation of fenbendazole, albendazole and levamisole in vitro. PLoS Negl Trop Dis 8:e2752

Part V

Advanced Technologies Expected for Application to Multidrug Efflux Transport Studies

Chapter 17

Single-Molecule Analysis of Membrane Transporter Activity by Means of a Microsystem

Rikiya Watanabe, Naoki Soga, Shin-ya Ohdate, and Hiroyuki Noji

Abstract

Emerging microtechnologies are aimed at developing a microsystem with densely packed array structure, i.e., an array with a femtoliter reaction chamber, for highly sensitive and quantitative biological assays. Here, we describe a novel femtoliter chamber array system (arrayed lipid bilayer chambers, ALBiC) that contains approximately a million femtoliter chambers, each sealed with a phospholipid bilayer membrane with extremely high efficiency (>90%). This novel platform enables detection of membrane transporter activity at the single-molecule level and thus expands the applicability of femtoliter chamber arrays to highly sensitive assays of transporters.

Key words Microsystem, Membrane transporter, Single-molecule analysis

1 Introduction

A variety of highly quantitative systems for single-molecule analysis of membrane transport proteins have been developed to elucidate the working mechanisms of these proteins [1, 2]. One of the most robust systems for transporter analysis is patch-clamp recording, which measures the flux of a charged substrate as an electric current. The latest method of patch-clamp recording is fully automated, thus enabling highly parallel analysis of transport activities. However, this method is not effective with transporters because they cannot generate sufficient electric current for detection ($>10^7$ molecules per second) due to low transport rates. Moreover, patch-clamp recording cannot detect a flux of electrically neutral substrates; therefore, a more versatile system for transporter analysis is urgently needed.

Recently, we addressed this issue by developing a microsystem based on arrayed lipid bilayer chambers (ALBiC) [3–5]; in this system, transport activity is measured on the basis of substrate accumulation or consumption in femtoliter chambers by means of optical methods. The ALBiC microsystems can perform highly

Akihito Yamaguchi and Kunihiko Nishino (eds.), *Bacterial Multidrug Exporters: Methods and Protocols*, Methods in Molecular Biology, vol. 1700, https://doi.org/10.1007/978-1-4939-7454-2_17, © Springer Science+Business Media, LLC 2018

sensitive and parallel analysis of membrane transport, allowing for single-molecule analysis of extremely weak transport activities (<10 molecules per second). Thus, this new method extends the applicability of the single-molecule transporter assay to active transporters. In this chapter, we describe protocols for single-molecule analysis of passive and active transporters by means of an ALBiC microsystem, in this case, α-hemolysin [6] and F_oF_1-ATP synthase, respectively [7–10].

2 Materials

Prepare all solutions using ultrapure water (generated by purifying deionized water to attain specific resistance of 18 MΩ cm at 25 °C) and analytical-grade reagents. Diligently follow all waste disposal regulations when disposing waste.

2.1 Microfabrication

1. A glass slide (24 × 32 mm; thickness 0.12–0.17 mm).
2. A hydrophobic carbon-fluoride polymer (CYTOP).
3. A positive photoresist.
4. A photo mask (mask pattern: $\phi = 3$ μm, interval = 6 μm).
5. A mask aligner.
6. An ultraviolet (UV) lamp controller.
7. Positive photoresist developer.
8. A reactive ion etching system.

2.2 Preparation of the Lipid Bilayer

1. 1,2-Dioleoyl-sn-glycero-3-phosphoethanolamine (DOPE).
2. 1,2-Dioleoyl-sn-glycero-3-phosphoglycerol (DOPG).
3. A glass block: it should have an access port for sample injection and be coated with CYTOP.
4. A spacer.
5. Assay buffer: 0.1 mM HEPES, 20 mM NaCl, and 2 mM $MgCl_2$; pH adjusted to 7.0 with NaOH.

2.3 Preparation of the Transporter

1. 1 mg/mL α-Hemolysin from *Staphylococcus* in assay buffer; store at −80 °C until use.
2. A plasmid encoding F_oF_1-ATP synthase from *Escherichia coli* (EF_oF_1): using a plasmid vector (pRA1000) encoding wild-type EF_oF_1, we inserted the His_3 tag at the C terminus of each c subunit; two helices in the C-terminal domain of the ε subunit were deleted by inserting a stop codon after ε-Asp91.
3. An EF_oF_1 expression system: the EF_oF_1-deficient RA1 strain of *E. coli* (*unc*$^-$/*cyo*$^-$) was transformed with the EF_oF_1 plasmid.
4. 40 mg/mL L-α-Phosphatidylcholine from soybean in assay buffer; frozen with liquid nitrogen and stored at −80 °C until use.

2.4 Imaging

1. A resonant scanning confocal system.
2. A laser with $\lambda_{ex} = 488$ nm.
3. A laser with $\lambda_{ex} = 561$ nm.
4. A 60× objective lens.
5. NIS Elements software.

2.5 The Single-Molecule Passive Transporter Assay

1. Alexa 488.

2.6 The Single-Molecule Active Transporter Assay

1. A fluorescent pH indicator, RhP-M: we obtained it from collaborators [11].
2. 5% (w/v) poly(ethylene glycol) (PEG with molecular weight of 2000) in assay buffer.
3. 200 mM ATP solution in deionized water whose pH was adjusted with NaOH to 7.0; store this solution at −30 °C until use; for the assays, dilute this solution to 200 μM with assay buffer.
4. 5 mM Nigericin in ethanol; stored at −30 °C until use; for assays, dilute this solution to 5 μM with assay buffer.

3 Methods

3.1 Microfabrication

All experimental procedures are to be performed in a clean room.

1. Immerse a glass slide in 10 N KOH and incubate overnight to clean up the surface of the slide.
2. Wash the glass slide with deionized water and dry the slide off with an air blower.
3. Place ~90 μL of the hydrophobic carbon-fluoride polymer (CYTOP) onto the center of the glass slide.
4. Spin-coat the glass slide with CYTOP (500 rpm for 10 s and then at 4000 rpm for 30 s) to achieve the coating thickness of 500 nm.
5. Bake the CYTOP-coated glass slide at 180 °C for 1 h on a hot plate.
6. Place ~90 μL of the positive photoresist onto the center of the CYTOP surface of the glass slide.
7. Spin-coat the CYTOP surface with the photoresist at 500 rpm for 10 s and then at 7500 rpm for 30 s.
8. Prebake the glass slide at 55 °C for 3 min and bake at 110 °C for 5 min on a hot plate.

9. Load the photo mask and the glass slide into the mask aligner. Expose the photoresist to UV light of 250 W for 7 s.
10. Immerse the photoresist in the positive photoresist developer and incubate for 5 min to develop the mask pattern on the CYTOP surface.
11. Rinse the glass slide with deionized water three times.
12. Etch the CYTOP layer with O_2 plasma by means of a reactive ion etching system for 15 min to fabricate the hole structures (femtoliter chambers: $\phi = 4$ μm, $h = 500$ nm) in the glass slide.
13. Wash the etched glass slide with acetone in the ultrasonic bath for 10 min to remove the positive photoresist from the CYTOP surface.
14. Rinse the etched glass slide with 2-propanol in the ultrasonic bath.
15. Rinse the etched glass slide with deionized water. Dry the slide off using the air blower.

3.2 Preparation of the Lipid Bilayer

1. Construct a flow cell ($V = \sim 20$ μL) by assembling the glass slide with the femtoliter chambers, spacer, and glass block (Fig. 1).
2. Inject ~20 μL of assay buffer containing the fluorescent dye into the flow cell and cool the whole device (*see* **Note 1** and Fig. 2).
3. Inject ~100 μL of chloroform solution of the lipid (containing 4 mg/mL of the lipid: a 1:1 [w/w] mixture of DOPE and DOPG) into the flow cell to flush the original aqueous solution (*see* **Note 2** and Fig. 2).
4. Inject ~100 μL of assay buffer into the flow cell to flush the lipid solution (*see* **Notes 3, 4** and Fig. 2).

3.3 Preparation of Proteoliposomes Containing the Active Transporter, EF_oF_1

1. Express and purify EF_oF_1 using the expression system as described previously [3, 12].
2. Wash crude soybean L-α-phosphatidylcholine with acetone [13, 14] and dissolve in assay buffer to the final concentration of 40 mg/mL (*see* **Note 5**).
3. Mix 100 μL of the purified 1 mg/mL EF_oF_1 solution with 1 mL of the 40 mg/mL liposome suspension (*see* **Note 6**).
4. Incubate this mixture for 1 h. Freeze with liquid nitrogen and store at −80 °C until use.

3.4 The Single-Molecule Passive-Transporter Assay

1. Fill the flow cell and femtoliter chambers with assay buffer containing 1 μM Alexa 488 as described in Subheading 3.2 (*see* **Note 7**).
2. Form lipid bilayers on the orifice of individual chambers as described in Subheading 3.2. Wash the flow cell with assay

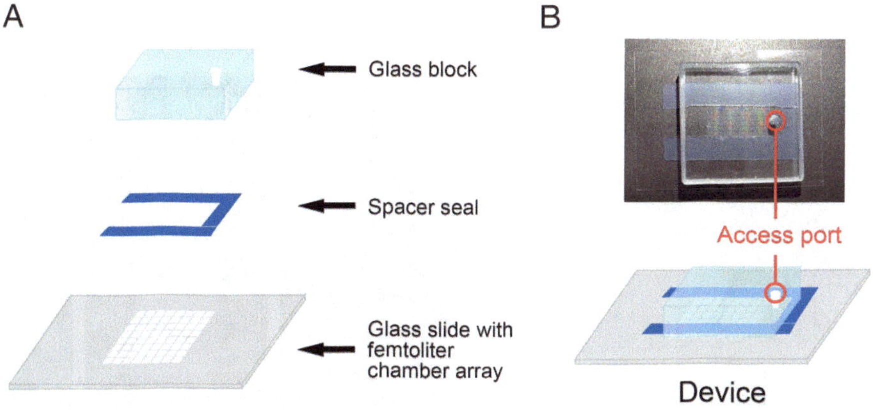

Fig. 1 The arrayed lipid bilayer chamber (ALBiC) microsystem. (**a**) Components of the ALBiC microsystem: *top*, a glass block with an access port for sample injection; *middle*, the spacer with one side open; *bottom*, the glass slide where more than 100,000 femtoliter chambers are fabricated. (**b**) A photograph (*top*) and illustration (*bottom*) of the ALBiC microsystem after assembly of the above components

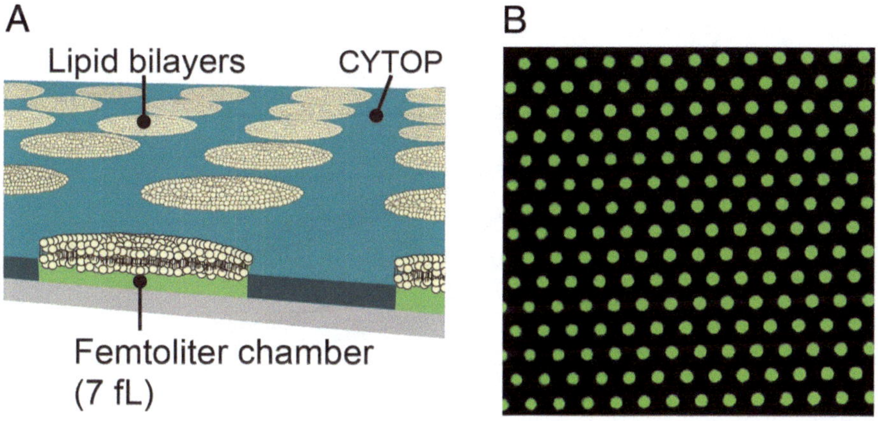

Fig. 2 Formation of a lipid bilayer on the orifice of femtoliter chambers. (**a**) A schematic diagram of the lipid bilayer formation on ALBiC. (**b**) A fluorescent image of femtoliter chambers containing 1 μM Alexa 488. Efficiency of the lipid bilayer formation is >90%

buffer to completely remove the residual Alexa 488 from the flow cell.

3. Start time-lapse recording of fluorescence, with the interval of 20 s, on a confocal-microscopy system equipped with the 60× objective lens and a laser ($\lambda_{ex} = 488$ nm).
4. Record the original fluorescence level for 5 min.
5. Inject ~100 μL of assay buffer containing 1 μg/mL α-hemolysin to start the passive transport reaction. α-Hemolysin is incorporated into the lipid bilayer membranes spontaneously (Fig. 3a, *see* **Note 8**).

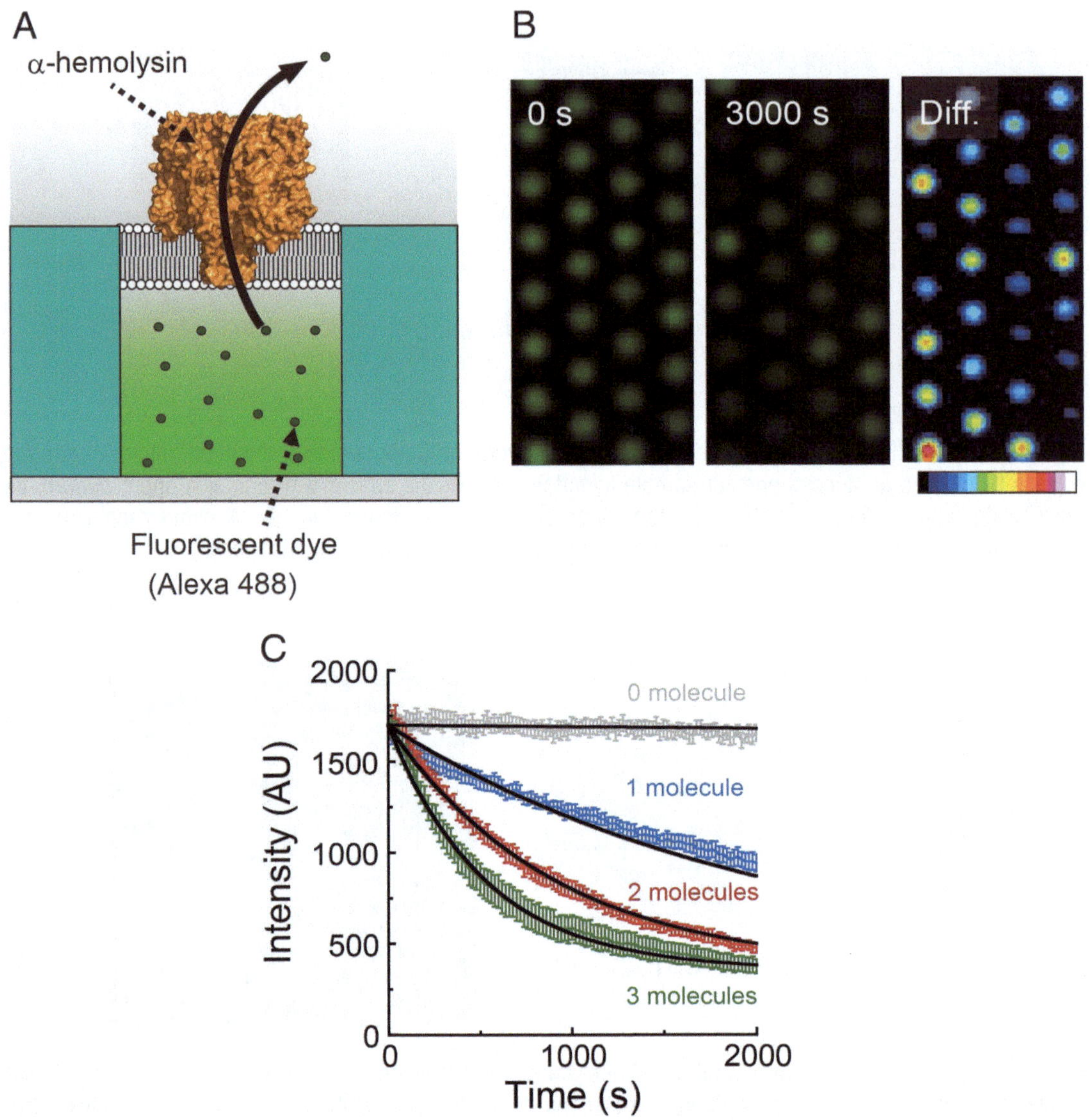

Fig. 3 Passive transport by α-hemolysin. (**a**) A schematic diagram of passive transport of Alexa 488 through the pore of α-hemolysin. The encapsulated fluorescent dyes, which are used as a passive-transport substrate, diffuse through the pore of the reconstituted α-hemolysin. (**b**) Fluorescent images of the passive transport of Alexa 488 by α-hemolysin. The images were obtained immediately after injection of 1 μg/mL α-hemolysin (*left*, 0 s) and 3000 s later (*middle*). The *right panel* shows the intensity difference between the images at 0 and 3000 s as a color gradient. (**c**) Time course of fluorescence decay of Alexa 488 because of the passive transport by 1 μg/mL α-hemolysin. *Gray*, *blue*, *red*, and *green* represent the activity of zero, one, two, or three molecules of α-hemolysin, respectively. *Solid lines* represent the fitting with single exponential decay according to the formula $y = C_1 \cdot \exp(-k \cdot t) + C_2$, where k is the rate constant of the passive transport. The error bars denote standard deviation

6. Acquire fluorescent time-lapse images to assess the decrease in the fluorescence intensity due to passive transport of Alexa 488 via the pore of α-hemolysin (Fig. 3b). Finish the recording at the end of the observation period (~1 h or longer).
7. Analyze fluorescence intensity of the chambers using the NIS Elements software.

3.5 The Single-Molecule Active-Transporter Assay

1. Fill the flow cell and femtoliter chambers with assay buffer containing 10 μM RhP-M and 1 μM Alexa 488 for detection of the pH change caused by the H^+-pump and for evaluation of the lipid bilayer formation, respectively.
2. Form lipid bilayers on the orifice of individual chambers as described in Subheading 3.2. Wash the flow cell with assay buffer to completely remove the residual RhP-M and Alexa 488 from the flow cell.
3. Dilute the proteoliposome solution 100-fold with assay buffer containing 5% (w/v) PEG. Inject ~100 μL of the diluted proteoliposome solution into the flow cell and incubate for 30 min to fuse proteoliposomes with the lipid bilayers (*see* **Note 9**).
4. Wash the flow cell with ~200 μL of assay buffer to remove proteoliposomes from the flow cell.
5. Start time-lapse recording of fluorescence, with the interval 100 s, on a confocal-microscopy system equipped with the 60× objective lens and the lasers ($\lambda_{ex} = 488$ and 561 nm).
6. Record the original fluorescence level for 5 min.
7. Inject ~100 μL of assay buffer containing 200 μM ATP to initiate the H^+-pump of EF_oF_1 (Fig. 4a).
8. Record fluorescent time-lapse images to assess the increase in the fluorescence intensity due to the H^+-pump activity of EF_oF_1 (Fig. 4b). At the end of the observation period (after ~2 h or longer), inject assay buffer containing 5 μM nigericin, an ionophore for H^+ (*see* **Note 10**).
9. After that, finish the recording. Analyze the fluorescence intensity of the chambers using the NIS Elements software (*see* **Note 11**).

4 Notes

1. The air bubbles that can form in the femtoliter chambers are not easy to remove. To remove the air bubbles from the femtoliter chambers, cool down the whole device on an ice block after injecting an aqueous solution.

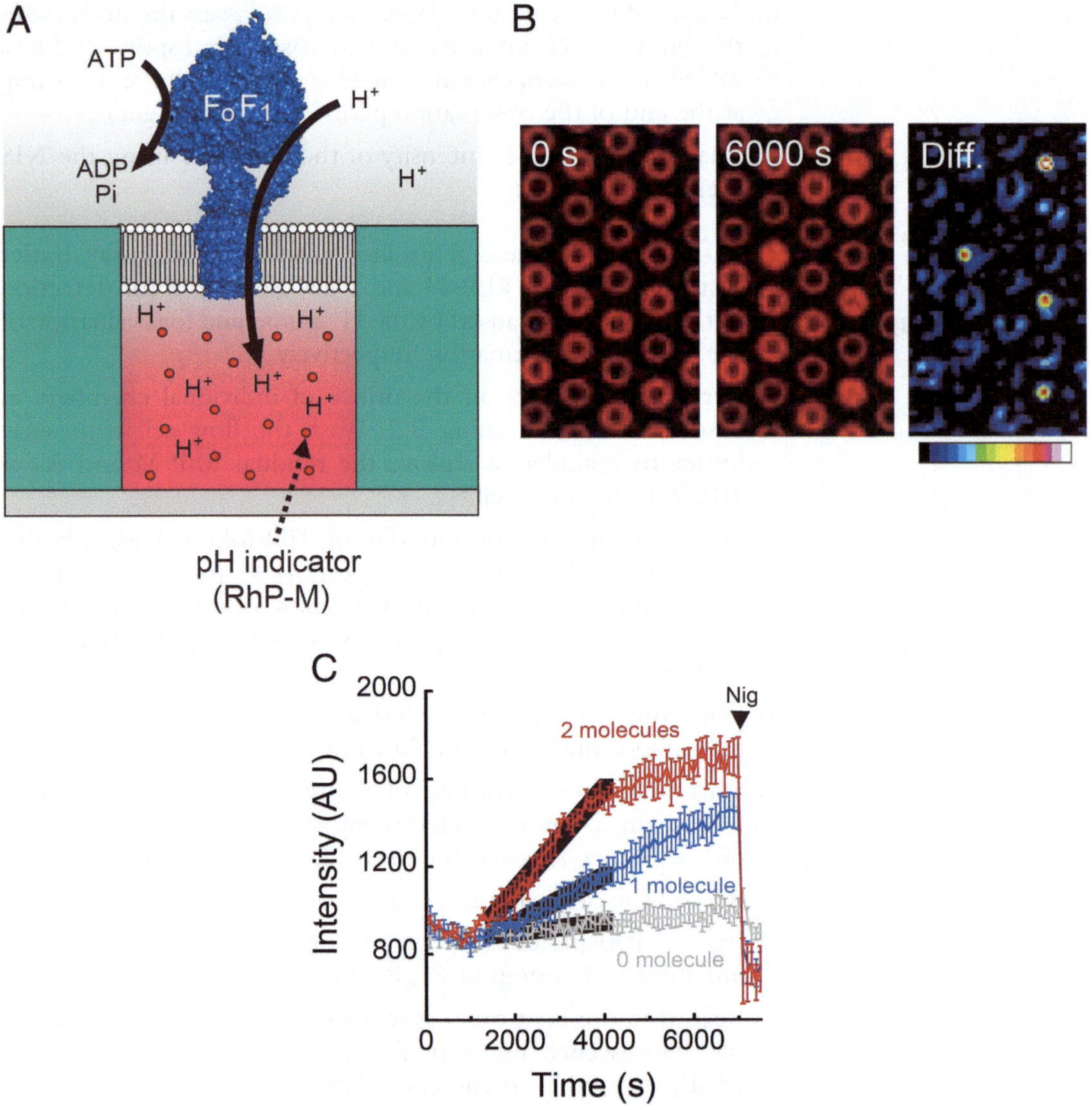

Fig. 4 Active transport of H^+ by EF_oF_1. (**a**) A schematic diagram of the H^+-pump of EF_oF_1. EF_oF_1 pumps H^+ into the chambers from outside using the energy of ATP hydrolysis. The pH change caused by EF_oF_1 can be detected with an encapsulated pH indicator (e.g., RhP-M). (**b**) Fluorescent images of the H^+-pump by EF_oF_1. The images were acquired immediately after the injection of 200 μM ATP (*left*, 0 s) and 6000 s later (*middle*). The *right panel* shows the intensity difference between the images at 0 and 6000 s as a color gradient. (**c**) Time course of the increase in RhP-M fluorescence due to the H^+-pump by EF_oF_1. *Gray*, *blue*, and *red* represent the activity of zero, one, or two molecules of EF_oF_1, respectively. *Black solid lines* represent the linear fitting from 1500 to 4000 s. The error bars denote standard deviation

2. The self-assembled lipid monolayer is formed at the interface of the aqueous solution and chloroform.
3. The residual lipids will form a lipid membrane on the orifice of the femtoliter chambers. Various organic solvents (e.g.,

chloroform, decane, or hexadecane) and lipid molecules (e.g., DOPE, DOPG, soybean lipids, or total *E. coli* lipids) can be used for preparation of the lipid bilayer.

4. The flow rate of assay buffer is critical for highly efficient lipid membrane formation. To achieve the high formation efficiency, gently inject final assay buffer with a constant flow rate (1–10 μL/s).
5. To attain homogeneous dispersion of the lipid in the suspension, incubate the mixture for 30 min with gentle stirring and disperse by brief ultrasonication with a tip-type sonicator for 30 s.
6. The weight ratio of EF_oF_1 to the liposomes is critical for control of the number of EF_oF_1 molecules incorporated into the lipid bilayers in the ALBiC microsystem.
7. A fluorescent dye, except for membrane-permeable dyes, such as calcein AM, can be used as the passive-transport substrate of α-hemolysin.
8. The concentration of α-hemolysin is crucial for control of the number of α-hemolysin molecules incorporated into the lipid bilayers in the ALBiC microsystem.
9. The proper concentration of the proteoliposomes in the fusion step is crucial for stability of the lipid bilayers in the ALBiC microsystem: a higher concentration of the proteoliposomes disrupts the lipid bilayers.
10. Due to nigericin, the H^+ gradient that was generated by EF_oF_1 will collapse, returning the active chambers to the original fluorescence intensity. This nigericin-induced fluorescence resetting ensures that the fluorescence increase represents the ATP-driven H^+-pump activity of EF_oF_1.
11. Some RhP-M molecules bound nonspecifically to the chamber wall due to its hydrophobicity (Fig. 4b). For quantitative analysis, analyze the fluorescent profile of RhP-M at the center of the chamber; omit the fluorescent signal localized along the chamber wall.

Acknowledgment

This work was supported by a Grant-in-Aid for Scientific Research No. 15H05591 and 15H01312 to R.W. from the Ministry of Education, Culture, Sports, Science and Technology, Japan, and by a Precursory Research for Embryonic Science (PRESTO) grant to R.W. from the Japan Science and Technology Agency.

References

1. Dunlop J, Bowlby M, Peri R, Vasilyev D, Arias R (2008) High-throughput electrophysiology: an emerging paradigm for ion-channel screening and physiology. Nat Rev Drug Discov 7:358–368
2. Zagnoni M (2012) Miniaturised technologies for the development of artificial lipid bilayer systems. Lab Chip 12:1026–1039
3. Watanabe R, Soga N, Fujita D, Tabata KV, Yamauchi L, Hyeon Kim S, Asanuma D, Kamiya M, Urano Y, Suga H, Noji H (2014) Arrayed lipid bilayer chambers allow single-molecule analysis of membrane transporter activity. Nat Commun 5:4519
4. Watanabe R, Soga N, Yamanaka T, Noji H (2014) High-throughput formation of lipid bilayer membrane arrays with an asymmetric lipid composition. Sci Rep 4:7076
5. Soga N, Watanabe R, Noji H (2015) Attolitre-sized lipid bilayer chamber array for rapid detection of single transporters. Sci Rep 5:11025
6. Song L, Hobaugh MR, Shustak C, Cheley S, Bayley H, Gouaux JE (1996) Structure of staphylococcal a-hemolysin, a heptameric transmembrane pore. Science 274:1859–1866
7. Junge W, Sielaff H, Engelbrecht S (2009) Torque generation and elastic power transmission in the rotary F_OF_1-ATPase. Nature 459:364–370
8. Watanabe R, Tabata KV, Iino R, Ueno H, Iwamoto M, Oiki S, Noji H (2013) Biased Brownian stepping rotation of F_oF_1-ATP synthase driven by proton motive force. Nat Commun 4:1631
9. Weber J (2010) Structural biology: toward the ATP synthase mechanism. Nat Chem Biol 6:794–795
10. Yoshida M, Muneyuki E, Hisabori T (2001) ATP synthase—a marvellous rotary engine of the cell. Nat Rev Mol Cell Biol 2:669–677
11. Asanuma D, Takaoka Y, Namiki S, Takikawa K, Kamiya M, Nagano T, Urano Y, Hirose K (2014) Acidic-pH-activatable fluorescence probes for visualizing exocytosis dynamics. Angew Chem Int Ed Engl 53:6085–6089
12. Iino R, Hasegawa R, Tabata KV, Noji H (2009) Mechanism of inhibition by C-terminal a-helices of the e subunit of Escherichia coli F_oF_1-ATP synthase. J Biol Chem 284:17457–17464
13. Soga N, Kinosita K Jr, Yoshida M, Suzuki T (2011) Efficient ATP synthesis by thermophilic Bacillus F_oF_1-ATP synthase. FEBS J 278:2647–2654
14. Soga N, Kinosita K Jr, Yoshida M, Suzuki T (2012) Kinetic equivalence of transmembrane pH and electrical potential differences in ATP synthesis. J Biol Chem 287:9633–9639

Chapter 18

Large-Scale Femtoliter Droplet Array for Single Cell Efflux Assay of Bacteria

Ryota Iino, Shouichi Sakakihara, Yoshimi Matsumoto, and Kunihiko Nishino

Abstract

Large-scale femtoliter droplet array as a platform for single cell efflux assay of bacteria is described. Device microfabrication, femtoliter droplet array formation and concomitant enclosure of single bacterial cells, fluorescence-based detection of efflux activity at the single cell level, and collection of single cells from droplet and subsequent gene analysis are described in detail.

Key words Microfabrication, Microdroplet array, Single cell analysis, Drug efflux, AcrAB-TolC, MexAB-OprM, *Escherichia coli*, *Salmonella enterica*, *Pseudomonas aeruginosa*, Optical microscopy, Fluorescence

1 Introduction

Active efflux of drugs such as antibiotics from the cell is one of the major mechanisms of multidrug resistance of bacteria [1]. The AcrAB-TolC system is a multicomponent efflux pump complex responsible for both the intrinsic and acquired drug resistance of gram-negative bacteria such as *Escherichia coli* and *Salmonella enterica* [2]. The AcrAB-TolC system recognizes and pumps out a wide variety of compounds including antibiotics, dyes, and detergents as substrates, driven by the electrochemical potential of proton across the inner membrane [3, 4]. MexAB-OprM and MexXY-OprM systems, a homolog of the AcrAB-TolC, elicit multidrug resistance in clinically isolated *Pseudomonas aeruginosa* [5].

Here, we describe a method assessing the activity of these multicomponent efflux pump systems at the single cell level [6, 7]. Our method uses a directly accessible femtoliter droplet array [8] and a fluorogenic substrate fluorescein-di-β-D-galactopyranoside (FDG) [9, 10], and simultaneously encloses individual bacterial cell and many FDG molecules into each droplet.

Akihito Yamaguchi and Kunihiko Nishino (eds.), *Bacterial Multidrug Exporters: Methods and Protocols*, Methods in Molecular Biology, vol. 1700, https://doi.org/10.1007/978-1-4939-7454-2_18, © Springer Science+Business Media, LLC 2018

Accumulation of a fluorescent dye fluorescein generated from the FDG in the cell and in the femtoliter droplet, observed under a fluorescence microscope, can be used as a marker of the efflux activity. Our method is rapid and simple, and potentially used for screening of efflux pump inhibitors and genes due to the accessibility to each droplet. In this protocol, preparation of the hydrophilic-in-hydrophobic micropatterned glass substrate, femtoliter droplet array formation concomitant with enclosure of the single cells of *E. coli* and FDG in each droplet, and fluorescence detection of the efflux activity are described in detail.

2 Materials

2.1 Microdevice Components

1. Electron beam resist (ZEP520A, ZEON).
2. Electron beam resist thinner (ZEP-A, ZEON).
3. Chromium mask blanks (2.5 in. in diameter, Clean Surface Technology Co.).
4. Spin coater (MS-A100, Mikasa).
5. Electron beam lithography system (JSM 6390, JEOL, and SPG 724, Sanyu Electron).
6. Electron beam resist developer (ZED-N50, ZEON).
7. Electron beam resist solvent (ZMD-B, ZEON).
8. Chromium etchant (Kanto Chemical).
9. Electron beam resist remover (ZDMAC, ZEON).
10. Microscope coverslips (30 mm in diameter, ~0.17 mm in thickness).
11. Fluorinated polymer CYTOP (9 wt%, type M, Asahi Glass).
12. Photoresist (AZ P4903, AZ Electronic Materials) (*see* **Note 1**).
13. Photoresist developer (AZ 300MIF developer, AZ Electronic Materials).
14. Mask aligner (ES410s, SAN-EI ELECTRIC).
15. Reactive ion etching instrument (RIE-10NR, Samco).
16. Disposable plastic petri dish (35 mm in diameter, Becton Dickinson).
17. Drill press (~20 mm in diameter).
18. Epoxy adhesive (Araldite AR-R30, NICHIBAN).
19. Knife (K-35, HOZAN).

2.2 Cells, Medium, Antibiotics, Inducer, Fluorogenic Substrates

1. *E. coli* strain MG1055 wild-type and its efflux-pump-gene deletion mutants Δ*acrB* (ΔB), Δ*tolC* (ΔC), Δ*acrB*Δ*tolC* (ΔBΔC).

2. The ΔBΔC mutant strain harboring the vector plasmid pMMB67HE recombined with the efflux pump genes *mexAB-oprM* from *P. aeruginosa* (pMMB67HE::*mexAB-oprM*).
3. Luria-Bertani medium (Becton Dickinson).
4. Ampicillin (Sigma-Aldrich).
5. Kanamycin (Sigma-Aldrich).
6. Isopropyl-β-D-galactopyranoside (IPTG; Sigma-Aldrich).
7. Fluorescein-di-β-D-galactopyranoside (FDG; Marker Gene Technologies).

2.3 Single-Bacterial Drug Efflux Assay Components

1. Fluorinated oil (Fluorinert FC40, SIGMA-ALDRICH) (*see* **Note 2**).
2. Air displacement pipette (Gilson).
3. Inverted microscope (BZ-8000; KEYENCE), equipped with an objective lens (CFI Plan Apo 20×, NA0.75; Nikon).

2.4 Single-Bacterial Drug Efflux Assay After Genetic Transformation

1. The ΔC mutant *E. coli* strain.
2. The vector plasmid pTH18kr recombined with the efflux pump gene *tolC* from *S. enterica* (pTH18kr::*tolC*).
3. Electroporation instrument (GenePulser Xcell; BioRad).

2.5 Cell Collection Components

1. Inverted microscope (Olympus IX71).
2. Micromanipulator (MNM-21, Narishige).
3. Capillary (75 μL, Drummond Scientific).
4. Puller (Model PC10, Narishige).
5. Microforge (MF900, Narishige).
6. Pressure controller (Femtojet, Eppendorf).

3 Methods

3.1 Microfabrication and Construction of the Device (Fig. 1)

Carry out all procedures in a yellow clean room.

3.1.1 Micropatterned Photomask Preparation

1. Dilute the electron beam resist to 1.4-fold (w/w) with a thinner (*see* **Note 3**). Place the diluted electron beam resist on chromium mask blanks and spincoat it with a spincoater using the following program:

   Slope 5 s.

   500 rpm 5 s.

   Slope 8 s.

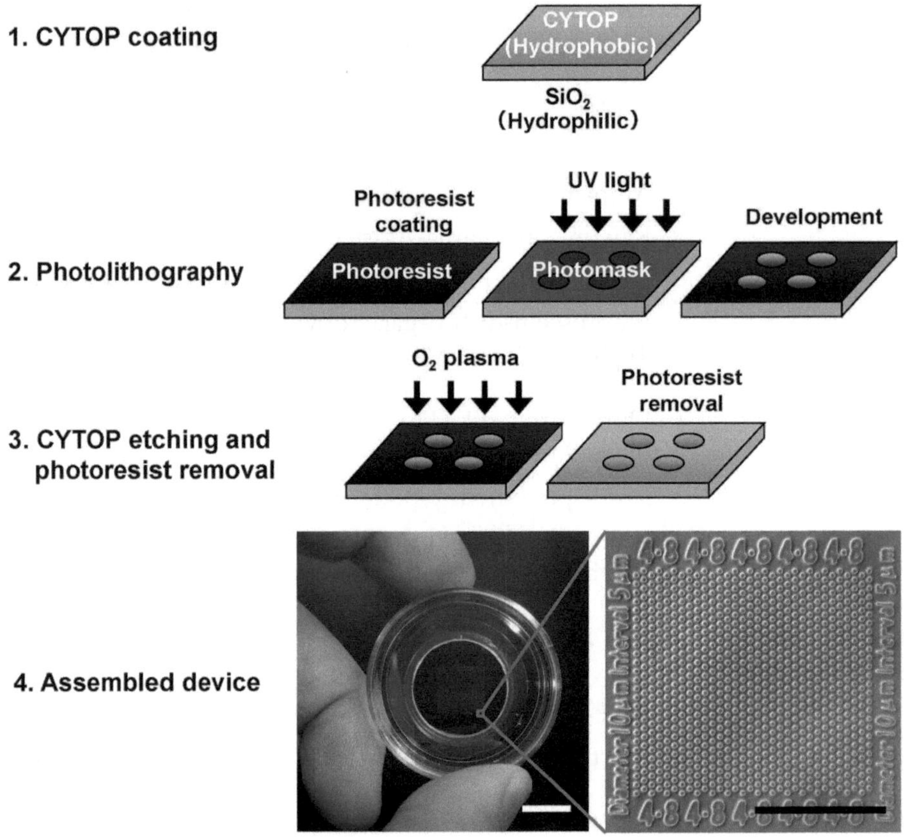

Fig. 1 (*1*, *3*, to) Schematics showing the procedure of microfabrication. (*4*, *Left*) Images of the assembled device. Scale bar, 10 mm. (*4*, *Right*) Microscopic image of the hydrophilic-in-hydrophobic micropatterned surface. Micropatterns are grouped into islands using numbers. This facilitates the identification of individual droplets and the cells enclosed in each droplet. Scale bar, 200 μm

3500 rpm 60 s.

Slope 5 s.

End.

2. Bake at 180 °C for 3 min.
3. Place the resist-coated chromium mask blanks in the electron beam lithography system and carry out lithography under the following conditions:

   Dose: 72 μC/cm^2.

   Voltage: 30 keV.

   Current: 1000 pA.
4. Immerse the chromium mask blanks in electron beam resist developer for 1 min.
5. Rinse well with electron beam resist solvent and then isopropyl alcohol, and dry with a blower.

6. Wet-etch the surface with the chromium etchant (*see* **Note 4**).
7. Remove the electron beam resist using a remover.

3.1.2 CYTOP Coating

1. Wash the microscope coverslips with ethanol and ultrapure water for 5 min each using a bath-type sonicator. Dip the coverslips in 10 N KOH and leave overnight at room temperature. Wash well with ultrapure water, dry on a heated plate at 180 °C, and cool to room temperature.
2. Place 75 μL of CYTOP on the coverslip, and spincoat using the following program:

   Slope 2 s.

   500 rpm 5 s.

   Slope 8 s.

   2000 rpm 30 s (*see* **Note 5**).

   Slope 5 s.

   End.
3. Pre-bake at 80 °C for 30 min (*see* **Note 6**).
4. Bake at 180 °C for 1 h.

3.1.3 Photolithography

1. Place the photoresist (~10 mm in diameter) at the center of the CYTOP-coated glass (*see* **Note 7**). Spincoat using the following program:

   Slope 2 s.

   500 rpm 5 s.

   Slope 8 s.

   4000 rpm 30 s.

   4500 rpm 1 s (*see* **Note 8**).

   Slope 5 s.

   End.
2. Bake at 55 °C for 3 min.
3. Bake at 110 °C for 5 min.
4. Tightly contact the photomask with the substrate using the mask aligner, and irradiate with UV light for 35 sec (*see* **Note 9**).
5. Immerse in developer for 6 min (*see* **Note 10**).
6. Rinse well with ultrapure water.

3.1.4 CYTOP Etching by Oxygen Plasma and Photoresist Removal

1. Place the photoresist-patterned coverslips into a reactive ion etching instrument, and dry-etch the CYTOP film with O_2 plasma using the following conditions:

   O_2: 50 sccm (standard cc/min).

Pressure: 10 Pa.

Power: 50 W.

Time: 30 min.

2. Rinse the etched coverslips three times with acetone for 1 min each using a bath-type sonicator (*see* **Note 11**).
3. Rinse the etched coverslips with isopropyl alcohol once, and dry with a blower.

3.1.5 Device Assembly

1. Punch a hole in the bottom of the disposable plastic petri dish using a drill press. Thoroughly remove burr with a knife, and wash with ultrapure water and ethanol for 5 min each in a bath-type sonicator. Dry at room temperature.
2. Coat the edge of the hole with the epoxy adhesive from the bottom of the petri dish, and completely cover the hole with the hydrophilic-in-hydrophobic micropatterned coverslip (*see* **Note 12**). The coverslip should be orientated such that the CYTOP-coated surface is in contact with the epoxy adhesive.
3. Allow the petri dish and the coverslip to completely adhere to one another overnight.

3.2 Single Cell Efflux Assay (Fig. 2)

First of all, we explain the principle of single cell efflux assay (Fig. 2). FDG is a fluorogenic substrate of β-galactosidase expressed in cytoplasm of the *E. coli*, and hydrolyzed into a fluorescent dye fluorescein. Both FDG and fluorescein are recognized and exported by the AcrAB-TolC system of the *E. coli*. In wild-type *E. coli* expressing high level of AcrAB-TolC, FDG is effectively pumped out before hydrolysis to fluorescein, and no fluorescence is observed (Fig. 2b and c, left). On the other hand, in the cells with weak efflux activity like ΔB and ΔC strains, FDG is imported into the cytoplasm and hydrolyzed to fluorescein. In ΔB cells, not only the cells, but also the droplets emit fluorescence (Fig. 2b and c, center) because the remaining minor RND efflux pumps slowly pump out the fluorescein. Although only a small amount of the fluorescein is pumped out, it can be easily detected because confined and accumulated in the femtoliter droplet. In ΔC cells, fluorescein accumulated in the cell (Fig. 2b and c, right) because TolC is a channel protein common to both the major and minor RND efflux pumps in *E. coli*. Therefore, form the fluorescence intensity and distribution of the fluorescein in the cells and droplets, efflux activity can be easily detected. Not only the AcrAB-TolC system, efflux activity of the MexAB-OprM and MexXY-OprM systems can be also specifically detected if expressed in ΔBΔC cells.

1. Inoculate overnight cultures of *E. coli* strains and incubate at 37 °C in the presence of IPTG (1 mM, to induce β-galactosidase) until the cultures reached a turbidity at

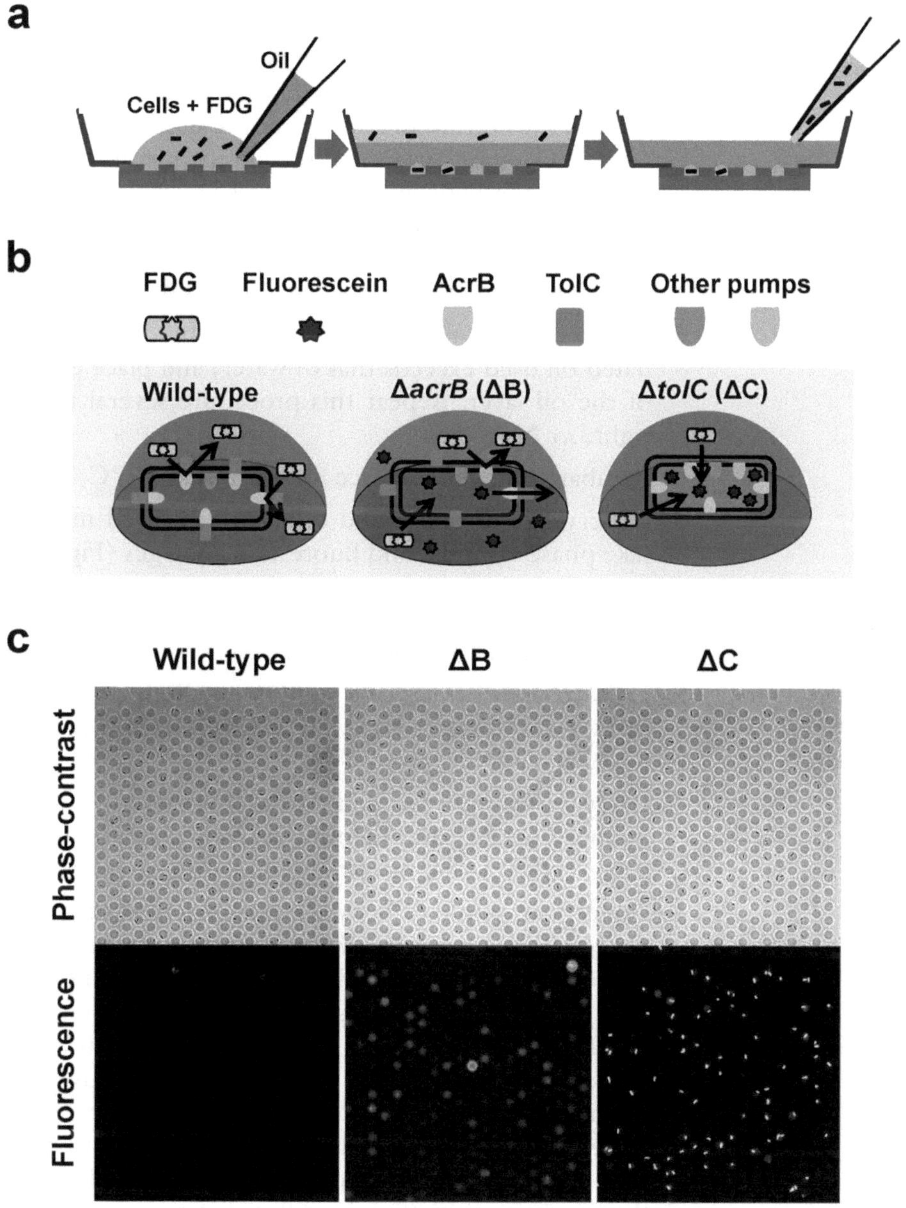

Fig. 2 (**a**) Procedure of droplet formation and cell enclosure. (**b**) Principle of single cell efflux assay. The detail is described in **Note 13**. (**c**) A representative assay. Phase-contrast (*top*) and fluorescence (*bottom*) images of the wild-type, ΔB, ΔC *E. coli* cells, in same image field each, are shown

600 nm of 0.6. For cultures of mutant strains harboring pMMB67HE::*mexAB-oprM*, add ampicillin (100 μg/mL) to the medium to ensure plasmid retention.

2. Mix 450 μL of bacterial culture with 50 μL of FDG solution (1 mg/mL).
3. Completely cover the micropatterned coverslip in the assembled device with the bacterial culture containing FDG. The

volume of the bacterial suspension on the micropatterned coverslip must be as low as possible (Fig. 2a left, *see* **Note 13**).

4. Using an air displacement pipette, introduce 1 mL fluorinated oil into the culture medium near the glass surface. The hydrophilic SiO_2 glass surfaces will retain the medium and the bacteria, while the hydrophobic surfaces are replaced with oil. As a result, femtoliter droplets (3×10^5 droplets per 1 cm^2 for a 10 μm diameter droplet) containing zero, one, or more bacteria will form (Fig. 2a center, *see* **Note 14**).
5. Remove excess medium on the oil (the density of the fluorinated oil used exceeds that of water) and place ethanol on top of the oil layer. Repeat this procedure several times (Fig. 2a right, *see* **Note 15**).
6. Incubate the whole device for 15 min at 37 °C.
7. Observe the droplets and cells with inverted microscope and take phase-contrast and fluorescence images (Fig. 2b and c).

3.3 Single Cell Efflux Assay After Genetic Transformation, Collection of Active Cell, Gene Analysis (Fig. 3)

1. Introduce pTH18kr::*tolC* into gene-competent ΔC cells by electroporation (*see* **Note 16**).
2. Culture transformed cells in SOC medium containing 50 μg/mL kanamycin and 1 mM IPTG to induce β-galactosidase at 37 °C for 3 h.
3. Carry out single-bacterial drug-efflux assay as described in Subheading 3.2, and identify droplets containing efflux active bacteria by phase-contrast and fluorescence microscopy.
4. Collect the single cells with a glass micropipette (inside aperture ~10 μm, prepared from a capillary using a puller and a microforge) under an optical microscope. Fill the micropipette with medium, set the pressure to a positive value (50–60 hPa) (*see* **Note 17**), immerse the micropipette into the oil layer approaching the droplet containing efflux-active cells. When the micropipette is close to the droplet, reduce the pressure to zero and allow the tip of the glass micropipette to make contact with the droplet. The droplet will spontaneously be drawn into the glass micropipette by capillary force.
5. Dip the glass micropipette into the medium in the test tube and culture the collected cell at 37 °C.
6. Transfer the contents of micropipette containing the single cell to the test tube containing 2 mL of LB medium supplemented with 50 μg/mL kanamycin and cultured overnight at 37 °C. Then, prepare grycerol stock and store at −80 °C.
7. Culture the cells again from the glycerol stock at a later date, extract the harbored plasmid. Carry out polymerase chain reaction using the extracted plasmid as a template. Analyze the amplified DNA fragment with agarose gel electrophoresis.

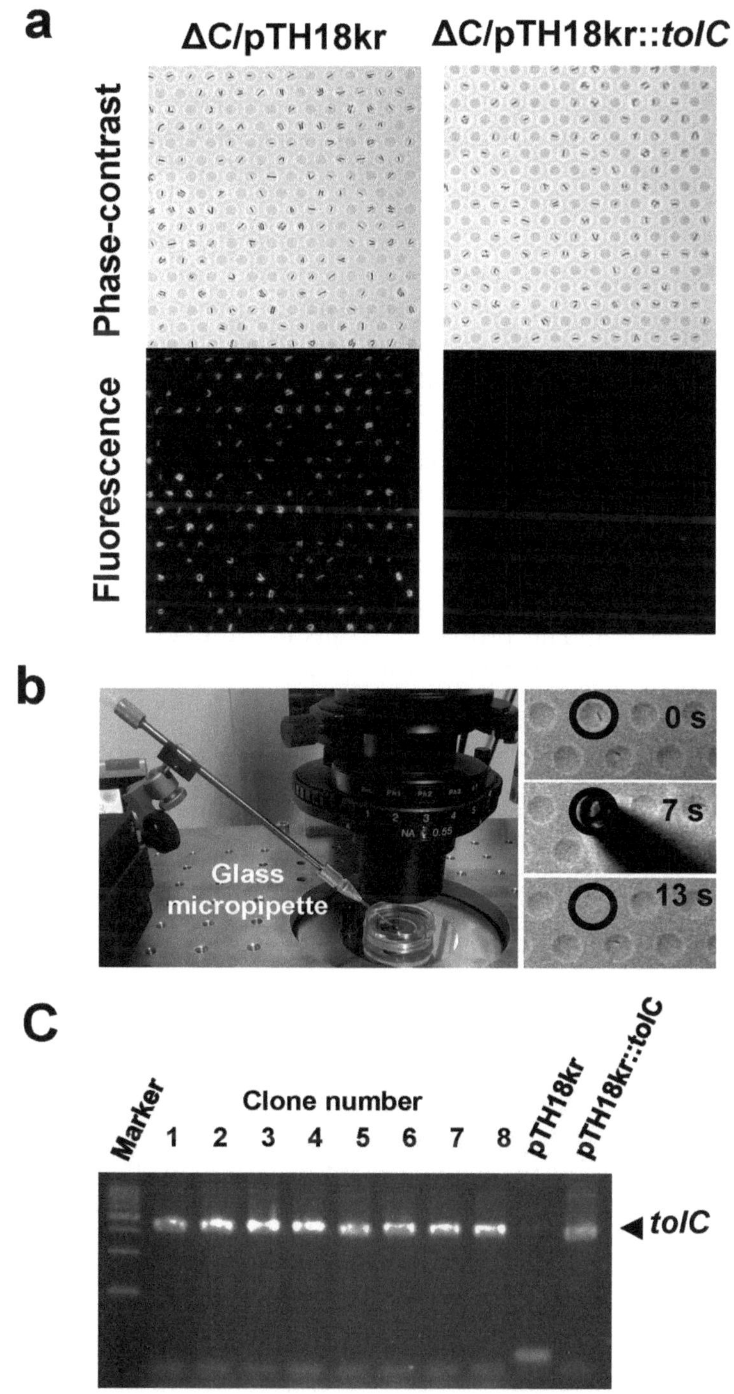

Fig. 3 (**a**) Phenotypic change after genetic transformation. Phase-contrast (*top*) and fluorescence (*bottom*) images of *E. coli* ΔC cells transformed with a control plasmid pTH18kr (*left*) or a plasmid expressing *tolC* from *S. enterica* (pTH18kr::*tolC*; right) (*bottom*). (**b**) Image of the micropipette used for droplet collection (*left*) and sequential images of droplet collection (*right*). (**c**) Example of gene analysis encoded in the plasmid by agarose gel electrophoresis

4 Notes

1. Photolithography must be carried out with a high viscosity photoresist, as the CYTOP-coated surface has very low friction and thus cannot be fully covered with a low-viscosity photoresist.
2. For the formation of femtoliter droplet arrays and for subsequent procedures, fluorinated oil with a density higher than that of water is favored.
3. Dilution of electron beam resist with thinner results in a thinner coat, which allows for shorter processing times in electron beam lithography.
4. Chromium can be wet-etched with an aqueous solution of ceric ammonium nitrate; however, the use of commercially available chromium etchant, which contains surfactants, is beneficial as the electron beam resist is generally hydrophobic and thus makes wet-etching with aqueous solutions difficult.
5. This procedure results in a CYTOP coat of ~1-μm thickness. However, if the speed used during the spincoat process is increased, a thinner coat will be formed.
6. The solvent used in the CYTOP solution has a boiling point of 180 °C. A pre-bake step at 80 °C results in slow evaporation of the solvent from the surface, and facilitates the formation of a uniform CYTOP layer.
7. Because the CYTOP-coated surface has very low friction, the placement of the photoresist at the center of the coverslip is important for uniform coating.
8. This process (spincoating at 4500 rpm, 1 s) is important for the removal of excess photoresist that has remained at the edge of the CYTOP-coated glass.
9. Repetition of this procedure will result in contamination of the photomask with photoresist, which will hinder the formation of the required tight contact with the substrate. If the procedure needs to be repeated, remove the photoresist on the photomask using gauze containing acetone, or wash the photomask in acetone using a bath-type sonicator.
10. The time required for development will vary depending on the temperature and the concentration of the developer. Completion of the developing process can be ascertained by observation under an optical microscope equipped with a yellow filter.
11. This procedure will remove photoresist. When photoresist is completely removed, the surface will repel acetone.
12. If the hole is not completely covered with epoxy adhesive and the coverslip, the culture medium will leak from the bottom of the petri dish in subsequent experiments.

13. Typically, 100–200 μL bacterial suspension is required. This low volume will make subsequent processes, including replacement of the medium on the surface of the substrate with oil, much easier.
14. Enclosure of the cells in the droplets is stochastic and is dependent on the cell density of the bacterial suspension. At an OD_{600} of 0.6, approximately 20–30% of the 10 μm droplets contain single cells. Increasing the droplet diameter to 20 or 30 μm increases the proportion of droplets containing multiple cells, but not that of droplets containing single cells. Therefore, we use 10 μm droplets, because the total number of droplets formed in a single device can be significantly increased.
15. This procedure is important for the complete removal of living cells from the medium on top of the oil layer. If living cells remain, they may grow on top of the oil layer and contaminate the collection process in Subheading 3.3.
16. The gene-competent cells have to be prepared without β-galactosidase induction to suppress leakage of β-galactosidase from the cell membranes permeabilized during the electroporation process.
17. At this pressure, medium slowly flows out from the glass micropipette into the water layer, preventing contamination. On the other hand, flow spontaneously stops in the oil layer due to the difference in the surface tensions of the water and the oil.

References

1. Nikaido H (2009) Multidrug resistance in bacteria. Annu Rev Biochem 78:119–146
2. Nishino K, Nikaido E, Yamaguchi A (2009) Regulation and physiological function of multidrug efflux pumps in *Escherichia coli* and *Salmonella*. Biochim Biophys Acta 1794:834–843
3. Nikaido H, Takatsuka Y (2009) Mechanisms of RND multidrug efflux pumps. Biochim Biophys Acta 1794:769–781
4. Yamaguchi A, Nakashima R, Sakurai K (2015) Structural basis of RND-type multidrug exporters. Front Microbiol 6:327
5. Lister PD, Wolter DJ, Hanson ND (2009) Antibacterial-resistant *Pseudomonas aeruginosa*: clinical impact and complex regulation of chromosomally encoded resistance mechanisms. Clin Microbiol Rev 22:582–610
6. Iino R, Hayama K, Amezawa H, Sakakihara S, Kim SH, Matsumono Y, Nishino K, Yamaguchi A, Noji H (2012) A single cell drug efflux assay in bacteria by using a directly accessible femtoliter droplet array. Lab Chip 12:3923–3929
7. Iino R, Matsumoto Y, Nishino K, Yamaguchi A, Noji H (2013) Design of a large-scale femtoliter droplet array for single cell analysis of drug-tolerant and drug-resistant bacteria. Front Microbiol 4:300
8. Sakakihara S, Araki S, Iino R, Noji H (2010) A single-molecule enzymatic assay in a directly accessible femtoliter droplet array. Lab Chip 10:3355–3362
9. Russo-Marie F, Roederer M, Sager B, Herzenberg LA, Kaiser D (1993) Beta-galactosidase activity in single differentiating bacterial cells. Proc Natl Acad Sci U S A 90:8194–8198
10. Matsumoto Y, Hayama K, Sakakihara S, Nishino K, Noji H, Iino R, Yamaguchi A (2011) Evaluation of multidrug efflux pump inhibitors by a new method using microfluidic channels. PLoS One 6:e18547

Chapter 19

Reconstitution and Transport Analysis of Eukaryotic Transporters in the Post-Genomic Era

Hiroshi Omote and Yoshinori Moriyama

Abstract

Measuring transport activity through reconstituted proteoliposomes is a key technique to resolve numerous problems found in the traditional methods. The system includes overexpression, purification, and reconstitution of transporters. Mixing of purified transporter with lipid and dilution below the critical micelle concentration result in rapid generation of proteoliposomes. Incubation of proteoliposomes in the presence of a driving force initiates substrate uptake. After starting the reaction, samples are passed through a gel filtration column to separate proteoliposomes from the reaction mixture. Here, we describe step-by-step procedures for such reconstitution assays.

Key words Transporter, Purification, Reconstitution, Proteoliposome, Driving force, Kinetics

1 Introduction

Membrane transporters are involved in various physiological processes through their transport activity. Measurement of transport activity is an essential step to determine their physiological function. Historically, transporter function has been measured using isolated membrane vesicles. After addition of radiolabeled substrate, membrane vesicles were separated from the reaction mixture by filtration. The transport activities of ion pumps, amino acid, sugar, and drug transporters have been analyzed using this system. At the end of the twentieth century, the development of molecular biology techniques enabled the use of cDNA for studying transporter function. Currently, heterologous expression systems with HEK cells, COS cells, and *Xenopus laevis* oocytes are widely used. However, these systems have multiple problems. For example, the presence of many intrinsic transporters on the membrane interferes with transport analysis. In addition, it is not possible to precisely control the intracellular conditions. While transport into the cell is easy to measure, export is difficult to evaluate.

Akihito Yamaguchi and Kunihiko Nishino (eds.), *Bacterial Multidrug Exporters: Methods and Protocols*, Methods in Molecular Biology, vol. 1700, https://doi.org/10.1007/978-1-4939-7454-2_19, © Springer Science+Business Media, LLC 2018

Reconstitution methods using purified transporters represent the best solution to resolve these problems. The reconstitution method was introduced in the 1970s, in which purified transporters were mixed with lipid and diluted below the critical micelle concentration [1–3]. By lowering the detergent concentration, micelles were decomposed and reorganized into liposomes. Along with liposome formation, protein in the detergent micelles is incorporated into the lipid bilayer of liposomes. The resultant proteoliposomes take up substrates with application of an appropriate driving force.

Transporter cDNAs can be obtained as a result of genome projects, and therefore purified transporters can be obtained through overexpression and purification. Reconstitution systems in the post-genomic era would be powerful tools to study the functions of any transporter of interest. In fact, this system has been successfully applied to analyze transporters of diverse species from mammals to plants and malaria parasites [3–7]. Here, we describe the procedures for reconstitution and transport assay using purified transporters.

1.1 Basics

Most secondary active transporters are reversible and transport their substrates in both directions. The driving force, such as H^+ and Na^+ gradients or membrane potential, determines the direction of transport for these transporters (Fig. 1) [6, 7]. This is important for reconstitution assay as we can control the direction of transport by application of a valid driving force. For example, MATE (Multidrug and Toxic Compound Extrusion) type drug transporter exports cationic drugs using counterflow of H^+ [8]. Although there is almost no pH gradient across the plasma membrane, the inside-positive membrane potential of the cell induces H^+ influx. Entry of H^+ drives extrusion of cationic drugs from the cell. In the reconstituted system, we generate an inside-acidic pH gradient by controlling the composition of the reconstitution buffer. This pH gradient stimulates uptake of substrates into the proteoliposomes by efflux of H^+ (Fig. 1). It should be emphasized that only half of the transporters in the proteoliposomes are oriented to the correct side. Nevertheless, essentially the same transport kinetics was observed in the system as observed using isolated vesicles [9].

2 Materials

2.1 Purified Transporters

Transporters are purified by either Ni-NTA column chromatography or other methods. Typical samples contain ~0.5 mg/ml of purified transporter in a buffer containing 20 mM MOPS-Tris (pH 7.0), 300 mM imidazole, 20% glycerol, and 0.1% decyl thiomaltoside [4, 9–15].

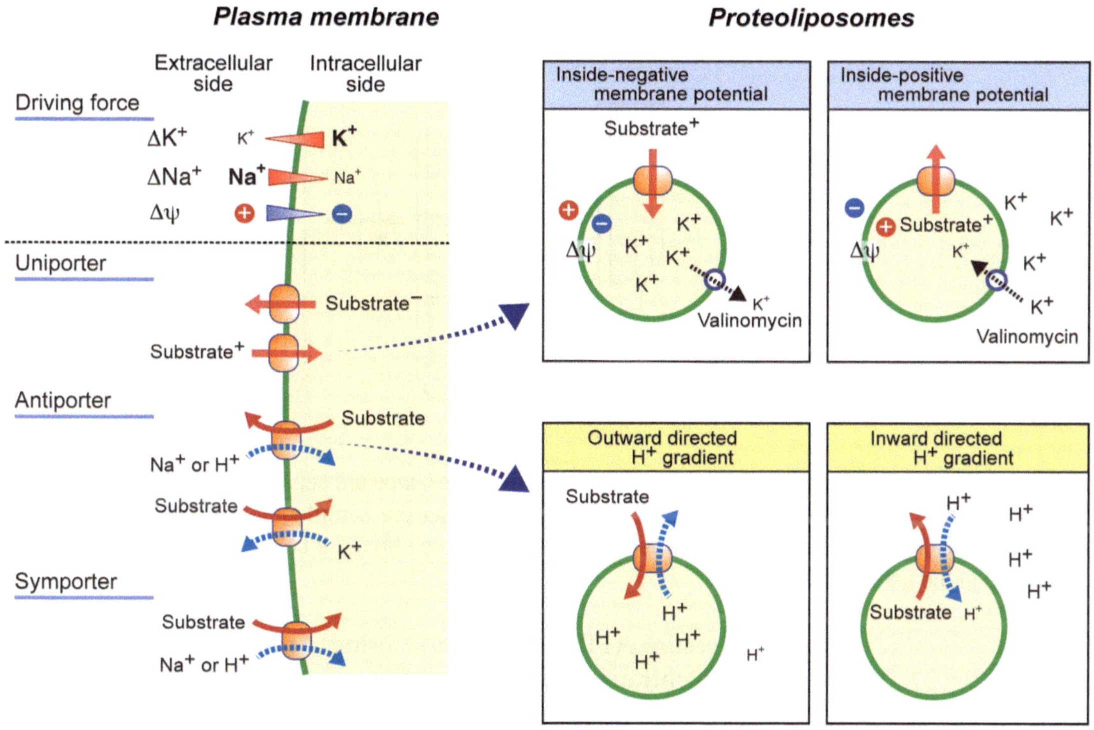

Fig. 1 Driving force of membrane transport. Membrane transporters use ion gradients and membrane potential as driving force. In the reconstituted system, inside negative and positive membrane potentials can be established by K^+ diffusion using K^+ ionophore. Ion gradient can be generated by changing ionic compositions of reconstitution buffer and reaction mixture

2.2 Lipids

Although various lipids can be used for transporter reconstitution, crude lipids give better results for transport assay.

1. Lipid buffer: 20 mM MOPS-NaOH (pH 7.0), 0.5 mM dithiothreitol.
2. Place 20 mg of soybean L-α-phosphatidylcholine (Sigma Type II-S) into a glass tube and add 2 ml of lipid buffer.
3. Seal the test tube with Parafilm.
4. Sonicate the lipid suspension using a bath-type sonicator until the solution becomes opalescent.
5. Divide into 200-μl aliquots and store at −80 °C until use.

2.3 ATP

ATP is a tetravalent anion and is decomposed under acidic conditions. To minimize nonenzymatic hydrolysis, ATP should be dissolved on ice and maintained at pH > 6.

1. Place 2 g of Tris-ATP into 20 ml of ice-cold water.
2. Adjust to pH 7.0 by adding 2 M Tris base with stirring (*see* **Note 1**).

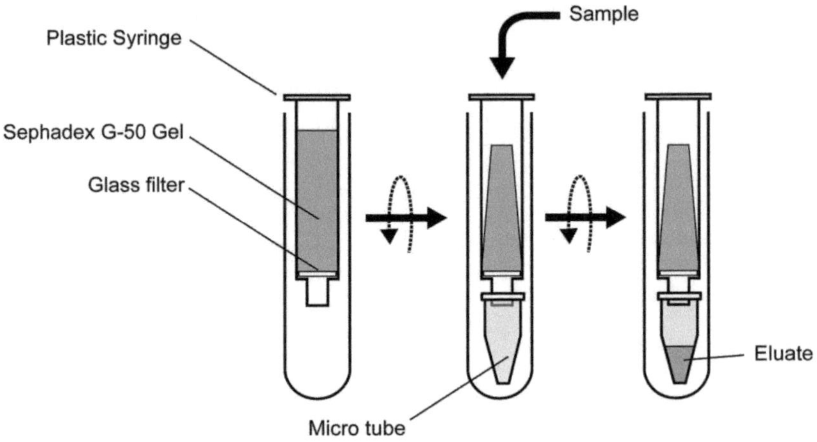

Fig. 2 Gel filtration by handmade spin column. Liposome and free ligand are separated by gel filtration. Plastic syringe is filled by Sephadex G-50 fine gel filtration medium. After pre-centrifugation, sample is applied and centrifuged again. Liposomes and transported substrates are eluted from the column whereas free substrate is trapped in the gel

3. Measure ATP concentration using a spectrophotometer. ATP concentration is calculated using $\varepsilon_{259} = 15{,}400$.
4. Prepare 100 mM ATP solution by dilution of solubilized ATP and store aliquots at −30 °C.

2.4 Spin Columns

We use gel filtration for separation of free radiolabeled substrates and proteoliposomes. This method can be used even for hydrophobic substrates, such as verapamil. We use handmade spin columns equipped with Sephadex G-50 fine gel that yield better separation than commercial desalting columns (~1/10,000 leakage) (Fig. 2).

Just before use, spin columns should be placed in a glass or plastic tube and centrifuged (750 × *g*, 2 min). After pre-spinning, transfer the column containing shrunken gel to a new tube, apply the sample, and centrifuge again. Proteoliposomes will be recovered in the eluate.

1. Swell 1 g of Sephadex G-50 Fine (GE Healthcare) in 20 ml of buffer containing 20 mM MOPS-Tris (pH 7.5), and keep for more than 3 h at room temperature (*see* **Notes 2** and **3**).
2. Cut Whatman glass GF/D filter with a cork borer.
3. Place glass filter in the bottom of 1-ml plastic syringe (6.5 × 70 mm) using a Pasteur pipette. Filter should be flat and placed horizontally.
4. Fill plastic syringe with 1 ml of Sephadex G-50 Fine gel. Ensure there are no bubbles in the column.
5. Wash out column with 1 ml of the same buffer.
6. Spin columns should be wrapped and placed at room temperature. Use within 1 day.

3 Methods

All procedures should be carried out on ice unless otherwise specified.

3.1 Co-Reconstitution with F-ATPase

The details of reconstitution were documented previously [9, 16]. Here, we describe reconstitution of purified vesicular glutamate transporter (VGLUT) with F-ATPase (Figs. 3 and 4). F-ATPase catalyzes ATP synthesis coupled with H^+ flow. This enzyme is reversible and transports H^+ using energy obtained from ATP hydrolysis [16]. F-ATPase is overexpressed and purified by glycerol density gradient centrifugation in the presence of 1% octylglucoside [16].

1. Thaw frozen transporter and lipid solutions, and place on ice.
2. Add 20 μg of purified VGLUT (in 1% octylglucoside) and 90 μg of purified F-ATPase (in 1% octylglucoside) to ice-cold microtubes containing 500 μg of lipids. The total volume is approximately 200–250 μl.

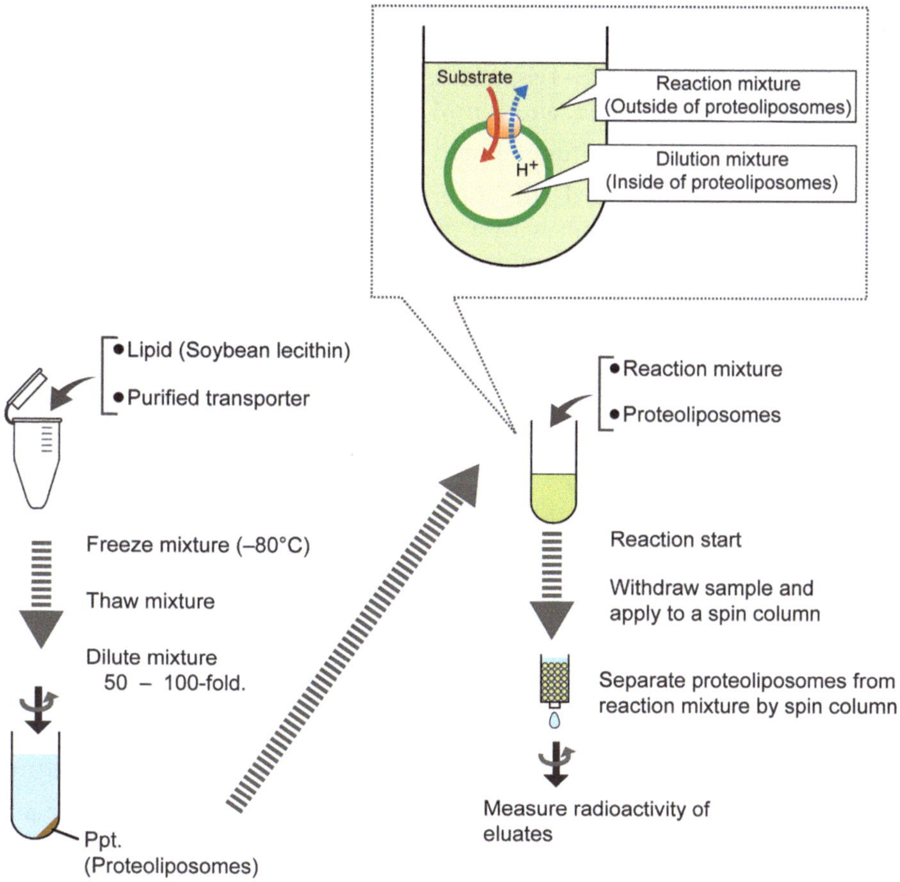

Fig. 3 Transport assay using reconstituted proteoliposomes and spin column. Flow diagram of transport assay is shown

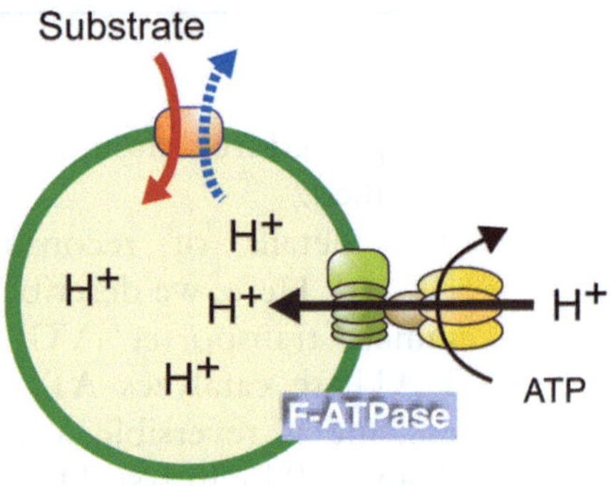

Fig. 4 Co-reconstitution with F-ATPase. F-ATPase hydrolyzes ATP and transports H^+ into proteoliposomes. This generates inside positive membrane potential and inside acidic H^+ gradient. These driving forces are powerful than other methods

3. After mixing, freeze the mixture by placing the tube at −80 °C for 15 min.
4. Quickly thaw the mixture by holding the tube in the hand and place on ice.
5. Dilute sample >60-fold by transferring the mixture to a centrifuge tube containing 20 ml of reconstitution buffer (20 mM MOPS-Tris, pH 7.0, 0.5 mM dithiothreitol, 0.1 M potassium acetate, and 5 mM magnesium acetate) (*see* **Notes 4** and **5**).
6. Centrifuge diluted sample for 60 min at 150,000 × *g*.
7. Discard the supernatant by decanting and remove residual solution with a pipette. A small precipitate will be present at the bottom of the tube.
8. Add 200 μl of reconstitution buffer and suspend using a glass homogenizer.
9. Transfer proteoliposomes to a microtube and keep on ice until use. Proteoliposomes should be used within 1 day.

3.2 Transport Assay Using Proteoliposomes Co-Reconstituted with F-ATPase

Reactions are carried out at 27 °C. Dependent on the situation, reactions can be initiated by addition of proteoliposomes, ATP, or substrate. The typical substrate concentration is 100 μM with specific activity of 0.5 MBq/μmol (Fig. 4).

1. Add 125 μl of reaction mixture (22.4 mM MOPS-Tris, pH 7.0, 112 mM potassium acetate, 5.6 mM magnesium acetate, and 4.8 mM KCl) to a glass test tube and incubate for 2 min at 27 °C (*see* **Notes 4** and **5**).
2. Add 10 μl of proteoliposomes to the reaction mixture.
3. Add 7.5 μl of 40 mM ATP to the reaction mixture and mix by vortexing.

4. Place a spin column in a test tube (15 × 105 mm) and centrifuge for 2 min at 2000 rpm (750 × *g*) just before use (*see* **Note 6**).
5. Three minutes after addition of ATP, add 7.5 μl of 2 mM [2,3-^{3}H] L-glutamate (0.5 MBq/μmol). The final volume of the reaction mixture is 150 μl.
6. One minute after starting the reaction, withdraw 130 μl of reaction mixture and apply to a pre-spun spin column.
7. Place the spin column in a new test tube containing a 500-μl microtube (*see* Fig. 2) (*see* **Note 7**).
8. Centrifuge the spin column for 2 min at 2000 rpm (750 × *g*).
9. Collect eluate and count radioactivity with a liquid scintillation counter.

3.3 Energetic Analysis of Transport with F-ATPase

F-ATPase generates an electrochemical gradient of H^+ composed of the H^+ gradient and membrane potential. To determine the contributions of these two components, H^+ and K^+ ionophores can be used. In this protocol, we include an appropriate agent to dissipate the H^+ gradient or membrane potential.

1. Add 125 μl of reaction mixture (22.4 mM MOPS-Tris, pH 7.0, 112 mM potassium acetate, 5.6 mM magnesium acetate, 4.8 mM KCl and 2.4 mM NH_4Cl) to a test tube and incubate for 2 min at 27 °C (*see* **Notes 8** and **9**).
2. Follow Subheading 3.2.

3.4 Reconstitution of Mouse MATE2

This section describes reconstitution and transport assay with the pH jump method. The experimental details were documented previously [17] (Fig. 3).

1. Thaw frozen transporter and lipid solutions and place them on ice.
2. Add 20–50 μg of purified mouse MATE2 (mMATE2) to an ice-cold microtube containing 550 μg of lipids.
3. After mixing the solution, freeze the mixture by placing the tube at −80 °C for 15 min.
4. Quickly thaw the mixture by holding the tube in the hand.
5. Dilute sample >60-fold by transferring the mixture to a centrifuge tube containing reconstitution buffer (20 mM MOPS-NaOH, pH 6.7, 0.1 M potassium acetate, and 5 mM magnesium acetate).
6. Centrifuge tubes for 60 min at 150,000 × *g*.
7. Discard supernatant by decanting and remove residual solution with a pipette.

8. Add 200 μl of reconstitution buffer and suspend using a glass homogenizer.
9. Transfer proteoliposomes to a microtube and keep on ice until use.

3.5 Transport Assay by pH Jump Method

This section presents the procedure for tetraethylammonium (TEA) transport by reconstituted mMATE2 [17]. The reaction is carried out at 27 °C, as described above. Transport reaction is started by transferring proteoliposomes prepared at pH 6.7 to the reaction mixture at pH 8.0 (*see* **Note 10**).

1. Add 140 μl of reaction mixture (20 mM tricine-NaOH, pH 8.0, 100 mM potassium acetate, 5 mM magnesium acetate, and 50 μM ^{14}C-TEA (0.5 MBq/μmol)) to a glass test tube and incubate for 2 min at 27 °C.
2. Centrifuge spin column for 2 min at 2000 rpm (750 × *g*).
3. Add 10 μl of proteoliposomes to the reaction mixture.
4. One minute after starting the reaction, withdraw 130 μl of reaction mixture and apply to a pre-spun spin column.
5. Place spin column in a new test tube containing a 500-μl microtube.
6. Centrifuge spin column for 2 min at 2000 rpm (750 × *g*).
7. Collect eluate and count radioactivity with a liquid scintillation counter.

3.6 Energetic Analysis in pH Jump Method

To determine the driving force of transport, the effects of each component of the electrochemical gradient on transport activity should be tested. As described above, most transporters are driven by H^+, Na^+, or K^+ gradient and membrane potential (Figs. 1 and 3). These driving forces can be applied by changing the compositions of reconstitution and reaction mixtures. The following table shows the components of the reconstitution and assay buffers (Table 1).

4 Notes

1. Solubilization of ATP will result in a decrease in pH as ATP is acid. Add Tris-base solution to maintain neutral pH during solubilization.
2. Gel swelling and spin column preparation should be carried out at room temperature. Do not cool.
3. Just before using, pre-spin the spin column to remove excess solution in the gel.
4. In this case, the compositions of reaction mixture and reconstitution buffer are the same.

Table 1
The components of the reconstitution and assay buffers

Substrate uptake driven by	Inside proteoliposomes (reconstitution buffer)	Outside proteoliposomes (assay buffer)	Additions
Sodium antiport	100 mM sodium acetate (or NaCl)	100 mM potassium acetate (or KCl)	None
Sodium symport	100 mM potassium acetate (or KCl)	100 mM sodium acetate (or NaCl)	None
H^+ antiport	20 mM MES-KOH, pH 6.0	20 mM Tricine-KOH, pH 8.0	None
H^+ symport	20 mM Tricine-KOH, pH 8.0	20 mM MES-KOH, pH 6.0	None
Inside-negative membrane potential	100 mM potassium acetate (or KCl)	100 mM sodium acetate (or NaCl)	2 μM valinomycin
Inside-positive membrane potential	100 mM sodium acetate (or NaCl)	100 mM potassium acetate (or KCl)	2 μM valinomycin

5. The formation of a membrane potential and ΔpH by F-ATPase are dependent on Cl^- concentration in the reaction mixture [9]. Under these conditions, most of electrochemical gradient of H^+ is composed of the membrane potential. If ΔpH is needed, the reaction mixture and dilution buffer must contain 100–150 mM KCl.
6. After pre-spinning, the bed volume of the gel is ~0.7 ml.
7. Place a 500-μl microtube at the bottom of a test tube after removing the cap.
8. To dissipate ΔpH, we use 2 μM nigericin (final concentration) or 2 mM NH_4Cl. To dissipate the membrane potential, we use 2 μM valinomycin. In addition, 1 μM CCCP or nigericin plus valinomycin are used to disrupt both ΔpH and membrane potential.
9. Valinomycin is hydrophobic. Dissolve valinomycin in dimethylsulfoxide (DMSO) and do not dilute with aqueous buffer.
10. In this case, intravesicular pH > 6.5 is used as MATE is inhibited by acidic pH.

References

1. Kasahara M, Hinkle PC (1977) Reconstitution and purification of the D-glucose transporter from human erythrocytes. J Biol Chem 252:7384–7390
2. Newman MJ, Foster DL, Wilson TH et al (1981) Purification and reconstitution of functional lactose carrier from *Escherichia coli*. J Biol Chem 256:11804–11808

3. Omote H, Miyaji T, Hiasa M et al (2016) Structure, function, and drug interactions of neurotransmitter transporters in the post-genome era. Annu Rev Pharmacol Toxicol 56:385–402
4. Juge N, Moriyama S, Miyaji T et al (2015) *Plasmodium falciparum* chloroquine resistance transporter is a H^+-coupled polyspecific nutrient and drug exporter. Proc Natl Acad Sci U S A 112:3356–3361
5. Miyaji T, Kuromori T, Takeuchi Y et al (2015) *AtPHT4;4* is a chloroplast-localized ascorbate transporter in *Arabidopsis*. Nat Commun 6:5928
6. Omote H, Miyaji T, Juge N et al (2011) Vesicular neurotransmitter transporter: bioenergetics and regulation of glutamate transport. Biochemistry 50:5558–5565
7. Omote H, Moriyama Y (2013) Vesicular neurotransmitter transporters: an approach for studying transporters with purified proteins. Physiology (Bethesda) 28:39–50
8. Otsuka M, Matsumoto T, Morimoto R et al (2005) A human transporter protein that mediates the final excretion step for toxic organic cations. Proc Natl Acad Sci U S A 102:17923–17928
9. Juge N, Yoshida Y, Yatsushiro S et al (2006) Vesicular glutamate transporter contains two independent transport machineries. J Biol Chem 281:39499–39506
10. Miyaji T, Echigo N, Hiasa M et al (2008) Identification of a vesicular aspartate transporter. Proc Natl Acad Sci U S A 105:11720–11724
11. Sawada K, Echigo N, Juge N et al (2008) Identification of a vesicular nucleotide transporter. Proc Natl Acad Sci U S A 105:5683–5686
12. Juge N, Gray JA, Omote H et al (2010) Metabolic control of vesicular glutamate transport and release. Neuron 68:99–112
13. Kato Y, Omote H, Miyaji T (2013) Inhibitors of ATP release inhibit vesicular nucleotide transporter. Biol Pharm Bull 36:1688–1691
14. Miyaji T, Sawada K, Omote H et al (2011) Divalent cation transport by vesicular nucleotide transporter. J Biol Chem 286:42881–42887
15. Leviatan S, Sawada K, Moriyama Y et al (2010) Combinatorial method for overexpression of membrane proteins in *Escherichia coli*. J Biol Chem 285:23548–23556
16. Moriyama Y, Iwamoto A, Hanada H et al (1991) One-step purification of *Escherichia coli* H^+-ATPase (F_0F_1) and its reconstitution into liposomes with neurotransmitter transporters. J Biol Chem 266:22141–22146
17. Komatsu T, Hiasa M, Miyaji T et al (2011) Characterization of the human MATE2 proton-coupled polyspecific organic cation exporter. Int J Biochem Cell Biol 43:913–918

Index

A

AcrA ... 25, 28, 76, 77, 114, 118, 122, 123, 126, 136, 137, 141, 142, 180, 181
AcrAB-TolC ... v, 3, 71, 180, 181, 331, 336
AcrABZ–TolC complex ... 76–79
AcrB ... v, 3, 25, 38, 76, 114, 147, 167, 180, 249, 254
Affinity binding assay ... 270, 273, 281, 282
Amphipol A8-35 ... 73, 79, 80
Anaerobic culture ... 263–265
Anaerobiosis ... 254
Antibiotics ... v, 3, 26, 83, 98, 113, 116, 168, 179, 180, 192, 240, 253, 254, 260, 272, 279, 286, 331, 332
  enhancement ... 295
  resistance ... 253
Antibody ... 4, 41, 53, 97–108, 149, 152, 156, 157, 160, 258, 264, 267, 280
Antimicrobial resistance ... 184, 221
Antiporter ... v, 98, 114, 180

B

Bacterial resistance ... 142, 240
BaeR ... 240, 241, 243, 244, 249, 250
β-Galactosidase activity assay ... 256, 263
Bicelles ... 85–88, 90–93
Binding free energy ... 185–187
Biological Warfare bacterial agents ... 293
Bodipy-FL-Maleimide ... 168–173, 175
BugBuster ... 257, 262
Burkholderia sp. ... 294

C

Channel ... v, 25, 59, 86, 93, 114, 119–121, 128–130, 143, 180, 184, 255
Co-crystallization ... 4, 51, 98
CPM assay ... 99, 100, 106, 107
CpxR ... 240, 241, 243, 244, 249, 250
Cryo
  EM analysis ... 71
  protection ... 31, 34
CYTOP ... 322–324, 332, 334, 336, 340

D

Designed ankyrin repeat proteins (DARPins) ... v, 3–22
Disulfide bonds ... 162
n-Dodecyl-β-D-maltoside (DDM) ... 9, 10, 12, 16, 27, 32, 38, 39, 42, 73, 78, 79, 81, 85–87, 100–102, 117, 123, 125, 130, 136, 139, 170, 172
Driving force ... 344, 345, 348, 350
Drug
  design ... 37
  discovery ... 232
  efflux ... 83, 85, 92, 113, 152, 157, 158, 221–234, 253, 266
  efflux assay ... 242, 248, 249, 333, 337
  resistance ... v, 25, 113, 142, 147, 158, 179, 221, 222, 239, 254, 270, 284, 331
  susceptibility assay ... 240, 241, 243
Dynamics ... vi, 4, 85, 92, 114, 122, 139, 167, 168, 180, 181, 184, 185, 195, 205

E

Efflux pump ... vi, 3, 59, 76, 83, 115, 137, 138, 168, 180, 221, 253, 269, 331
Electron cryo-microscopy (cryo-EM) ... vi, 71
EmrE ... 84, 91, 92
Epitope tag ... 255, 256, 259, 260
Escherichia coli ... v, 4, 26, 38, 41, 42, 53, 59, 76, 86, 98, 113, 147, 167, 180, 239, 255, 270, 322, 331
EvgA ... 240, 241, 243, 244, 248–250
Expression ... vi, 5, 26, 41, 60, 73, 85, 97, 124, 149, 222, 239, 253, 269, 322, 343

F

Fab fragment ... 98, 104–108
Flow cytometry ... 293–316
Fluorescein-di-β-D-galactopyranoside (FDG) ... 331–333, 336, 337

Akihito Yamaguchi and Kunihiko Nishino (eds.), *Bacterial Multidrug Exporters: Methods and Protocols*, Methods in Molecular Biology, vol. 1700, https://doi.org/10.1007/978-1-4939-7454-2, © Springer Science+Business Media, LLC 2018

Fluorescence ... 51, 99, 101, 102, 108, 135–138, 142, 143, 161, 162, 242, 249, 325–329, 332, 336–339
Fluorescence microscopy ... 332
Fluorescent
- assay ... 167
- imaging ... 170, 325, 326, 328
- spectrophotometer ... 169, 172, 223, 242, 249, 256, 258, 259, 274, 296, 346

Force field ... 182–185, 187, 189, 195, 199, 203
Francisella sp. ... 294, 296
Free energy calculations ... 185–187
Functionally rotating mechanism ... 147, 148

G

Gel filtration chromatography ... 10, 17, 79
Gel-shift assay ... 273, 280, 289
Gram-negative bacteria ... v, 3, 25, 113, 114, 147, 179, 232, 259, 331

H

Heavy metal efflux pump ... 59–70
Highly sensitive bioassay ... 322
High-resolution structure determination ... v, 3–22
High throughput bioassay ... vi, 51
Homology modeling ... 51, 180, 182, 188–190, 195, 207
Hybridoma ... 99, 100, 102–104

I

IMAC ... 73, 78, 79
Imaging ... 107, 170, 323
In-frame gene deletion ... 270, 271, 273, 287, 289
Inhibitors ... v, 25–35, 37, 39, 43, 49, 53, 73, 78, 79, 117, 118, 120, 135, 149, 180, 186, 189, 192, 245, 280, 332
In vitro ... 40, 43, 46, 47, 53, 54, 114, 123, 143, 285

K

Kinetics ... 114, 134, 141, 344

L

Ligand path ... 167–175
Lipid bilayer formation ... 325
Lipidic cubic phase (LCP) ... 41, 51, 54, 55, 99, 106, 108

M

Macrocyclic peptides ... 38, 43, 46–49, 51, 53, 54
Major facilitator superfamily (MFS) ... v, 97, 98
Membrane
- channel ... vi, 25, 59, 119–121, 128–130, 143, 255
- fusion protein (MFP) ... 25, 59, 113–115, 126–128, 180
- protein ... v, 3, 4, 17, 41, 49, 51, 55, 59, 60, 72, 84, 90, 92, 98, 114, 125, 127–129, 139, 140, 155, 157, 167, 168, 184, 185, 188, 195, 255, 259
  - expression ... 41
  - purification ... 41, 138
- transport ... 83, 322, 345
- transporter ... v, vi, 59, 116–118, 124–126, 253, 321, 343

Methanethiosulfonate (MTS) ... 153, 161–163
MexAB-OprM ... 136, 138, 331, 333, 336, 337
Microdevice ... 332
Microfabrication ... 322–324, 333–336
Micromanipulator ... 333
Micropipette ... 337, 339, 341
MicroPlanter ... 241, 244, 245
Miller Units ... 262, 267
Molecular
- databases ... 184
- docking ... 180, 182, 183, 188, 192–194
- dynamics simulations ... vi, 181–185, 188, 189, 192–207, 209
- modeling ... 179–209
- properties ... 179

Multidrug
- efflux ... v, vi, 3, 71, 167, 180, 250, 253–267
- exporter ... v, vi, 26, 37, 39, 43, 49, 53, 84, 148, 239–250
- resistance (MDR) ... v, 37, 83, 98, 113, 179, 183, 240, 250, 254, 331
- and toxic compound extrusion (MATE) ... v, 37–55, 344, 349–351

Multipoint inoculator ... 244, 245

N

NMR
- pedagogy ... 83
- spectroscopy ... vi, 83–94, 184

O

ONPG (2-Nitrophenyl β-D-galactopyranoside) ... 256, 261
Optical microscopy ... 337, 340
Oriented solid-state NMR (O-SSNMR) ... 83, 84, 87

P

Parametrization ... 184
Peristaltic pump mechanism ... 4, 281, 288
Physiological substrates ... 266
Polarization spin exchange at the magic angle (PISEMA) ... 85, 88–93
Protein
  crystallography ... 3, 167
  expression ... 26, 33, 73, 126, 271, 279, 280
  purification ... 9, 10, 28, 29, 34, 78, 79, 117, 125, 132, 169, 172, 272, 273, 277–281, 289
  stabilization ... 85
Proteoliposome ... 87, 101, 102, 114–116, 121–123, 130–143, 324, 325, 329, 344, 346–351
Proton
  gradient ... 114–116, 137, 138, 143
  motive force (PMF) ... vi, 4, 163
  relay ... 152, 160
Pseudomonas aeruginosa ... 114, 115, 138, 142, 180, 265, 331, 333
Purification ... v, 4, 6, 9, 10, 14, 16–18, 22, 26–29, 34, 38, 40–42, 61–65, 70–81, 98, 99, 101, 105, 116–121, 124–130, 132, 138, 169, 172, 224, 225, 241, 243, 258, 272–274, 277–281, 287, 289, 344
PyMol Caver plug-in ... 168
Pyrococcus furiosus ... 38, 41

Q

qRT-PCRs ... 270–272, 285, 287

R

Random nonstandard Peptide Integrated Discovery (RaPID) system ... 39, 40, 43–48
Reactive ion etching ... 322, 324, 332, 334
Real-time quantitative RT-PCR ... 242, 243, 245
Reconstitution ... vi, 86, 101, 113–143, 343–351
Regulator ... vi, 240–243, 255, 266, 269
Resistance-nodulation-cell division (RND) ... v, vi, 3, 26, 32, 38, 59, 114, 115, 137, 147–164, 179, 180, 182, 184, 336
  family ... 3, 38, 59, 148
  transporters ... 25, 35, 114, 148, 179, 181, 188, 189, 193, 195
Response regulator ... 240–243

S

Salmonella enterica ... 331, 333, 339
Seeding ... 29, 31, 34
Single
  cell analysis ... 331–341
  molecule analysis ... vi, 321–329
Site-directed-mutagenesis ... vi, 72, 76, 77, 92, 151, 153, 154, 168, 169, 171
Size exclusion chromatography (SEC) ... 4, 32, 34, 99, 104–107
Solid-state NMR ... 86, 93
Solubilization ... 9, 16, 17, 100, 101, 119, 125, 128, 139, 140, 172, 350
Staphylococcus aureus ... 269–290
Structural analysis ... 41
Structure ... v, 3, 32, 37, 59, 71, 83, 98, 114, 147, 167, 180, 222, 253, 269, 324
Substrate path ... 167
Sucrose Monododecanoate (Dodecanoyl sucrose/DDS) ... 26

T

Three-dimensional (3D) structure ... 71, 98, 182, 186, 189, 190
TolC ... v, 3, 25, 71, 114, 119–121, 123, 128, 129, 135, 138, 142, 180, 181, 255, 263, 331, 333, 336, 337, 339
Transporter ... v, 3, 25, 37, 59, 83, 97, 113, 147, 167, 180, 253, 321, 343
Tripartite
  assembly ... 72
  multidrug transporter ... v, vi, 76
Two-component signal transduction system ... 239

V

Vapor diffusion ... 29, 49, 51, 54, 55, 67

X

X-ray crystallography ... 3, 49, 54, 59, 83, 92
Xenobiotic exporter ... 37

MIX
Papier aus verantwortungsvollen Quellen
Paper from responsible sources
FSC® C105338
FSC
www.fsc.org

If you have any concerns about our products,
you can contact us on
ProductSafety@springernature.com

In case Publisher is established outside the EU,
the EU authorized representative is:
Springer Nature Customer Service Center GmbH
Europaplatz 3, 69115 Heidelberg, Germany

Printed by Libri Plureos GmbH
in Hamburg, Germany